The NASA Kepler Mission

AMERICAN ASTRONOMICAL SOCIETY

IOP | ebooks™

AAS Editor in Chief

Ethan Vishniac, Johns Hopkins University, Maryland, USA

About the program:

AAS-IOP Astronomy ebooks is the official book program of the American Astronomical Society (AAS), and aims to share in depth the most fascinating areas of astronomy, astrophysics, solar physics and planetary science. The program includes publications in the following topics:

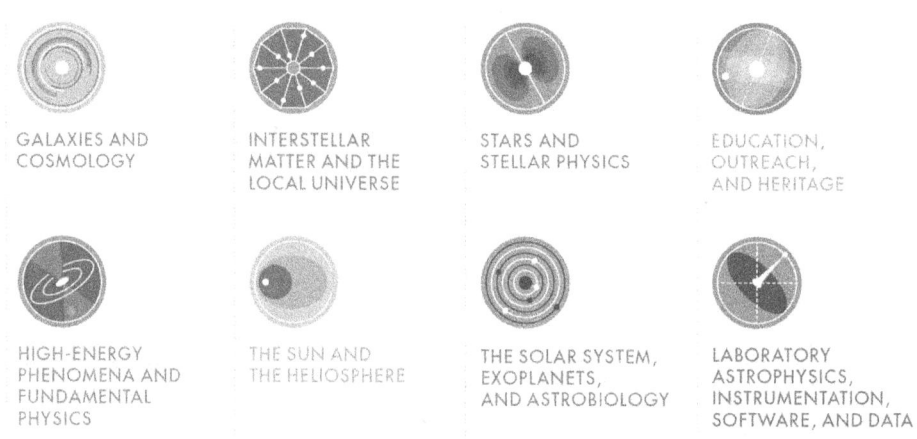

GALAXIES AND COSMOLOGY

INTERSTELLAR MATTER AND THE LOCAL UNIVERSE

STARS AND STELLAR PHYSICS

EDUCATION, OUTREACH, AND HERITAGE

HIGH-ENERGY PHENOMENA AND FUNDAMENTAL PHYSICS

THE SUN AND THE HELIOSPHERE

THE SOLAR SYSTEM, EXOPLANETS, AND ASTROBIOLOGY

LABORATORY ASTROPHYSICS, INSTRUMENTATION, SOFTWARE, AND DATA

Books in the program range in level from short introductory texts on fast-moving areas, graduate and upper-level undergraduate textbooks, research monographs and practical handbooks.

For a complete list of published and forthcoming titles, please visit iopscience.org/books/aas.

About the American Astronomical Society

The American Astronomical Society (aas.org), established 1899, is the major organization of professional astronomers in North America. The membership (~7,000) also includes physicists, mathematicians, geologists, engineers and others whose research interests lie within the broad spectrum of subjects now comprising the contemporary astronomical sciences. The mission of the Society is to enhance and share humanity's scientific understanding of the universe.

The NASA Kepler Mission

Edited by
Steve B Howell
NASA Ames Research Center, Moffett Field, CA, 94035, USA

IOP Publishing, Bristol, UK

Multimedia content is available for this book from http://iopscience.iop.org/book/978-0-7503-2296-6.

ISBN   978-0-7503-2296-6 (ebook)
ISBN   978-0-7503-2294-2 (print)
ISBN   978-0-7503-2297-3 (myPrint)
ISBN   978-0-7503-2295-9 (mobi)

DOI   10.1088/2514-3433/ab9823

Version: 20200901

AAS–IOP Astronomy
ISSN 2514-3433 (online)
ISSN 2515-141X (print)

British Library Cataloguing-in-Publication Data: A catalogue record for this book is available from the British Library.

Published by IOP Publishing, wholly owned by The Institute of Physics, London

IOP Publishing, Temple Circus, Temple Way, Bristol, BS1 6HG, UK

US Office: IOP Publishing, Inc., 190 North Independence Mall West, Suite 601, Philadelphia, PA 19106, USA

Cover image credit: NASA/Ames Research Center/W. Stenzel/D. Rutter

For my parents who accepted and supported my scientific questioning, exploration, and experimentation. And for my family, children, loved ones, and friends who put up with it.

Contents

Preface

The Kepler Mission opened up a new chapter in astronomy; the study of thousands of planets orbiting stars other than the Sun (i.e., exoplanets). By continuously monitoring the brightness of 170,000 stars simultaneously at a photometric precision sufficient to detect Earth-size planets it showed that the Galaxy is full of planets; there are more planets than stars. Many of these planets are small and possibly rocky like the Earth and some are found orbiting in the habitable zone. When the Mission was first proposed in 1992, humankind knew of no exoplanets. Today we face the opposite situation; a deluge of information about exoplanets orbiting every type of star. Although several types of search methods have contributed to our new knowledge, the photometric method pioneered by the Kepler Mission provided the breakthrough needed to detect and characterize thousands of small planets. To accomplish this breakthrough, the Kepler team had to develop an instrument with a photometric precision never-before attained. To detect the 84 ppm change in the brightness of a star when an Earth-size planet transits a solar-like star, required development of an instrument with both extreme precision and a very wide field-of-view (FOV). The effort to prove that such an instrument could be designed began in 1983 and ended when the Kepler Mission finished operations in 2017. Over its lifetime, Kepler detected over 5000 planets and planetary candidates.

The Kepler mission emblazoned as a corn maze. Shown here in an aerial view of the Kepler Mission-themed corn maze at the Dell'Osso Family Farm in Lathrop, California during the fall of 2011. The maze was one of seven "Space Farms" around the country that honored NASA and the 50th anniversary of human spaceflight. Credit: The MAiZE Inc.

This book provides a comprehensive discussion of the development of the instrument and spacecraft as well as thorough discussions of many of the discoveries accomplished by the Kepler Mission. In particular, the chapters on the development of the instrument and spacecraft detail the design and operation of the thermal, optical, detector, power, and guidance systems. Dozens of photos illustrate the assembly of the telescope and spacecraft. To monitor the thousands of stars needed to obtain statistically significant results, a Schmidt-type telescope with a FOV ten thousand times that of the Hubble Telescope was built. Because its FOV covers such a large area of the sky (over 100 deg^2), the focal surface is strongly curved requiring individual sapphire lenses to be inserted over each of the 21 detector modules in order to obtain sharp images. Because slight movements of the star images on the detectors could overwhelm the small signals from planets transiting their stars, the spacecraft was designed to track so precisely that the image of each star stayed on the same pixels for months at a time. Even so, the orbital motion of the spacecraft causes the star field to expand and shrink (velocity aberration) so that provision had to be made to account for the motion. After four years of monitoring the single star field in the Cygnus–Lyra constellations, the loss of two of the reaction wheels that controlled the pointing precision terminated observations of the original star field.

The book's authors describe how the scientists and engineers at Ames and Ball rose to the challenge of designing a productive mission based on the operation of the two remaining wheels. The new mission, K2, successfully conducted both searches for short-period exoplanets and astrophysical observations in 19 different fields of view before the spacecraft ran out of fuel.

The many types of planets discovered are fully covered in the chapters on exoplanet discoveries. Examples are provided of; terrestrial worlds, circumbinary planets, and planets so close to their star that they are disintegrating. Many of these planets are part of multi-planet systems; some have as many as six planets. Surprising discoveries include planets larger than the largest planet in the solar system, many planets with sizes that imply that they may be water worlds, and planetary systems that are completely unlike our solar system. Also included in the exoplanet chapters are the data showing transits; both the raw data and the data after the analyses have been performed to remove the systematic noise. Examples demonstrate that before the analyses are completed, the transits were often not recognizable, but afterwards, the signals of all but the smallest planets are apparent. The authors note that when data from ground-based radial velocity measurements of the planet's mass are available, the planet's density can be estimated and a determination made of whether the planet is a rocky planet rather than a gas or ice giant planet. Another method of determining planet mass is to note the changes in the time of the transits. Because of gravitation attraction between the planets in nearby orbits, the pattern of changes in transit time are used to determine planet masses when a sufficient number of transits are available that show measurable changes. Due to the large number of exoplanets discovered, the results can be used to estimate size distributions and the occurrence frequency of planets orbiting different types of stars. These results are also discussed in the exoplanet chapters. Of particular interest is a figure showing the position of a variety of planets found in and near the HZ relative to the temperature of the stars they orbit and the heat flux they receive. Also shown are several planets that, like Venus, are slightly too close to their stars to keep from losing all their water by the "runaway greenhouse" process.

One of the most difficult aspects of interpreting the data is the determination of whether the detection is of an actual planet or an astrophysical phenomenon that mimics planetary transits. Situations such as a small star grazing a large star in a binary star system, or a binary star in the background of the target star often produce signals that cause confusion. The authors discuss the statistical analysis of the data and observational techniques used with both ground-based and space-based observatories to distinguish the source of the detected signals. Although the detection and characterization of planets and planetary systems were the prime goals of the Kepler Mission, the results of astrophysical investigations were also of high priority. Because fluctuations in star brightness provide information about waves on the surface and the interior of stars, Kepler measurements were used to deduce the size of the star and thereby the size of the planet. Estimates of the stellar age and internal structure were also obtained. Based on the measured ages of stars with planetary systems, the Kepler results found that rocky planets were being formed when the first stars in our galaxy were being formed; i.e., exoplanets are plentiful both in space and in time.

The authors devote three chapters to the results from astrophysical investigations, observations of solar system objects and to extragalactic studies. Because Kepler monitored over 170,000 stars, information on the variability of stars of every age and type were recorded. Asteroseismologists investigated the interior structure and rotation of red giant stars as they evolved from burning hydrogen in their shells to burning helium in their cores. From the ages and distributions of giant stars, a history of the development of our galaxy could be deduced. Several supernova were observed. For these, the very beginning of the explosion event was observed as well as their decay. Results for solar system objects including planets, moons, and minor objects are also discussed. In summary the book's authors have provided the reader with a plethora of information about the development of the Kepler and K2 missions and of the many exciting results.

William J. Borucki, PI of the Kepler Mission
Mountain View, CA
2019 December 18

Acknowledgments

David Lee Summers would like to thank Jeff Young and Dr Mike Brotherton for their assistance tracking down books and stories featuring known exoplanets.

Csaba Kiss, László Molnár, and András Pál would like to acknowledge members of the Hungarian K2-TESS Solar System group: Attila Bódi, Zsófia Bognár, Anikó Farkas-Takács, Ottó Hanyecz, Csilla Kalup, Viktória Kecskemthy, Lászlo Kiss, Gábor Marton, Emese Plachy, Krisztián Sárneczky, Róbert Szabó Gyula M. Szabó Róbert Szakáts, and József Vinkó as well as our international affiliates in the "Small Bodies Near and Far" project, whose collaborative work made these results possible. We also thank the Kepler/K2 Guest Observer Office team for their efforts to allocate special pixel masks to track moving targets. We thank the hospitality of the Veszprm Regional Centre of the Hungarian Academy of Sciences (MTA VEAB), and the University of Pannonia, Veszprm, for hosting a series of workshops dedicated to the Solar System data. Our works have been supported by the Hungarian National Research, Development and Innovation Office (NKFIH) grants (including projects K-119517, K-125015, GINOP-2.3.2-15-2016-00003); by the Hungarian Academy of Sciences through their Lendlet Program (projects L-2009, LP2012-31, and LP2018-7/2019), Premium Postdoctoral Research Program, and János Bolyai Research Scholarships.

Armin Rest and Peter Garnavich thank Brad Tucker, Ed Shaya, Rob Olling, Richard Mushotzky and the "Kepler ExtraGalactic Survey" (KEGS) team. KEGS was partially funded by NASA K2 grant #NNX17AI64G, #80NSSC18K0301, #80NSSC18K0302, #80NSSC19K0112. AR and PG further acknowledge Geert Barentsen, Jessie Dotson, Gautham Narayan, Ashley Villar, Ryan Ridden-Harper, Dan Kasen, Steve Margheim, Alfredo Zenteno, and many others that recognized the importance of a K2 Supernova Experiment and worked to make it successful.

AV was supported by NASA through the Sagan Fellowship Program executed by the NASA Exoplanet Science Institute under contract with the California Institute of Technology/Jet Propulsion Laboratory.

PS acknowledges support from NASA grant HST-GO14912 and NSF grant AST-1514737.

SEM would like to thank Jeff Coughlin for looking up facts and find a few figures related to Kepler exoplanet vetting. I want to thank everyone who worked tirelessly to make the Kepler mission the amazing success that it was. You continue to be an inspiration to me. I would also like to thank Pádraig and Róisín for their sacrifice of missing several Mommy-told bedtime stories so that I could find the time to finish my sections of this book. Finally I want to thank Fergal Mullally for the endless conversations that help so much in getting my thoughts together.

KLS gratefully acknowledges the support of the National Aeronautics and Space Administration through Einstein Postdoctoral Fellowship Award Number PF7-180168, issued by the Chandra X-ray Observatory Center, which is operated by the Smithsonian Astrophysical Observatory for and on behalf of the National Aeronautics Space Administration under contract NAS8-03060. KLS also thanks the research group of Gordon Richards at Drexel University, as well as the endlessly helpful staff of the K2 Guest Observer office at NASA Ames Research Center, all of whom provided crucial input to understanding the effects of Kepler/K2 systematics on AGN science.

AMC acknowledges thoughtful suggestions from Trevor David on the cluster exoplanet and eclipsing binary sections of this book. She would also like to thank SBH, the NASA Postdoctoral Program, and the Kepler/K2 Guest Observer office for making her an official part of these fantastic space missions.

RAS gratefully acknowledges NASA K2 Grant NNX15AC97G, and would like to thank the Kepler Mission Team for their outstanding work, and the K2C9 Microlensing Science Group for their contributions.

RAM would like to thank Douglas Gies, Steve Howell, and the NASA Postdoctoral Program for involving her in the groundbreaking work of Kepler. RAM also acknowledges the Kepler Eclipsing Binary Working Group for their efforts in detecting and cataloging EBs, as well as leading the analysis and interpretation of Kepler data.

Calen B. Henderson acknowledges the truly heroic efforts made by the K2 Guest Observer Office as well as the full staff of engineers involved for facilitating all aspects of the microlensing campaign, including the selection of the Microlensing Science Team, the late addition of targeted postage stamps for each campaign half, the spacecraft reorientation, and the impressive recovery from Emergency Mode. I also thank all ground-based observers who contributed their time and effort—Mike Lund, Masha Kleshcheva, Wei Zhu, Dun Wang, Rahul Patel, Savannah Jacklin, Joey Rodriguez, Rachel Street, Giuseppe D'Ago, and Virginie Batista in particular.

George H. Jacoby wishes to thank Orsola De Marco, Todd Hillwig, Alison Crocker, Josh Dey, Jonathan Hurowitz, and Anne Marie Cody for assistance with the science and data analysis discussed in his report.

SBH would like to thank all the co-authors for their tireless efforts in writing this volume.

There were hundreds of engineers, scientists, computer programmers, and administrators that together made the NASA Kepler mission a success, I wish to applaud them all. Sally Seebode supported me throughout the writing of this book even when I was grumpy, whining, or elated. I appreciated it always. I would like to thank Sarah Armstrong and Leigh Jenkins at IOP for their support of this book from the beginning and their help in making it become a reality. I had the pleasure of a conversation with David and Sarah Lardner, possibly long lost relatives of Dionysius himself. Their help in tracking down the Lardner portrait as well as additional related information was greatly appreciated. I want to thank Dawn Gelino for suggesting me and this topic at the start of the process. Finally, thanks to Mojgan Momeni Hassan-harati for help with some technical aspects of the book production.

Editor biography

Steve B Howell

Dr. Steve B. Howell is currently a senior research scientist at the NASA Ames Research Center in Mountain View, California. He was formerly the head of the Space Science and Astrobiology Division and the project scientist for NASA's premier exoplanet finding missions: Kepler and K2.

Steve received his PhD in Astrophysics from the University of Amsterdam and has worked in many aspects of astronomy including pioneering the use of charge-coupled devices (CCDs) in astronomy, building new technology instruments for ground and space-based telescopes, university teaching and research, and public outreach and education. Steve is also involved in informal and formal scientific education for kids to adults including his participation in SuperChefs, a non-profit organization teaching food science and healthy eating to young adults. Dr. Howell has written over 800 scientific publications, numerous popular and technical articles, and has authored and edited ten books on astronomy and astronomical instrumentation. Combining the latest scientific discoveries about exoplanets, fundamental physics principles, and modern food chemistry, Steve and collaborator Bill Yosses (former White House chef) have developed a series of educational and highly entertaining scientific and public outreach presentations on the physics of food and healthy eating as well as incorporating them into the classes at the Harvard cooking school and the Los Angeles Unified School District. In early 2020, Steve and Grammy award winning musician Henri Scars Struck debuted their exoplanet-related music installation, Beyond Me, in New York City. Steve has edited and written two science fiction books related to exoplanets and space travel. He lives in the San Francisco Bay area with his partner Sally and enjoys scientific challenges, the great outdoors, vegetarian gourmet cooking, and playing blues music. And yes, he still considers Pluto to be a planet.

List of Contributors

William Borucki
NASA Ames Research Center, CA, USA

Steve B. Howell
NASA Ames Research Center, CA, USA

John R. Troeltzsch
Ball Aerospace, Broomfield, CO, USA

Jessie Christiansen
NASA Exoplanet Archive, USA

Stephen Kane
University of California, Riverside, CA, USA

Susan Mullally
Space Telescope Science Institute, Baltimore, MD, USA

David Ciardi
NASA Exoplanet Archive, USA

Calen Henderson
NASA Exoplanet Archive, USA

Andrew Vanderburg
The University of Texas at Austin, Austin, TX, USA

Fergal Mullally
SETI Institute, Mountain View, CA, USA

Rachel Street
Las Cumbres Observatory, USA

William Chaplin
University of Birmingham, Edgbaston, Birmingham, UK

J. J. Hermes
Boston University, Boston, MA, USA

Ann Marie Cody
NASA Ames Research Center, CA, USA

Elliott Horch
Southern Connecticut State University, New Haven, CT, USA

Rachel Matson
US Naval Observatory, Washington, DC, USA

Steven Kawaler
Iowa State University, Ames, IA, USA

Csaba Kiss
Konkoly Observatory, Budapest, Hungary

Laszlo Molnar
Konkoly Observatory, Budapest, Hungary

Andras Pal
Konkoly Observatory, Budapest, Hungary

Krista Smith
Stanford University, Stanford, CA, USA

Armin Rest
Space Telescope Science Institute, Baltimore, MD, USA

Peter Garnavich
University of Notre Dame, Notre Dame, IN, USA

The NASA Kepler Mission

Steve B Howell

Chapter 1

Prelude

I was lucky enough to be the Kepler project scientist and then the K2 project scientist during the majority of the eight years of flight operation. The role of the project scientist is to be an interface between the mission team, NASA, and the scientific community as well as being the science leader and taking responsibility for producing the best and highest quality science from the mission and making it available to the community. Working together with project management, budget decisions, scientific direction, and requests and proposals for funding were also part of the job. The Kepler mission lasted for four years until hardware failures of two reaction wheels occurred. At that time, the K2 mission was born from the ashes, using non-standard, innovative, and never before attempted spacecraft operations. The story of the spacecraft, its mission and rebirth, and many science accomplishments are chronicled in this book. Many individual people were a part of this story from the early concepts to the final deliveries to NASA data archives. Many pages exist in the scientific and engineering literature, in past proposals, in software and hardware documentation, in notebooks and emails, and personal correspondence that trace the phenomenal work and years of effort that went in to the Kepler mission. It is impossible to cover all of the contributions and all of the science in any one volume, so realize, as you read through these pages, we present only the tip of the iceberg.

When I started to think about ultrahigh-precision photometry with the Kepler mission, about a decade or so prior to selection and even longer before launch, I never realized what a changeling the mission would be. I had been involved in a number of previous NASA missions and large-scale telescope projects, but none were as paradigm and life changing as Kepler. When first proposed in 1992, a mission named FRESIP (FRequency of Earth-Size Inner Planets), and in subsequent proposals to NASA's Discovery Program, including a name change to Kepler, the frequency of exoplanets, particularly small, Earth-size, potentially rocky orbs, was unknown. Proposals to fund and fly Kepler were often met with skepticism as to the scientific success as well as the technical ability. Eventually, the proposal submitted in 2000 won, and the mission was accepted for flight. Now the real work began. The next seven years were a whirlwind for the science team, many engineers, and software experts, as the spacecraft, telescope, and the all-important photometer were developed, built, and integrated, culminating in the 2009 March launch.

The Kepler mission extends far beyond scientific discovery. It reaches deep within our souls as humans and brings to mind thoughts of alien life. Are we alone? What is our place in the Universe? While Earth-size planets were discovered, some residing in the habitable zones of stars, none of them are worlds that we will be traveling to and living upon any time soon. A realization that the Earth is a special place and we need to take care of it is a humbling yet powerful message. The impact of Kepler and its scientific search to find planets orbiting other stars was quickly summed up by us Earthlings as the "Search for Life." This immediately brought religious, philosophical, and human feelings to the fore and I too was amazed and privileged to interact in many ways with members of the world community on such topics.

Planets orbiting stars other than our Sun, exoplanets, is not a new idea or even a new scientific pursuit. Sometime in 460–370 BC, Democritus pined, "There are innumerable worlds of different sizes. Some of the worlds have no animal or vegetable life nor any water." Bruno, around 1598 proclaimed, "The universe contains an infinite number of inhabited worlds populated by other intelligent beings." He was burned at the stake for his beliefs. Fortunately, we exoplanet scientists are not as threatened today. Astronomy texts written in the 19th Century had great public appeal. The details of the heavens and some knowledge of what it means has always captured our imagination as was particularly true of Lardner's *Handbook of Natural Philosophy and Astronomy* published in 1858 (Figure 1.1). In a section describing why stars vary, Lardner writes, "Periodical dimming or total disappearance of the star, may arise from transits of the star by its attendant planets." But he goes further, perhaps thinking of Bruno, and says "These various conjectures are merely conjectures, scarcely deserving the name of hypothesis or theory." Given this popular opinion of the time that other planets might be mere conjecture and not really exist, exoplanets remained in shadow for the next ten decades or so, finally getting a reality check in 1992, when radio astronomers Aleksander Wolszczan and Dale Frail announced the discovery of two planets orbiting the pulsar PSR 1257+12.

The Kepler and K2 missions were filled with technological and scientific advancements and discoveries, many of which will be highlighted through this volume. The two figures below (Figures 1.2 and 1.3) present artist renderings of early concepts for the spacecraft and telescope. We will read about the engineering that went into the creation of the most precise imaging camera ever made, the

Figure 1.1. Dionysius Lardner (1793–1859) found his father's law practice distasteful and instead dedicated himself to a life of scientific discovery. In the mid-19th Century, he postulated exoplanets as a possible cause of stellar variability. Credit: Dionysius Lardner. Lithograph by D. Maclise [A. Croquis], 1832. Credit: Wellcome Collection. Attribution 4.0 International (CC BY 4.0).

amazing story of spacecraft failure and its Phoenix-like renaissance. Exoplanet discoveries by the thousands, alien solar systems to boggle the mind, and planets that even theorists could not dream up. The story does not end there however. The astrophysical discoveries of Kepler and K2 may be even more astounding. Unique observations of stars, solar systems objects, and far away galaxies round out the paradigm changing scientific discoveries of these two NASA missions. Due to its ecliptic viewing, specific K2 observational campaigns were developed for micro-lensing and supernovae surveys. Along the way, social and science community changes occurred as well. The fully open data policies and world-wide community involvement led to a boon in science results and publications as well as new, creative methods of collaboration, promoting similar expectations for future space missions.

This volume is truly a work of insight and retrospection. The writers were all an integral part of the Kepler and K2 missions, spending years of their careers and lives, living and breathing Kepler. You could not ask for a more learned and informative group of essayists. I am deeply indebted to them all for their

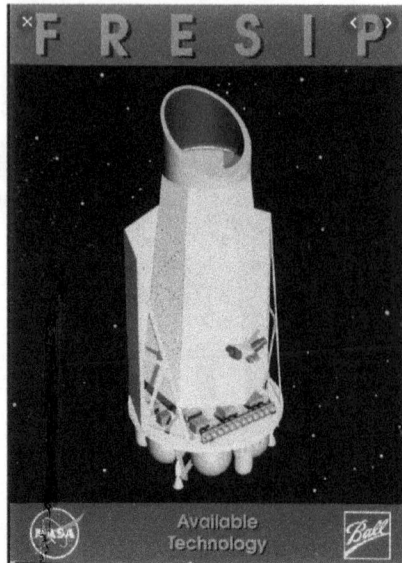

Figure 1.2. A early artist concept of the FRESIP spacecraft. Credit: NASA/Dominic Hart.

Figure 1.3. Artist view of the Kepler spacecraft showing the articulated antenna for data download to Earth. This antenna was later traded for a less costly body-mounted antenna giving rise to the need for brief stopping of data collection to allow spacecraft maneuvers to point the fixed antenna toward Earth. Credit: Kepler Telescope Project/NASA.

agreement to help forge this volume and to their dedication to its completion. Hundreds of additional engineers, scientists, and software experts played equally important parts in the creation, formation, development, and operation of the Kepler spacecraft and the Kepler and K2 missions. There is nothing as far reaching and as life-changing as being a part of a space mission and of all that I have been privileged to be a part of. This mission stands alone for its scientific achievement and wide-ranging influence on humanity.

Steve B. Howell, Project Scientist, Kepler and K2 missions
Redwood City, CA
2020 January 16

Chapter 2

The NASA Kepler and K2 Missions

John Troeltzsch and Steve B Howell

2.1 Introduction

Are we alone? This is likely humankind's greatest question. An answer has been sought by us Earthlings since the beginning of time using many forms of inquiry such as spiritualism, mysticism, mathematics, and looking toward the night sky. Astronomers have taken their own path, that of observational searches throughout the Universe aimed at seeking signs of intelligent life among the stars.

Arguably the most impactful human experiment to search for alien worlds orbiting other suns, planets possessing similar traits to our own Earth including the potential to harbor life, has been the NASA Kepler and K2 missions. No other scientific experiment, space telescope, or in-depth study has been capable of finding and characterizing planets outside of our solar system that are potentially similar in size, rocky composition, water content, and surface temperature as our Earth.

Additionally, the Kepler and K2 missions have changed the scientific horizon in many areas of stellar astrophysics, time-series observations, planetary science, computational analysis, solar system science, and extragalactic systems. These topics will all be expounded upon in the following pages. To read many press results, view many images, and see videos made specifically during and for the Kepler and K2 missions, visit the official NASA Kepler website.[1]

2.2 Kepler and K2 Mission Overview

The Kepler Space telescope was launched on 2009 March 6 but the story of this NASA Discovery Mission starts long before that. In 1983, Bill Borucki, working at the NASA Ames Research Center, began to develop a mission concept and hone photometric techniques that would be precise enough to detect the minute drop in light from a far away star caused when a planet the size of Earth crossed or transited

---

[1] https://www.nasa.gov/mission_pages/kepler/main/index.html.

in front of it. Bill was a pioneer in this area and thought, by many, to be performing work that would never come to fruition. After a number of attempts proposing to NASA that including many changes in the mission time line, concept, and goals, the Kepler mission was finally approved in 2001 as NASA's 10th Discovery mission. The long saga of the Kepler mission, including eight years of proposal submission before acceptance, is told in detail by Bill himself in Borucki (2016).

Kepler was a very simple experiment—it had one instrument and performed one type of science, high-precision photometry over a wide field of view. The focal plane of the photometer was the largest ever flown in space, so large in fact that data collected on the 10 by 10 degree imager could not all be transmitted back to Earth. Only about 1% of the total collected image could be routinely returned to Earth, yielding 9 GB of science data per month. These data consisted of observations obtained every 30 minutes for approximately 170,000 stars in the field of view and, for a smaller number (512 stars), 1 minute samples.[2] Measurements of the light output of a star over time are called time-series observations or light-curves and Kepler provided scientists with the longest, best sampled, highest precision data sets even obtained. These were the science products that changed the world.

Kepler's time-series light-curves present relative photometry of extreme high precision, better precision than astronomers had ever obtained before. Figure 2.1 illustrates how Kepler detected exoplanet transits. The Kepler mission was designed to perform this one very simple, yet incredibly powerful type of observation. The mission goal was well established and very specific—determine the frequency of Earth-size and larger planets orbiting in the habitable zone of stars similar to our Sun. The information obtained for each detected planet consisted of its radius, orbital period, and, by using Kepler's laws of planetary motion, the distance at which the planet orbits its host star. Since transits only occur if the planets' orbit is aligned along the line of sight of the observer (spacecraft), the small probability of such a random occurrence led to the need to monitor at least 100,000 stars at once.

These and other requirements produced a final design of the Kepler space telescope having a wide field of view, a Schmidt telescope design, no shutter or filters, a large focal plane array of charge-coupled devices (CCDs) detectors, and a heliocentric orbit. The mission was slated to have a four year lifetime. The photometer consisted of 21 2048 × 2048 custom made CCDs able to image 105 deg^2 of sky at once in the optical bandpass of 400–850 nm (see Borucki 2016).

The selection of the Kepler field of view, Figure 2.2, was based on the need to observe a large number of stars simultaneously, but not too many so as to be crowded and blended together. Additional constraints required that the field of view be above 55° of ecliptic latitude (to avoid sunlight scattering into the telescope) and the need to avoid very bright stars (which would cause scattered light across the focal plane leading to a high background and large saturated areas on the CCDs).

---

[2] The CCD arrays were actually read out every 7 s and co-added on-board the spacecraft to produce the 1 minute and 30 minute final postage stamp images, which were then sent back to Earth.

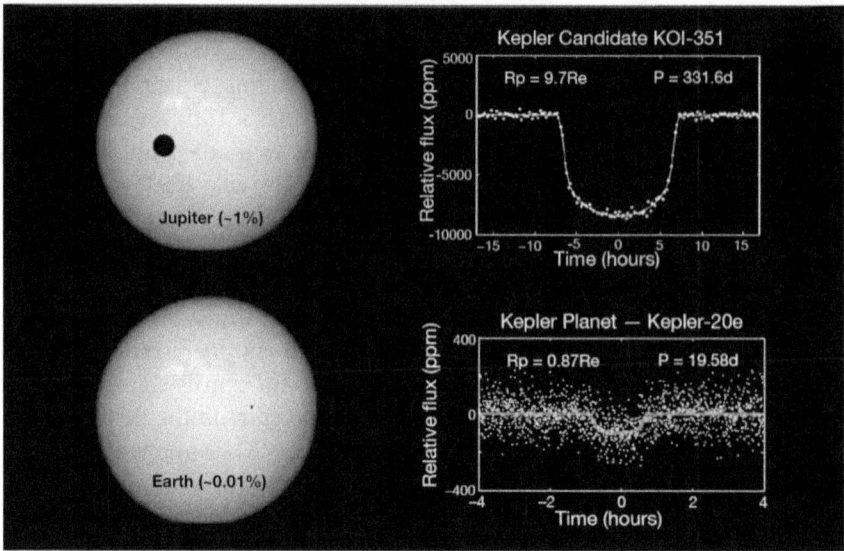

Figure 2.1. Kepler observations of exoplanet transits. On the left are shown artist concepts of what the planets Jupiter and the Earth would look like if we could observe them transiting our Sun. The Earth covers an area of only 0.01% that of the Sun. The right-hand panels show actual Kepler transit measurements for a Jupiter and an Earth-size planet. The planet's radius and orbital period are given. Note that the Jupiter-size planet's transit dip, near 1% of the starlight, is easily detected while that of the Earth-sized body is difficult, requiring multiple transits to be observed and data binned (red and orange) to aid the eye and reveal the transit. Credit: Howell/Stencil/NASA.

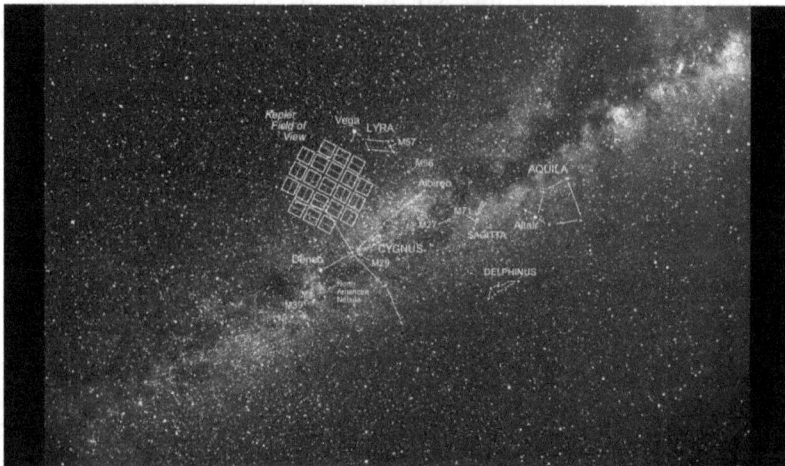

Figure 2.2. The Kepler mission field of view located primarily in the constellation of Cygnus, the Swan. The field offered the right star density and stellar type as well as accessibility for follow-up observations from the ground. Credit: Carter Roberts.

Ground-based telescopes, needed for confirmation and characterization via follow-up observations, were more plentiful and available in the northern hemisphere as well as the fact that the Cygnus field offered many Sun-like stars.

On 2009 April 8, Kepler recorded its first-light image (Figure 2.3). The image shows the light collected by each of the 42 CCDs in the focal plane as well as the rotational symmetry inherent in the imager design. The gaps between CCDs are large enough to contain an image of the full Moon—thus giving an idea of the very large amount of sky observed by Kepler. The image shows the variation in star density across the field of view, the lower left being closest to the Galactic plane.

The Kepler mission was designed and operated primarily for its goal of exoplanet discovery and characterization. From the beginning, a robust guest observer program was also in place, allowing astronomers from around the world to propose for and obtain high precision light-curves of a wide variety of sources. This philosophy greatly extended the science output of the mission as well as providing paradigm changing results across all of astrophysics. As we will see throughout this book, especially for the K2 mission, the world-wide community rallied to propose, obtain, and use the high precision Kepler spacecraft results to attack many outstanding astrophysical, planetary science, and extragalactic questions as well as provide new, exciting never before seen results.

After ∼3.5 years of operation, the Kepler spacecraft suffered failures of two of its four reaction wheels, hardware critical to the pointing ability of the space telescope.

Figure 2.3. First-light image obtained in 2009 April from the Kepler imager. The image shows the 42 individual CCDs and the varying star density across the field of view. Three open clusters are also visible in the image. Credit: NASA/Ames/J. Jenkins.

In late 2013, innovative work by the engineers at Ball Aerospace and scientists at NASA Ames, as well as interested members within the community, developed and demonstrated a novel, unique, and fascinating new mission design. This led to the NASA K2 mission[3] (Howell et al. 2014), which surveyed the ecliptic for an additional four years until running out of fuel in 2019 September. K2 changed the scientific paradigm again making observations of a variety of sources not available to the original Kepler mission, the majority of which were proposed by the world-wide scientific community.

Unlike the Kepler mission and due to its unique operation mode, K2 was constrained to observe a continuous set of approximately 80 day pointings along the ecliptic or zodiac in the sky. Figure 2.4 illustrates such fields of view, ever changing around the sky as the spacecraft orbited the Sun. Using the pressure of Sun light on the solar panels,[4] a pressure equal to the weight of two mosquitoes landing on a square meter size trampoline, the Kepler spacecraft could balance the momentum

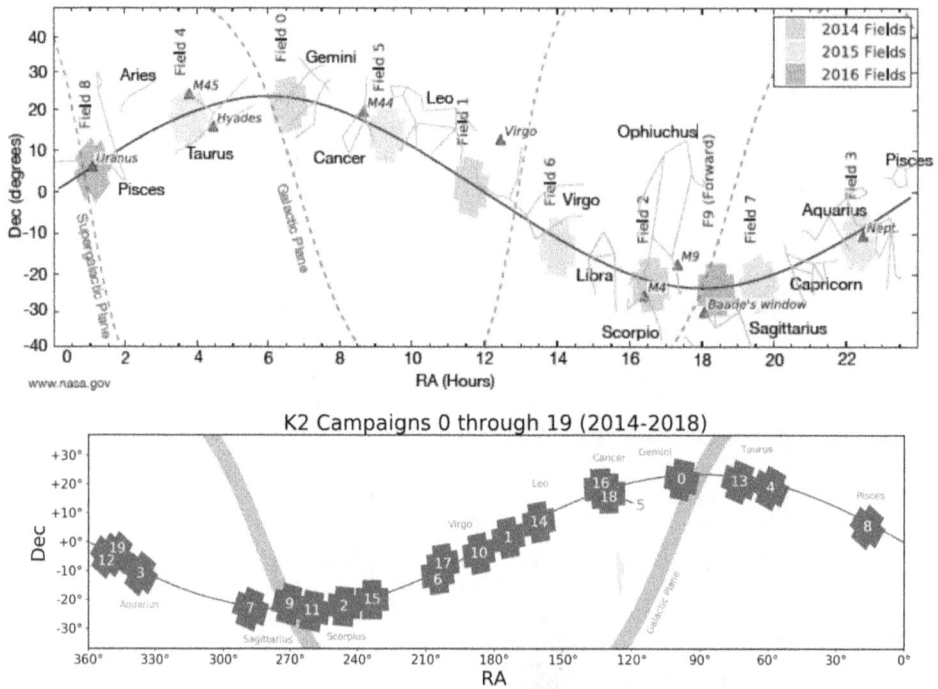

Figure 2.4. K2 fields of view along the zodiac or ecliptic. Top: Campaigns to be performed in the first two years of K2 operation. The constellations along the zodiac are shown as well as the location of the galactic and supergalactic planes. Bottom: Locations of the 19 separate K2 pointing fields along the ecliptic as observed during its four year mission. Credit: Howell/Mullally/NASA.

---

[3] A clever name hinting at the use of only 2 reaction wheels to operate and point the spacecraft or perhaps the second Kepler mission, but in fact, named after the second highest but harder and more interesting to climb mountain peak, K2.

[4] Note, not the solar wind, but actual pressure exerted by solar photons.

available from the two remaining working reaction wheels and accurately point the telescope on the sky. K2 charted new territory in how to operate a spacecraft and in its scientific discoveries.

The K2 mission was born of necessity and was the first and likely will be the only, NASA astrophysics mission proposed and accepted which, at the time, was not 100% functional nor did most of its science goals have proof they could be accomplished. However, the promise of its unique and novel operation, the legacy of what it would discover based on the Kepler mission results, and the wide variety and unprecedented new science it could deliver were seen as so important and new, K2 was an easy sell and an immediate success (Howell et al. 2014). K2 had key science goals of exoplanet discovery, focusing on two new areas: nearby low-mass, faint red stars and close bright stars, areas of discovery space that were not available to Kepler. Additionally, K2 would take advantage of its ecliptic pointing to observe numerous bright open star clusters ranging in age from 2 to over 4000 million years, star-forming regions toward the galactic center, red giant stars across the Galaxy, extragalactic and solar system objects as well as provide the first dedicated space-based campaigns aimed at observations of supernovae and microlensing events.

Overall, K2 observations reached nearly the same fantastic photometric precision as Kepler did and covered 19 separate sky pointings (campaigns) throughout the Ecliptic during its four year mission. The scientific results obtained using the data collected by the NASA Kepler and K2 missions, have had a profound and enormous impact on many areas of astrophysics. These results span nearly every area of observational science and will be highlighted throughout this book within the following chapters. But first, let us take a look at the building and operation of the Kepler spacecraft.

2.3 The Kepler Spacecraft

2.3.1 Introduction

The Kepler spacecraft was built and assembled at a Ball Aerospace facility in Boulder, Colorado. Figure 2.5 shows the fully assembled spacecraft awaiting delivery to Cape Kennedy. The development and construction of the spacecraft involved over 2,000 Ball Aerospace employees working more than 1,350,000 hours over 7 years. Kepler also required the dedication and support of employees from more than 40 companies throughout the US and UK that provided components and subsystems for the mission.

NASA Discovery missions are critical to its portfolio because they tend to return great science for a relatively small investment, but these programs also face significant challenges due to the way they are selected and often funded. Teams proposing a Discovery mission must maximize the science return, minimize the technical risk, and fit within a rigid cost cap that is based more on historical cost models than current market pricing. As a Discovery mission, the Kepler program faced and met these challenges, but it was never easy.

Kepler is unique among NASA missions due to its singular science focus and the need to stare at the same stars for many years. Determining the statistical prevalence

Figure 2.5. Kepler spacecraft fully assembled in high bay at Ball Aerospace. The orange material is thermal blanketing and note the cleanroom-suited workers as well as the telescope cover in place. Credit: NASA/JPL/ Ball Aerospace.

of Earth-size planets requires obtaining photometric data from more than 100,000 stars. In order to make these measurements, Kepler would need extreme stability over a 100 deg^2 field of view. This requirement led to the need to launch a large space telescope and minimize external pointing and thermal disturbances to previously unproven levels. These challenges were met by using a heliocentric orbit and by minimizing any changes to the temperature of the photometer due to re-pointing of the spacecraft. The Earth trailing, heliocentric orbit chosen for Kepler was first pioneered by the Spitzer Space Telescope and includes the fundamental challenges of increasing distance of the spacecraft from Earth and the need for autonomy to keep the observatory safe during long unattended periods of operation.

Figure 2.6 presents a schematic view and identifies the major components of the observatory. There are two main elements that make up the flight segment or observatory. The right-hand side of the figure shows the photometer element, which consists of a 1 m class telescope with a 96 megapixel camera located in the center of the telescope barrel. The remainder of the figure shows the spacecraft element and its large solar panel, which serves to both provide electrical power and to shade the telescope from the Sun.

Much has been written about the early development of the Kepler concept from a revolutionary idea through the NASA mission selection process. This section of Kepler's story starts with the formal development of the spacecraft and its systems.

2.3.2 Preliminary Design and Systems Engineering

When the Kepler team set out to design and build the spacecraft in 2002 October, there were no existing blueprints. NASA and Ball had extensive experience building

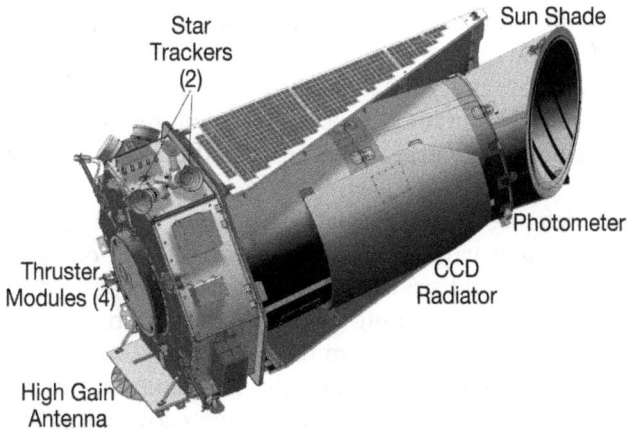

Figure 2.6. Major components identified for the Kepler spacecraft. Credit: NASA/JPL-Caltech.

and operating deep space missions but no one had ever tried to build a deep space photometer to measure the minute signals of an exoplanet transit. The team's job was to build a one of a kind machine for finding exoplanets and we started from scratch.

The bedrock of the Kepler Spacecraft is systems engineering. The systems engineering effort began before NASA selected the mission and completed 60 days after launch when the spacecraft was commissioned and shown to operate to specification. The process started by accurately and completely specifying the high-level science requirements needed to accomplish the mission. These requirements were dissected, analyzed, and propagated through many lower levels to define specific, key parameters and requirements for everything from mechanical tolerances between the photometer and the spacecraft to how much radio frequency interference the detector electronics could tolerate and still measure a transit. Through the accurate and succinct definition of a full set of Kepler requirements, the systems engineering team would pave the way for all the work to come.

The driving requirements for the spacecraft were as follows:

1. Compatible with a Delta 2 launch vehicle.
2. Operate in a heliocentric orbit.
3. Acquire photometric data throughout the operational lifetime through 100,000 apertures.
4. Continuously observe the same portion of the celestial sphere.
5. Operate with the optical axis pointed at the scientifically specified Kepler field.
6. Use Ka and X band radio frequencies to communicate with a 34 m Deep Space Network Antenna on Earth.
7. Have no credible single point of failure that could cause loss of mission.
8. Focus the photometer to provide a point-spread function adequate for the target photometric aperture.
9. Command CCD integration times from 2.5 to 8 s.

10. Provide fault protection to ensure Kepler can attain a safe state upon detection of a fault.
11. Provide adequate margin against predicted performance for mass, power, battery storage, computer processing, photometric precision, and data storage including any degradation expected during the mission.

In parallel with defining the requirements, a team of expert analysts modeled and predicted how the system would perform through the various environments and operating conditions it would see during the mission. These models covered every possible aspect of the system including predicting the structural strength and modal frequencies of the structure to make sure the spacecraft would survive launch. This covered everything from the bolts holding the spacecraft together to the adhesive holding the Schmidt corrector in place. Other models covered the rate of molecular evaporation in vacuum from composite structures, adhesives, and lubricants that could affect the optical throughput of the telescope or the shape and length of the optical truss. Software and firmware requirements were modeled to verify that the various computer processors and electronics circuits in the spacecraft could adequately execute the algorithms needed to operate the spacecraft. Models were established to cover the effects of radiation on electrical parts accounting for the suns solar cycle and the thermal performance degradation of the paint and insulting blankets on the outside of the spacecraft that would be exposed to the solar wind. The analysts and systems engineers allocated budgets and margins to be measured and tracked during the fabrication and assembly phase to ensure the assembled spacecraft would be able to meet all requirements and detect exoplanets.

Once the requirements and models were in place, the systems engineers defined a verification method for every requirement. Requirements were to be verified by test, analysis, demonstration, or similarity to document adherence to specifications and performance. Constant adjustments were made to move surplus margin from one area into areas where the design was excessively challenging. Tests were planned and designed to make sure the desired verification measurement could be made and what requirements had to be placed on the test equipment. Setting the requirements and then defining the criteria for successful verification is a key aspect in avoiding the desire to take short cuts later when time and money start to run out.

The hard work of the systems engineers enabled the leaders of the various subsystems to start defining which elements of the system could be purchased from other companies and which had to be designed and built by Ball. This, in turn, provided the schedule information needed to prioritize the tasks that would take the longest and thus had to be started early. At the top of the list were the many, many CCDs that would be needed to build a 96 megapixel camera and the large primary mirror and Schmidt corrector.

The preliminary design phase of the program culminated in a preliminary design review (PDR) in 2004 October. At this point an independent NASA review board provided their assessment, recommendations for moving forward, and a green light to proceed. Following the review the momentum of the team significantly increased as staff was added and the team worked to place 41 major orders with other companies for hardware needed to build Kepler.

In 2005 February NASA announced a serious agency budget issue due to challenges on another Discovery mission and cut the Kepler FY05 funding by 35M dollars. This occurred just as Ball completed its major push to ramp up the program. Although the funding cut came with a revised cost and a new 2008 June launch date, the loss of momentum and disruption of the programmatic plan significantly impacted the program. The decision was made to focus all the available resources on the more difficult photometer element and halt work on the other spacecraft subsystems. On the positive side, this allowed the team to focus on the challenges of the photometer and solid progress was made on CCD detector production.

2.3.3 Detailed Design and Reliability

All large space missions have a major design review, the critical design review (CDR), to establish the detailed design is complete and mature enough to move into major production. The period leading up to the Kepler CDR required hundreds of engineers, all working in unison, to complete individual designs, but more importantly, to verify that their design would properly connect to the adjacent design, and the adjacent, adjacent design .... The importance of all of these interfaces is paramount to successful integration of the system once all the hardware and software is complete. The role of the systems engineers and the subsystem leaders during this effort is to document and triple check that nothing is missed. This involves verifying the bolt patterns of an electronic box agree with those on the support panel where it will mount. Checking that each wire in an electrical connector and its associated harness is defined, its route specified, and that it matches the destination connector on the next electronics box. The effects of the heat generated by an electrical box on an adjacent optical mount must be considered. Detailed plans are established for an orderly and safe assembly of the spacecraft. All these details are documented and reviewed during the CDR before NASA will authorize a project to move forward. The Kepler program had 4089 plans, drawings, and procedures for building the spacecraft.

Kepler, like most space missions, required very high reliability. Once a spacecraft is launched from Earth into an orbit around the Sun there is currently no way to physically fix it. A space astronomical observatory must perform at a level of reliability much higher than an equivalent observatory on Earth. This reliability is provided by space qualifying each part of the spacecraft. Sometimes this is done by using a previously qualified parts other times a part must be physically stressed by the expected environment (vibration, thermal, vacuum, radiation) during development to establish the qualification. Each electrical part on the spacecraft has an allowable temperature range that must be checked, and a worst-case analysis is performed on all electronics circuits to verify the circuit will work properly after aging during its life in space. The CDR covered the Kepler design but as importantly it covered the predicted reliability of the design.

Another element of high reliability is ensuring the high-quality workmanship. Every fastener on the spacecraft has a required torque specified to make sure it is tight, and many fasteners receive a drop of adhesive to make sure they don't shake lose during the vibration of launch. Constant care is required to prevent a dirty glove from contaminating an optic or to keep static electricity from damaging a sensitive electrical component. The rules governing how the spacecraft will be assembled and the quality checks to verify the rules are followed are all written down in the detailed design drawings. The Kepler mission assurance team worked side by side with the engineers and technicians designing and building the observatory. They provided an independent check on every task to make sure the hardware was properly designed and assembled and kept safe at all times.

Thermal Design

Contributed by Denny Teusch, Thermal Engineer, Ball Aerospace.

The thermal subsystem is commonly referred to as the thermal control subsystem (TCS). The purpose of the Kepler TCS is to provide thermal control of both the photometer and spacecraft. In addition, the TCS provides monitoring sensors to access the thermal health and status of Kepler.

The TCS is responsible for maintaining Kepler component temperatures within allowable flight temperature (AFT) limits, minimizing rate-of-change of photometer temperatures, managing temperature gradients across and through optics, maintaining stable CCD temperatures during operation, providing heating of all components during lower power states, providing heating for photometer annealing, and providing temperature sensors for sensing of spacecraft and photometer critical temperatures.

The Kepler system has two main thermal control areas; the photometer and the spacecraft. The photometer must be kept much colder than the spacecraft and it must also have very small temperature fluctuations during the mission. The spacecraft temperatures are maintained much closer to nominal ambient conditions. Figure 2.7 provides an overview of the techniques used to maintain and control the temperature of the photometer. The photometer is mounted to the spacecraft with titanium flexures to minimize heat transfer and the solar array and thermal blankets shield the photometer from direct solar heating. This isolation from the spacecraft and the photometer's view to cold deep space maintains the photometer at cold stable temperatures. Heat that is generated by the focal plane array (FPA—the CCD imaging camera) inside the photometer is transferred to the external FPA radiator via heat pipes. Another set of heat pipes is situated immediately behind the focal plane to minimize temperature gradients. The camera electronics box walls radiate the electronics dissipated heat within the cold telescope cavity, which maintains the telescope and optics temperatures at working temperature levels.

Heat that is generated by spacecraft electrical boxes serves to maintain the electronics at flight-allowable temperatures while the bus and exposed electronics box surfaces serve as radiators of excess heat. Redundant heaters and heater controllers are used to maintain survival temperatures in the event of a failure.

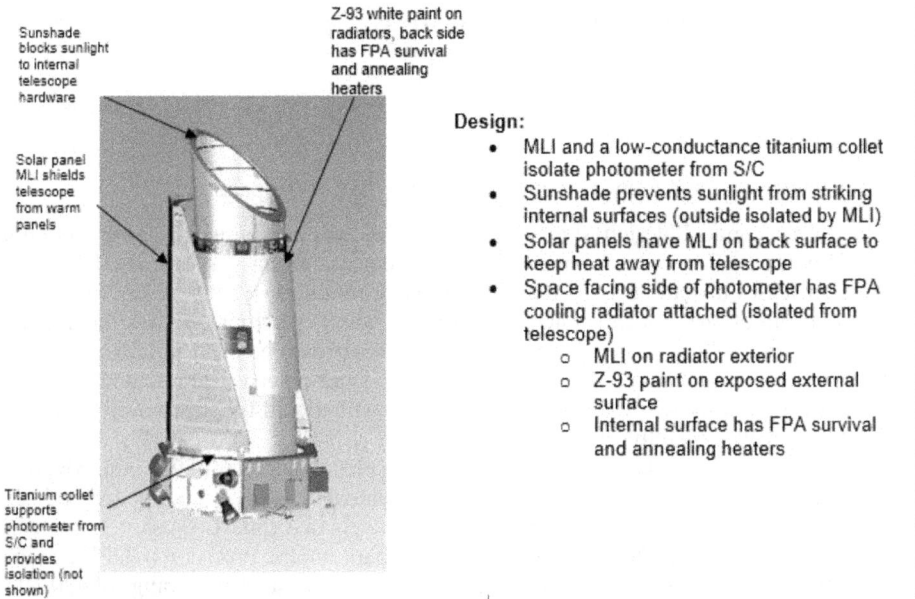

Sunshade blocks sunlight to internal telescope hardware

Z-93 white paint on radiators, back side has FPA survival and annealing heaters

Solar panel MLI shields telescope from warm panels

Titanium collet supports photometer from S/C and provides isolation (not shown)

Design:
- MLI and a low-conductance titanium collet isolate photometer from S/C
- Sunshade prevents sunlight from striking internal surfaces (outside isolated by MLI)
- Solar panels have MLI on back surface to keep heat away from telescope
- Space facing side of photometer has FPA cooling radiator attached (isolated from telescope)
  - MLI on radiator exterior
  - Z-93 paint on exposed external surface
  - Internal surface has FPA survival and annealing heaters

Figure 2.7. Kepler's outer surfaces are designed to protect against the extremes of direct sunlight and deep space. Figure courtesy of Ball Aerospace.

Two types of planned maneuvers affect the spacecraft thermal state: (1) Kepler rotates 90° every 3 months to keep the solar arrays pointed to the Sun and the radiator and photometer's boresight pointed away from the Sun and; (2) every 30 days, the spacecraft points the antenna toward Earth for approximately 24 hours for science downlink. In both of these cases, a period of thermal settling will be required before the spacecraft can re-establish fine-point and commence re-taking science data. This settling time is needed for the star trackers to stabilize adequately for correct pointing.

In 2006 January, the team was deep into preparations for CDR. In parallel with the technical work, a current cost and schedule analysis was performed including past and future expenditures. This detailed programmatic baseline review determined some serious cost concerns and the team was forced to focus on how to revise the plan to meet the technical requirements and the cost limitations. A number of trade offs and descopes were identified and implemented in order to get the program to fit within the budget constraints. Ultimately a program rebaseline was required which involved additional funding and a launch delay to 2009 February. Kepler passed its CDR in 2006 October. The program finished out 2006 making good technical progress manufacturing the observatory but the team had spent more than one-half the year in reviews and replans, which seriously eroded overall efficiency and morale.

2.3.4 Photometer Optics and Focal Plane Array

Optical Design

Contributed by Chris Stewart, Optical Engineer, Ball Aerospace.

The Kepler optical design shown in Figure 2.8 was an on-axis, classic Schmidt telescope consisting of a Schmidt corrector element, primary mirror, and 25 field flattener lenses including the fine guidance sensor modules (FGSs). A key functional requirement of the optical system was that it was a photometer, not an imager, spreading the stellar energy over several pixels to improve the differential photometry precision by making the system less sensitive to inter-pixel variations and pointing jitter. An optical system that provided uniform point-spread functions over the entire field of view was desirable.

The telescope had a large collection entrance aperture diameter of 950 mm, covering a large 13.9° field of view to meet the requirement to provide >100 deg^2 of sky coverage, making it the largest Schmidt Telescope in orbit. Low shot noise was necessary to achieving the combined differential photometric precision (CDPP), so the spectral bandpass was as broad as possible and covered a wavelength range of 420–915 nm. The short wavelength cutoff was chosen to avoid CaII and H&K lines below 400 nm, which are known to be the most variable portion of the solar spectrum. The resulting F-number for the system was 1.473 on-axis with an effective focal length of 1400 mm. The telescope fed 42 1044 × 2200 pixel science CCDs and four 512 × 512 pixel FGS CCDs over a curved focal plane array made up of 27 micron pixels and 13 micron pixels respectively.

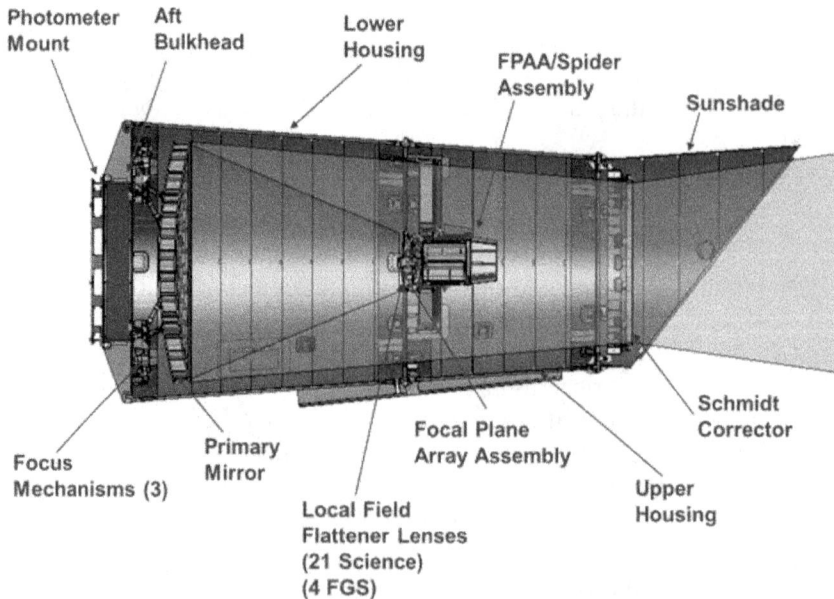

Figure 2.8. Cutaway view of the Kepler optical elements. Figure courtesy of Ball Aerospace.

The aspheric Schmidt corrector was manufactured from Fused Silica 7980 due to its high radiation resistance, low coefficient of thermal expansion (CTE), and ability to be fabricated at the 1 m diameter required. The primary mirror was manufactured from ULE Corning Titanium Silicate 7972 due to the low CTE required to cover the operational temperature range of 222.5–255.3 K, which precluded the use of light-weight metal mirrors. The primary mirror substrate also utilized a flight qualified FRIT bonded mirror blank construction to yield the desired 88% lightweighting for this 1.45 m diameter optic. The local field flattener lenses, which provided local flat field imaging over each pair of science CCDs, were fabricated out of sapphire where the high density of this material provided efficient radiation shielding for each CCD in the required 7 krad (Si) radiation environment.

The optics were maintained in alignment with a graphite cyanate-ester structure that provided the thermal-mechanical stability required between the Schmidt corrector, primary mirror, and focal plane assembly. The mirror focus mechanisms and the redundant, one-time deployed aperture cover release mechanism were the only mechanisms on the photometer.

Meter class plus optical components are generally high-risk, long-lead items so procuring them early in the program lifecycle was essential to receiving these critically important components in time to support successful integration/alignment of the photometer system. Detailed technical and programmatic oversight of the suppliers for the key photometer components, including the optics, structure, and mechanisms, was required to successfully fabricate, test, and deliver these critical components for integration. A number of technical and programmatic issues were encountered during the fabrication of these components/subassemblies, resulting in cost and schedule impacts to the program, but all issues and challenges were successfully navigated and resolved by the team to yield a photometer system that met the demanding requirements of the mission.

Integration and test of a unique sensor the size of Kepler required the development of specialized opto-mechanical ground support equipment (GSE) and specialized test equipment (STE). A specialized GSE was produced to support integration/bonding of the 1 m diameter Schmidt corrector assembly. A specialized GSE was also produced to support integration/alignment of the primary mirror assembly onto the aft-bulkhead assembly, and then through optical wavefront testing at ambient and operational cryogenic temperatures during thermal vac testing. This specialized GSE supported key performance testing before and after vibration testing to verify subassembly performance after exposure to launch levels. STE was developed to support dust cover latch mechanism testing to minimize the effects of gravity during ground testing, and verify focus mechanism assembly performance.

For integration and build-up of the photometer system, a custom multi-story alignment tower was built to support metrology methods used to integrate/align the primary mirror assembly, photometer structure, Schmidt corrector assembly, and focal plane array assembly. At the integrated photometer level, generalized vertical collimator assembly (VCA) STE was designed and built to support Kepler performance testing at ambient temperature and pressure and at flight cryogenic temperatures and pressure during thermal vacuum testing. The VCA was designed to support Kepler testing over the full field of view, and also designed to support testing of future large aperture systems up to approximately 1.3 m in diameter.

In the spring of 2007 the program was faced with programmatic challenges getting the optics coated and technical issues completing the manufacturing of the focal plane assembly. The optical coating issues were driven by delays due to US government military programs at the chosen mirror coating supplier, which resulted in Ball having to switch coating vendors to maintain the launch window. This change required quickly finding two new companies (one for the primary mirror and one for the Schmidt corrector) that had the skills and facilities needed to safely and accurately coat 1 m class optics. Ball then worked with these companies to develop the procedures and handling equipment needed to unload the optics, prepare them for coating, install them into large coating chambers, and deposit the required coating. This effort resulted in successfully coating the optics and delivering them to Ball in just 8 months to maintain the schedule for launch. Figure 2.9 shows the coated primary mirror and Figure 2.10 shows the completed Schmidt corrector lens.

The manufacture and coating of large optics is well established for ground-based observatories, but Kepler's optics required unique designs to survive the violent ride to space and the extreme temperatures the photometer would experience on-orbit. In addition, the difference in the coefficient of thermal expansion between glass, the flexure, and the composite structure holding the optic would result in breaking the glass if not accommodated. The solution was to precisely glue mounting pads to the optic and then connect the pads to carefully designed flexures, which allow movement during launch but return to precise locations after launch. Figure 2.11 shows a strut and pad on the bottom of the primary mirror. Figure 2.12 illustrates the complete suspension system for the primary mirror.

Figure 2.13 shows bond pads mounted to the Schmidt lens. These pads attach to 10 flexures and a composite mounting ring to form the complete Schmidt corrector

Figure 2.9. The coated primary mirror is inspected and cleaned following deposition of the enhanced silver coating. Photo courtesy of Ball Aerospace.

Figure 2.10. Inspection of the figured Schmidt corrector plate before installation of the bond pads. Photo courtesy of Ball Aerospace.

Figure 2.11. The glass to metal bonding of the primary mirror to the strut is characteristic of optical bonding for space missions. Photo courtesy of Ball Aerospace.

Figure 2.12. The primary mirror is suspended on six struts to isolate the optic from the spacecraft structure during launch. Photo courtesy of Ball Aerospace.

Figure 2.13. Ten titanium bond pads are bonded to the glass lens to allow attachment to the photometer structure. Photo courtesy of Ball Aerospace.

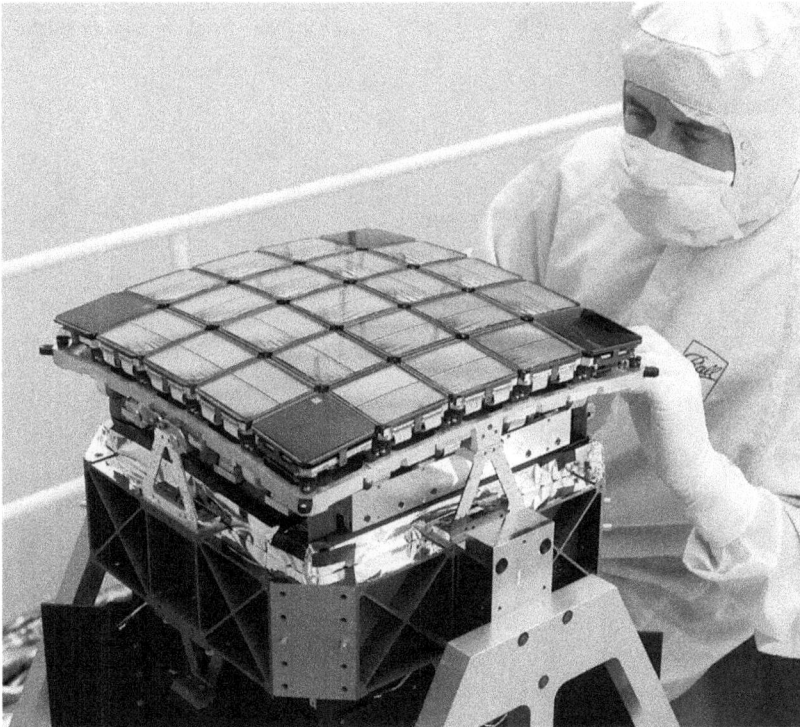

Figure 2.14. The Kepler camera contains >22,000 individual parts. Credit: NASA Ames and Ball Aerospace.

assembly. In all cases great care was required to clean the surface and establish the correct thickness of adhesive to obtain the required strength.

The primary challenges faced by the focal plane assembly were the result of underestimating the complexity of building, testing, and integrating 42 CCDs and 10 electronics boards into a 1.5 cubic foot assembly (Figure 2.14). There were few changes to the design as work progressed but there were a number of physical issues related to the detectors and field flattener lenses that had to be resolved and the time required to complete the assembly was underestimated thus driving up the cost. The Kepler focal plane assembly is the largest, most complex, and expensive camera ever launched into interplanetary space and was far more challenging than had been assumed in the original proposal.

Focal Plane Assembly

Contributed by Chris Miller, Focal Plane Engineer, Ball Aerospace.

The Kepler focal plane array assembly (FPAA) is housed within the photometer telescope, and consists of the four items shown in Figure 2.15: the focal plane array (FPA), the local detector electronics (LDE), the spider assembly, and the radiator.

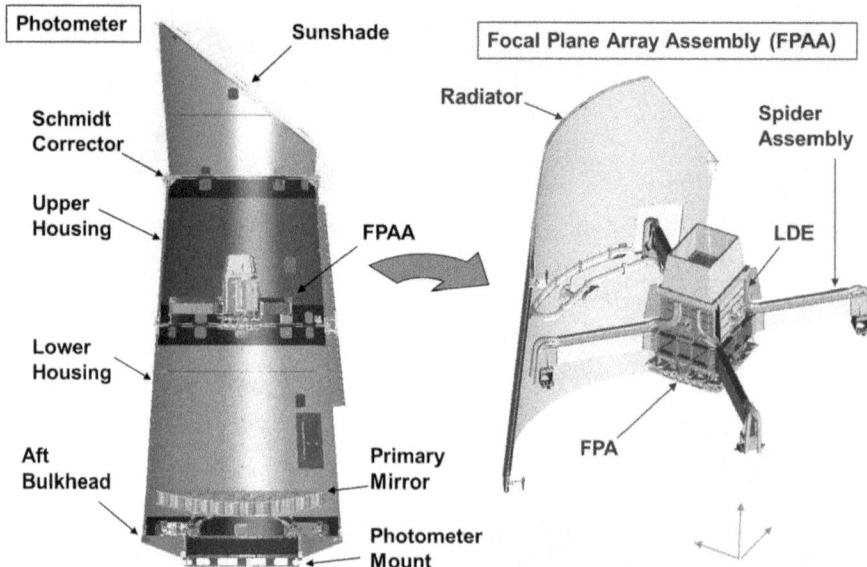

Figure 2.15. The focal plane array assembly contains four sub-assemblies and is located in the core of the photometer. Figure courtesy of Ball Aerospace.

The FPAA architecture was driven by the early selection of a wide field of view Schmidt optical system, which required locating the focal surface between the Schmidt corrector and primary mirror. With this optical configuration, the focal plane would need to be supported in the middle of the telescope by means of a spider support structure. In addition, the focal surface would need to be spherical and not planar, which introduced considerable complexity into the hardware design and manufacturing.

As with the Kepler photometer design, the FPA design was driven by maximizing the signal-to-noise ratio (SNR). For the CCD detectors, signal was maximized through a very large detector surface area provided by 21 individual detector modules, and by choosing detectors with large charge storage (well) depth. Noise was minimized by cooling the detectors to −95°C in order to minimize spurious generation of charge in the form of dark current, and by means of a carefully architected readout scheme.

The LDE architecture reviewed in the Kepler proposal was re-engineered in 2003 to a multiplexed acquisition system and the LDE was relocated inside the telescope moving behind the FPA to perform analog-to-digital conversion in close proximity to the CCDs. While this presented significant packaging challenges, it simplified the overall design, offered greatly improved noise performance, and reduced mass and power.

In order to maximize the amount of signal (photons) collected by the telescope, it was also extremely important to minimize the optical obscuration of the FPA, LDE, and spider assembly. This required the LDE to be placed in the "shadow" of the FPA, and all cabling and heat pipes to be routed in the "shadow" of the four spider support legs.

Because stable pointing and tracking of the photometer would be critical to the Kepler mission, a high precision, high resolution fine guidance system was necessary. Co-locating the fine guidance sensors (FGS) on the same focal surface as the science detectors was optimal, but they needed to have large physical separation in order to provide the required resolution, so placing a guidance sensor in each of the four corners of the array as shown in Figure 2.16 provided a good solution.

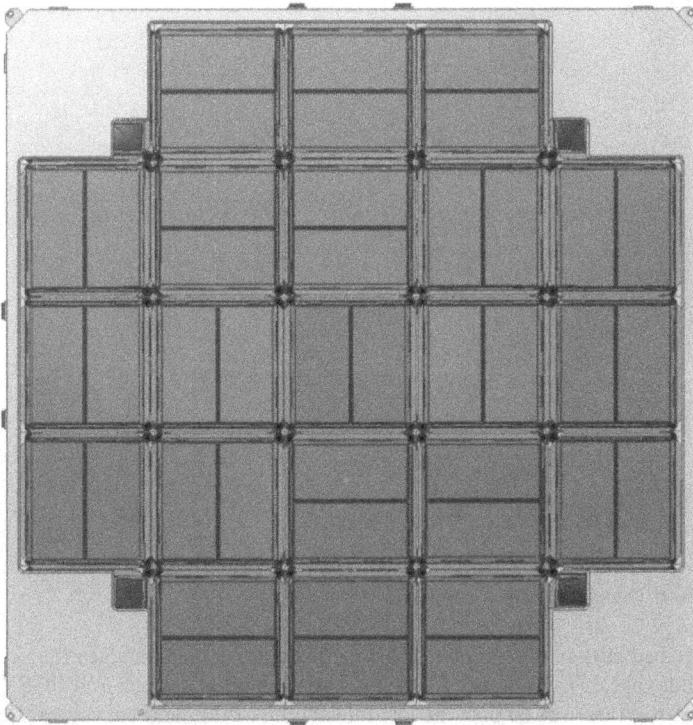

Figure 2.16. The focal plane assembly includes 21 science CCD modules and 4 fine guidance CCDs in the corners to form a "compound eye" camera system. Figure courtesy of Ball Aerospace.

Selection of the science CCDs involved a number of trades that considered photo-response uniformity, quantum efficiency, charge/well depth, and read noise, but also manufacturability, due to the large number of science-grade devices that would be required. The latter consideration drove selection of commercially available devices with mature technology that could be expected to have high yields. Because there was risk associated with producing large numbers of high-grade CCDs, we did a parallel procurement with two vendors to provide engineering-grade detectors, and then down-selected to one vendor for flight production.

The very large area of the focal surface, combined with its spherical shape, dictated the use of multiple detectors positioned in a faceted array. The optical system provides an 101° FOV with <11% optical vignetting in a circle with a diameter of 0.34 m. The arrangement of 21 science modules and 4 FGS modules as shown in Figure 2.16 provides complete detector coverage within this area, and also provides four-fold rotational symmetry so that the inactive areas are in the same location when the telescope is rotated 90° about the optical axis every 91 days during the mission.

The design of the science modules was driven by a number of requirements, including the ability to replace a module that becomes defective after installation into the FPA (Figure 2.17). Because repeating a precision FPA alignment process after such a replacement would be lengthy and costly, the modules and supporting structure were built to a tightly controlled set of tolerances, enabling "plug-and-play" approach.

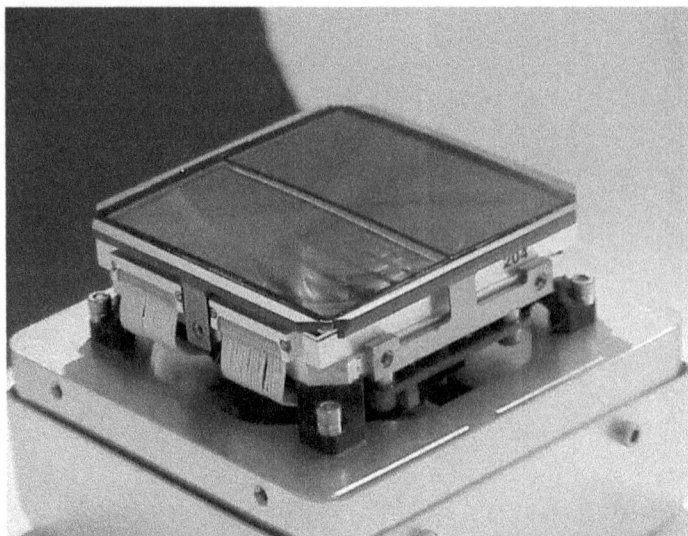

Figure 2.17. The CCD detectors were packaged in sets of 2 with a field flattener lens to form a modular assembly. Photo courtesy of Ball Aerospace.

As described earlier, the Kepler telescope provides a focal surface that is spherical with a radius of 1.394 m (Figure 2.18), but the individual science and FGS CCDs are planar. The solution was to provide local flattening of the field curvature by means of refractive lenses mounted above each set of CCDs. The field flattener lenses are supported by a four-piece titanium frame that has low-reflective coating applied to its illuminated surfaces to provide stray light mitigation. In addition to anti-reflective coating, the lenses also have a dark mask coating applied to selected areas, which prevents illumination of reflective surfaces on the CCDs and module components.

In addition to maintaining the CCD detectors at −95° C, the spatial temperature gradient across the entire array of modules must be held to within 2° C, and the temporal stability must be within 0.5° over 90 days. Keeping these requirements across such a large focal array drove our use of a thermal spreader with four individual ethane spreader heat pipes to equalize temperatures both spatially and temporally. Another set of two high-efficiency ethane heat pipes transport heat from the spreader to the radiator that is mounted to the exterior of the telescope.

The local detector electronics configuration is shown in Figure 2.19. Electrical Component Assemblies (ECA1-ECA5) are acquisition/driver board pairs that perform sequential integration and readout of the 21 science modules, while also supporting continuous 10 Hz readout of 4 fine guidance sensor modules. At one end of each ECA driver Printed Wire Assembly (PWA) is a 264 pin *hypertronics* connector that is the electrical interface to five FGS and/or science modules. At the opposite end, each ECA acquisition PWA connects to its neighbors through the rigid-flex motherboard. ECA1 and ECA5 each service two FGS and three science modules, while ECA2, ECA3, and ECA4 each service five science modules. ECA3 contains the 48 MHz master clock that provides all science and FGS CCD control, while ECA1, 2, 4, and 5 are slave controllers. The LDE power supply, and ECA digital control and I/O data are fully block redundant, while the CCD data acquisition/control and engineering data are all single-string.

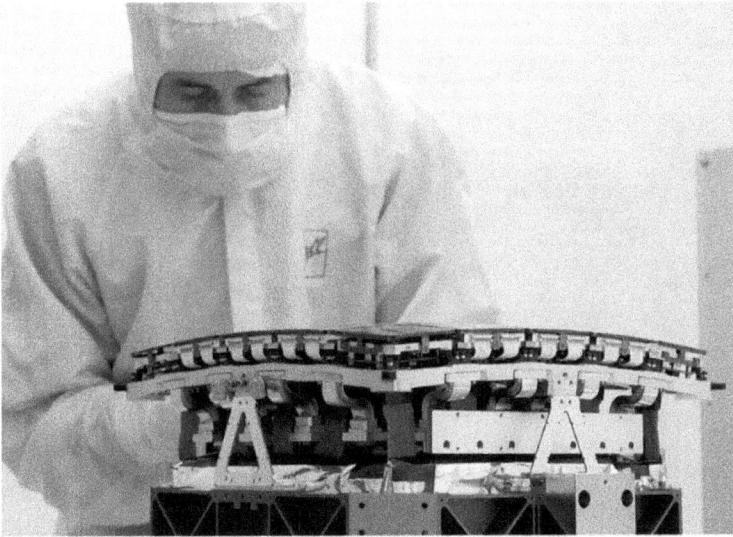

Figure 2.18. The focal plane assembly incorporated an optically precise curvature to match the wavefront coming from the telescope. Photo courtesy of Ball Aerospace.

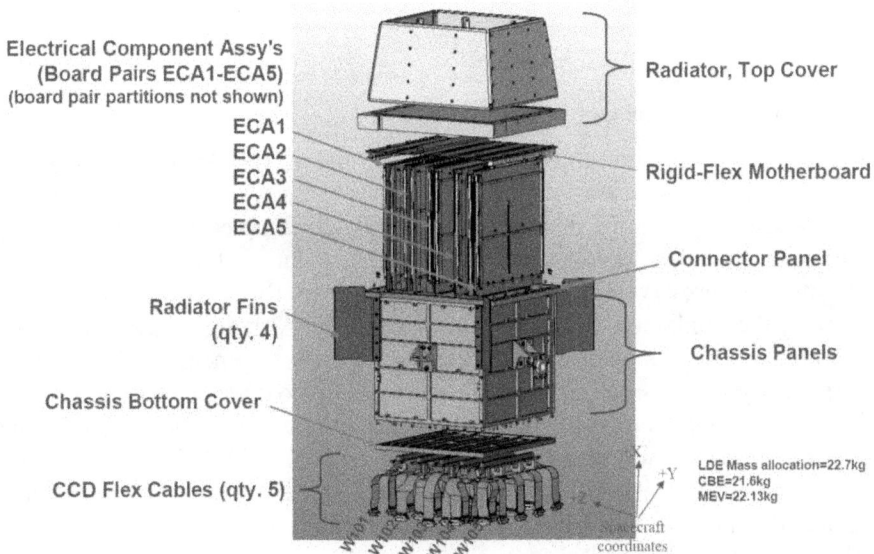

Electrical Component Assy's
(Board Pairs ECA1-ECA5)
(board pair partitions not shown)
ECA1
ECA2
ECA3
ECA4
ECA5

Radiator, Top Cover

Rigid-Flex Motherboard

Connector Panel

Radiator Fins
(qty. 4)

Chassis Panels

Chassis Bottom Cover

CCD Flex Cables (qty. 5)

LDE Mass allocation=22.7kg
CBE=21.6kg
MEV=22.13kg

spacecraft
coordinates

Figure 2.19. The local detector electronics controls the CCDs and converts the analog electrical signals from the CCDs into digital numbers for computer processing. Figure courtesy of Ball Aerospace.

Demanding thermal performance also drove the design of the cabling between the FPA and the LDE. Because the LDE is maintained at a relatively warm temperature compared to the FPA, use of standard copper conductors in this cabling would result in very high conducted heat loads that would raise the CCD operating temperature. Our solution (shown in Figure 2.20) was to use laminated flex circuit with conductors

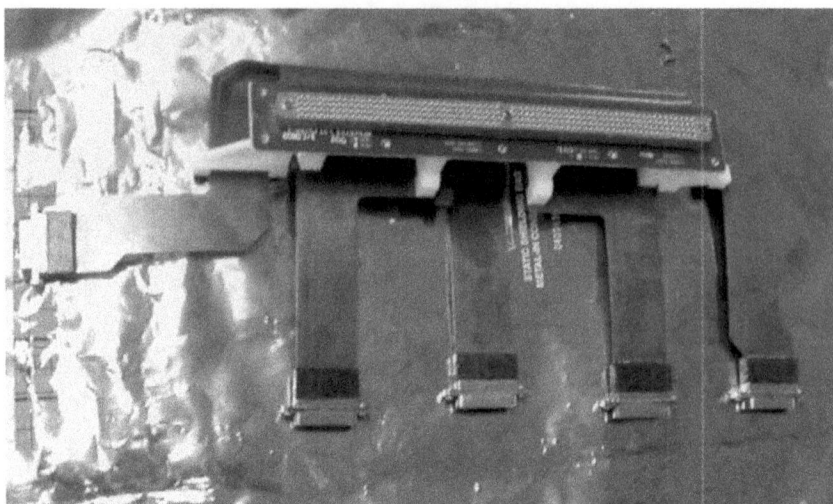

Figure 2.20. One of five flexible cables connecting the detectors to the electronics. Photo courtesy of Ball Aerospace.

made of a 55% copper–45% nickel alloy known as Constantan®, which has a thermal conductivity only 5% that of pure copper.

2.3.5 2007 Program Replan

As a result of the mounting cost and schedule pressure NASA began discussing canceling Kepler if the team did not find a way to meet our obligations with the existing budget. The team's devotion to the mission coupled with an unwavering funding limit forced us to sit down and craft a solution that would satisfy the constraints and preserve the mission. Elemental to this process was a critical revisiting of the fundamental, top-level requirements. Although the Kepler requirements were amazingly stable over the course of the development, we had expanded the mission beyond the minimum requirements. The goal was to create a plan that optimized the strengths of each Kepler team member and returned to the basics of the mission. This plan was approved by NASA in 2007 July. NASA's insistence on the team delivering to its previous commitments under threat of cancellation had fundamentally altered the team's approach and set the stage for a successful completion of the program.

The 2007 programmatic replan required the program to complete development and launch within the allocated, existing resources, which required the Kepler team to make several fundamental changes. Two of these changes directly impacted the technical work on the spacecraft. First, the mission duration was reduced from 4 to 3.5 years, which was the minimum duration that met the top-level requirement to detect and determine the statistical prevalence of Earth-size planets around solar like stars in the habitable zone. This change was made with the assumption the mission could be extended if it was successful during the 3.5 year baseline. This change did not require any modifications to the spacecraft design and the decision was made not to reduce any of the on-board consumables like fuel. Second, the raw photometric performance (RPP) test of the photometer was eliminated from the test plan. The RPP test was a complicated, thermal vacuum, optical test that had been in development for 4 years. It was designed to directly validate Kepler could detect Earth-size planets utilizing a transit simulator placed over the flight photometer in one of Ball's large thermal vacuum chambers. The ground support equipment for the test was considered a significant cost and schedule risk and there were doubts that the test would be sensitive enough to work and might even result in inaccurate or ambiguous results. Lab testing using the non-flight single string transit verification hardware at Ames had demonstrated the overall Kepler concept and the program determined it would be an acceptable risk to descope the test of the flight segment. In addition, there was a series of rigorous environmental and performance tests planned for the photometer to verify it would function correctly in space.

2.3.6 Photometer Assembly, Integration, and Test

Assembly of the photometer began once the telescope barrels, mirror assemblies, and focal plan assembly were complete. The focal plan assembly was the last piece needed for assembly and it was delivered to the integration team in 2007 October. Assembly of the photometer began on 2007 November 14 with the installation of the FPAA onto the lower telescope barrel (Figure 2.21). This was the first of series of high stakes operations involving the overhead crane to move and position very high value hardware over other equally expensive items and then hold things in place while the technicians made the flight connections. There was a conference room with a window overlooking the cleanroom where the FPAA installation took place. The window was standing room only as we watched the delicate operations in the cleanroom from above. On the next day the lower telescope barrel with the FPAA installed was lifted and lowered over the primary mirror assembly. The primary mirror was carefully inspected (Figure 2.22) since this would be the last time we had access to it. Figure 2.23 shows the lower telescope hovering over the primary mirror during the installation process. At this point the team took a break for Thanksgiving.

On December 6, the team of technicians gathered in the cleanroom again and prepared for the installation of the upper telescope house and Schmidt corrector on to the lower telescope (Figure 2.24). On the following day the fully assembled optical system was moved into an alignment tower shown in the background of Figure 2.24 and light was put through the end to end system for the first time. Throughout the design phase engineers had created mechanical drawings and optical models for the photometer and all of its sub-assemblies but the only way to know if an optical system is performing correctly is to shine light through it. Over the next 3 weeks the optical team took over and carefully measured and adjusted the pieces of the telescope and by December 25 they were convinced it worked as predicted, at least at room temperature. In just 6 weeks the Kepler team had assembled, aligned, and tested the photometer and it was working with no significant issues.

Figure 2.21. The completed FPAA is positioned with the CCDs facing the primary mirror in the center of the telescope lower barrel. The electronics and wires for power and signals are positioned to avoid stray light. Photo courtesy of Ball Aerospace.

Next, the photometer had to demonstrate its performance under vacuum and at its planned operating temperature. During the time the photometer was being assembled an independent team was installing and certifying a large optical test system called the vertical collimator assembly (VCA) into the largest thermal vacuum chamber at Ball at that time, called Brutus. Lessons from the Hubble Space Telescope optical error in 1990 taught space telescope engineers the importance of having a completely independent way to verify optical performance. Since the VCA was designed and built by a different team at Ball than the team that assembled and aligned the photometer, we were assured the independence required to avoid systemic errors in our test equipment. The picture in Figure 2.25 was taken on 2008 January 8 with the photometer nestled in the VCA in the Brutus vacuum chamber at Ball. Thermal vacuum testing involves running a series of performance tests on a space system in high vacuum while adjusting the surrounding temperature from the minimum to the maximum predicted operating temperature. The test temperatures are chosen to include margin so the conditions on orbit will always be within the extremes of the testing on Earth. The photometer successfully completed its thermal vacuum test and went on to vibration testing. During the vibration test the photometer was placed on a large shaker table and exposed to the predicted launch vibrations in three orthogonal directions. After being "shaked and baked" the photometer was moved back into the cleanroom to verify it still performed to specification after environmental testing. It passed these tests and was delivered for integration to the spacecraft in 2008 April.

Figure 2.22. The completed primary mirror assembly is inspected before installation into the photometer. Photo courtesy of Ball Aerospace.

2.3.7 Spacecraft Assembly, Integration, and Test

The Kepler spacecraft design was derived from previous Ball designs with a number of key modifications to support the mission. Ball has built spacecraft for Earth orbit since the late 1950s but the Deep Impact spacecraft launched in 2005 was the first Ball system designed to operate away from the comforts of Earth orbit. Operating in deep space has the benefits of a very stable thermal and disturbance environment. It also requires communicating at large distances and being able to autonomously keep the system safe in the event of a failure or a false alarm. Kepler used many of the systems and designs developed for Deep Impact along with experience gained by NASA on the Spitzer Space Telescope.

It would be inefficient for an aerospace company to build all the subsystems of a spacecraft from scratch due to the special knowledge and equipment required in each area. As a result many of the items on a spacecraft are procured through a complex supply chain. When it is time to start a new spacecraft, Ball sends out

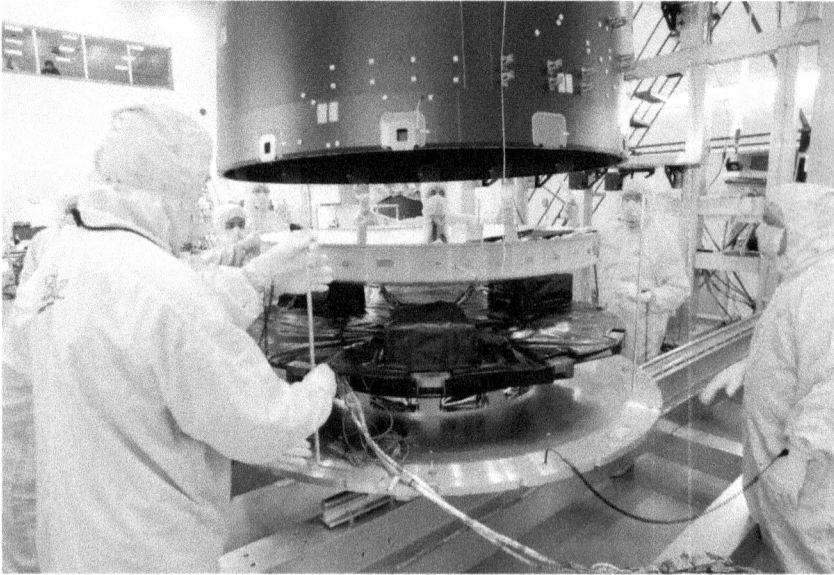

Figure 2.23. The lower telescope including the FPAA is lowered over the primary mirror. Photo courtesy of Ball Aerospace.

requests for proposals to a set of proven suppliers and then evaluates the products and prices provided. Throughout the process NASA reviews and offers advice based on their expertise and experiences on other missions. For Kepler, the spacecraft supply chain involved working with 20 companies throughout the US. In order to manage this large effort Ball formed small teams, organized by discipline, to specify, order, design, oversee manufacturing and test, and integrate the required subsystems.

The Kepler spacecraft consisted of the following subsystems (Figure 2.26).
1. Avionics system including flight computers, data storage, and precision clock electronics.
2. Flight software including command and control, data processing, and fault protection.
3. Attitude control system including star trackers, Sun sensors, a rate sensor, and reaction wheels.
4. Reaction control system (RCS) including fuel tank, valves, thrusters, pressure sensors, and associated plumbing.
5. Power system including the solar array and a battery
6. Telecommunications including a radio, amplifiers, and antennas.

The spacecraft bottom deck provides for the precise connection of four major elements of the mission. On the bottom side, the deck connects Kepler to the launch vehicle adapter ring (Figure 2.27) and thus to the rocket. On the top side, the deck

Figure 2.24. The upper telescope including the Schmidt corrector is moved across the cleanroom for installation onto the lower telescope. Photo courtesy of Ball Aerospace.

mates to the bottom of the photometer and provides the structure to hold the entire photometer to the spacecraft. The edges of the bottom deck support the side panels of the spacecraft bus structure and the top deck and solar array assembly. Finally, the bottom deck supports the reaction control system deck. This single assembly is responsible for supporting and connecting every major element of Kepler.

Figure 2.25. The photometer is readied for thermal vacuum testing to verify predicted optical and thermal performance in space. Photo courtesy of Ball Aerospace.

Unlike other deep space spacecraft that require powerful thrusters to enter orbit around distant worlds, the Kepler "propulsion" system was responsible for maneuvering the spacecraft during critical events like release from the launch vehicle, maintaining attitude during anomalies resulting in safemode, and dissipating the energy accumulated by the reaction wheels during science operations. This subsystem was named the reaction control system (RCS) since it would never significantly propel a change to the velocity of the spacecraft, it only changes the orientation of the vehicle. The RCS consists of four pairs of mono-propellant thrusters, plumbing and valves, and a tank to store the hydrozine fuel for the

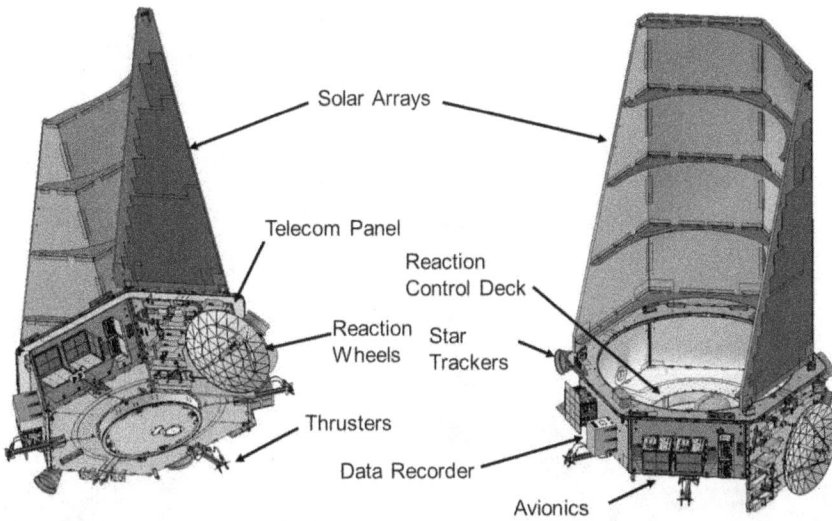

Figure 2.26. Subsystems are mounted onto a spacecraft structure to provide all the functionality needed to support the photometer during the mission. Figure courtesy of Ball Aerospace.

Figure 2.27. The bottom deck of the spacecraft is match machined to the launch vehicle adaptor ring to precision tolerances to assure clean separation during launch. Photo courtesy of Ball Aerospace.

mission. The RCS was assembled and tested in a separate cleanroom at Ball in parallel with other work on the spacecraft. This allowed for the unique assembly steps like welding to progress without impacting other parts of Kepler's construction (Figure 2.28). When the RCS was completed we faced the logistical challenge of moving it to the cleanroom where the rest of the spacecraft was being assembled.

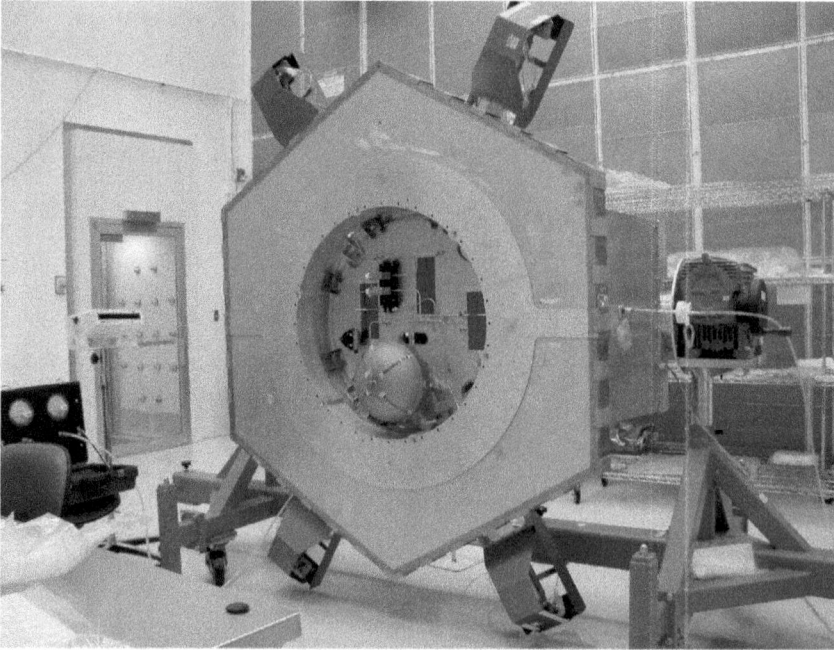

Figure 2.28. The reaction control system fuel tank, valves, thrusters, and plumbing lines were integrated into a structural deck to ease assembly and testing. Photo courtesy of Ball Aerospace.

Due to some unique aspects of our facility, we rolled the structure through the heart of the factory to a small loading dock. There it was loaded onto a large flatbed truck for a 100 foot trip (Figure 2.29) to another loading dock where it was unloaded and moved to the spacecraft cleanroom. Boulder, CO has been known for years for holding the worlds shortest Saint Patrick's day parade. The movement of the RCS that day reminded us of the worlds shortest aerospace parade. Soon afterwards the RCS and the spacecraft bus were mated moving us one step closer to the final spacecraft configuration.

The Kepler reaction wheels were chosen in 2004 during the spacecraft design phase based on the Kepler attitude control requirements and a heritage review. The Kepler design team determined the required performance characteristics for the reaction wheel assemblies (RWAs) based on the size of the spacecraft, mission concept of operations, and required mission life. Heritage reviews, also known as inheritance reviews, are performed for all spacecraft subsystems and components where an existing product and design is planned to be used. These reviews cover the history and previous uses of the device including any changes in the design or intended use for the upcoming mission and include the supplier, Ball, and NASA engineers.

Kepler required four RWAs to provide a mechanism to point the spacecraft during science operations and data downlink and since only three wheels are required to control the spacecraft a 4th wheel was included to provide redundancy

Figure 2.29. The reaction control system being transported across the parking lot. Photo courtesy of Ball Aerospace.

in the event one of the wheels failed. The Kepler reaction wheels are the only part of the spacecraft designed to move during the nominal mission.

In 2005, John Troeltzsch recalls being approached by Bill Borucki at Ball with concerns about the RWAs. Bill asked about some troubling news concerning RWA failures on Japan's Hayabusa mission. NASA was beginning to get concerned about the number of recent failures and an effort was started to record, track, and analyze all issues related to RWAs that might impact missions currently under construction. A RWA failure on NASA's TIMED in 2007 created a critical turning point in the level of concern. The Kepler RWAs had already been delivered and installed on the spacecraft but in early 2008 the decision was made to remove them and send them back to the manufacturer for detailed inspection. A team of US experts was convened to review the design and inspection results. The decision was made to rebuild the RWAs with some parts being replaced by new parts and some design modifications to improve dynamic margins. The four wheels were reassembled and tested and returned to Ball for installation on the spacecraft. There were some lingering concerns but based on the evidence and analysis on hand, NASA's tiger team cleared the wheels for flight. The combination of the rebuild, a 3.5 year mission, and the redundancy inherent in a four wheel design was deemed by NASA as acceptable for launch.

Electrical Power Design

Contributed by Roger Lapthorne, Power System Engineer, Ball Aerospace.

The single-fault tolerant Kepler electrical power and distribution subsystem (EPDS) consisted of the following elements:
1. Solar array (4 non-coplanar facets, 7.3 m^2 in total area, 2860 triple-junction GaAs solar cells).
2. Battery (16 strings, 24 A hr nameplate capacity, 128 small-format Li-ion cells).
3. Bus control assembly (BCA; forms the essential bus, sub-buses, single-point ground).
4. Power control boards, or PCBs (control the flow of power from the solar array to the BCA).
5. Power distribution boards, or PDBs (control the flow of power from the BCA to the loads).

There were several engineering challenges faced by the design team, starting with the solar array (Figure 2.30). The solar array not only generated power for the Kepler spacecraft, it also served to block sunlight from directly striking the Kepler telescope—a key feature that allowed the telescope components to maintain the consistent low temperature regime required to meet science objectives. And that was not enough—because the solar array also was not allowed to radiate waste heat from the backside (as most solar arrays are designed to do). In order to prevent this from happening, the backside of the solar array was covered with multi-layer insulation. The instruments were kept very cool, but the solar array got very warm as a result.

The solar array incorporated a couple of atypical design features to mitigate the impacts of the unique thermal environment:
1. Typically, as much of a solar array's Sun-side as possible is covered in solar cells. In the case of Kepler, some of the area that would have otherwise been a "hot zone" was covered with optical surface reflectors (OSRs)—basically, small mirrors that reflected incoming solar energy away from Kepler.
2. Each solar cell in a solar array is typically covered with a small glass sheet to improve performance and robustness. Kepler elected to use coverglass that included a pair of specialized coatings (UVR and IRR) to help keep the cells cooler than would have otherwise been possible.

Another challenge posed by the solar array was the unique physical configuration—four rigid-panel facets each pointing in a different direction. Kepler needed to continuously "stare" at the same star field for three months at a time. As a result, the amount of sunlight impinging on each facet changed slowly over time, day in and day out. This process called for a hard look at exactly how the solar cells would best be "segmented" (a segment is a set of parallel solar cell strings), and how they would be controlled by the flight software to provide a smooth power profile. That hard look at the solar array segmentation in turn necessitated creation of an analysis tool to investigate and predict the solar array's performance, assuming different segmentation configurations and/or operational scenarios. In addition, the wrap around shape of the composite array structure that held the individual solar cells proved to be quite challenging to manufacture and transport (Figure 2.30). The wide temperature range of the solar array structure challenged the ability to glue the structure together and the unusual shape required development of custom fixtures to hold it in place. By contrast,

Figure 2.30. The solar array being moved for installation with the photometer. Photo courtesy of Ball Aerospace.

the Kepler battery did not figure to have many design challenges. This was mainly because the battery had very little "battery work" to do during the Kepler mission. Since Kepler's solar array necessarily had to point sunward continuously throughout the mission, the battery was never required to discharge (except to support the brief launch and ascent phase of the mission). Instead, the battery served only to clamp the bus voltage at the desired operating point (~32 V), and to absorb the occasional power transient resulting from load changes on the bus. The Kepler battery was manufactured in Oxford, England and consisted of strings of lithium ion laptop batteries. This battery was one of the first Li-ion batteries to be used in space and was the first to be used on a deep space mission.

That aside, there was an important lesson learned regarding the battery early in the Kepler mission. Kepler's fault detection and safing system detected and responded to a couple battery over-voltage (OV) conditions during certain commissioning operations. These OV events were not harmful to the battery or the spacecraft, but they were a nuisance that could have stretched out the commissioning phase. These OV events resulted from the unforeseen conflict between two subsystems, each just trying to "do their job":

1. The EPDS adopted a conservative approach with regards to battery energy. The battery charging algorithm was set up to keep the battery fully topped-off, maximizing the energy available to respond to contingencies. This approach, however, put the battery voltage as near to the OV detection threshold as it could possibly be.

2. The ADCS was required to maneuver the spacecraft more often and more quickly during commissioning ops than it would during mission ops. This resulted in the Sun passing over the solar array facets more quickly than had been anticipated—which in turn resulted in solar array power output capability rapidly increasing over short periods of time.

The timeframes involved were such that the battery charge control algorithm in flight software (running at 1 Hz) was unable to keep up with the rapidly changing capability of the solar array. This resulted in the battery having to briefly absorb excess current from the solar array, which in turn led to a transient OV detection and response. The solution—the battery charge control algorithm was re-tuned to charge the battery to a lower (albeit less conservative) voltage, allowing for more headroom between the battery voltage and the OV threshold voltage.

By all measures, the EPDS flawlessly supported the entirety of the Kepler science mission. The Kepler mission was followed by the K2 mission, which asked the EPDS to perform in a way in which it was not originally designed. From the EPDS perspective, there was one primary difference between Kepler and K2. During the Kepler mission, the Sun moved over the solar array facets from side to side; during the K2 mission, the Sun passed over the solar array facets from top to bottom. (Note— because of this relative motion of the Sun over the solar array facets, two of the facets would actually never see the Sun during the K2 mission, they were always pointed 90° off-Sun).

Using the design/analysis tool previously created to support the original design of the solar array, it was determined that the solar array could indeed support K2 with little to no additional risks. The only caveat for K2 was that mission planners had to adhere to a maximum solar array tip angle. The max tip angle was not an impediment to the science objectives and as a result, the EPDS flawlessly supported the entirety of the K2 science mission just as it had done for the original Kepler mission.

Attitude Determination and Control Design

Contributed by Dustin Putnam, Attitude Control Engineer, Ball Aerospace.

The differential photometric requirements for the Kepler mission placed strict require-ments on the pointing accuracy and stability of the telescope. The astronomers not only needed the point-spread function (PSF) of each star to remain on a fixed set of pixels, but they also required the PSF stay in the same regions of those pixels to minimize the effect of intra-pixel nonuniformity. Line-of-sight pointing stability was paramount to keeping the CDPP as high as possible: pointing stability over time spans longer than 30 minute was limited to 9 mas per axis. The required pointing stability over shorter time spans (higher temporal frequencies) was given by a power spectral density curve.

A spacecraft ordinarily combines output from a single or set of external star cameras and an inertial reference unit to create high accuracy attitude and rate estimates. The requirements for Kepler were beyond the capability of star trackers in use at that time. The Kepler solution was to place four small CCDs on the corners of the focal plane array assembly (FPAA) that could capture light from stars near the science field and use those stars to maintain an ultra-stable pointing reference. Utilizing the light gathering power of the telescope's 95 cm aperture allowed guide stars as faint as 14th visual magnitude to be used in the attitude measurements. Figure 2.31 shows a cleanroom photo of the FPAA with an FGS CCD highlighted.

These four CCDs were collectively called the Kepler fine guidance sensor (FGS). Each CCD was a 535 × 550 array with pixel resolution of 1.92 arcsec. The local detector electronics would read-out square blocks of pixels, each block capturing the light from a single guide star, at a rate of 10 Hz. Flight software algorithms used the distribution of light in each block to create a star centroid measurement, and multiple centroids would be combined to create an attitude estimate. The attitude estimates were back differenced and low pass filtered to create an angular rate estimate. The flight software (FSW) could process up to 20 guide stars per CCD each 10 Hz frame, but only 10 stars/CCD were used during science data collection. Adding more stars does not necessarily make the measurement better if the additional stars are extremely dim; a total of 40 stars was determined to provide the required level of attitude measurement stability.

Figure 2.32 shows the cross-boresight line of sight motion power spectral density (PSD) and its reverse cumulative summation. The PSD stays well below the short time-period requirement (red line) and shows that high frequency motion only contributes about 8 mas of motion per axis. The spikes in the PSD curve are the result of periodic cross-talk between the FGS and the science data during readouts.

Flight software algorithms compared the FGS attitude and rate measurements to desired quantities and used a classical PID (proportional + integral + derivative) control law to synthesize torque commands that were sent to a set of reaction wheels which provided the actuation. Kepler was equipped with four Goodrich TW-16B200 reaction wheels, each of which could provide up to 200 mN m of torque and store up to 16 N m s of momentum.

Solar radiation pressure (SRP) produces the dominant disturbance torque on the spacecraft. The vehicle's center of mass was offset approximately 1 m from the center-of-pressure, creating an SRP torque of about 50 N m. This near constant external torque caused the reaction wheel rotors to slowly accelerate over time. Rotor speeds

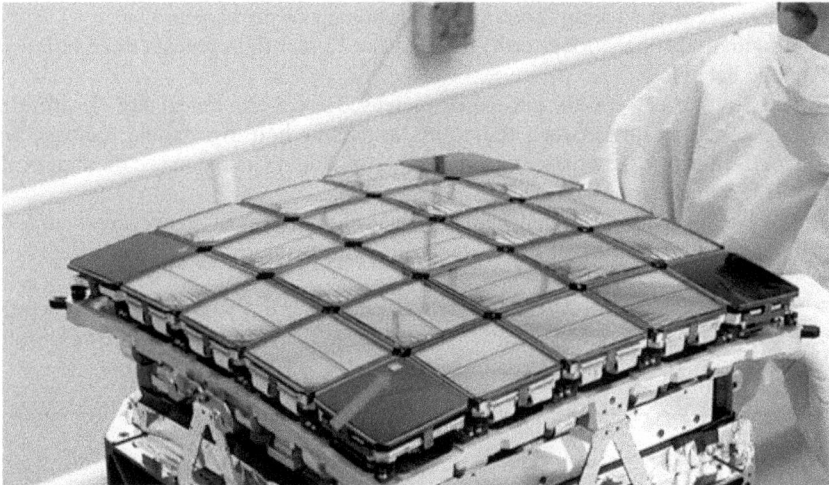

Figure 2.31. The Kepler focal plane array assembly is equipped with four small field-of-view CCDs, one in each corner, that provided an ultra-stable attitude measurement. Credit: NASA Ames and Ball Aerospace.

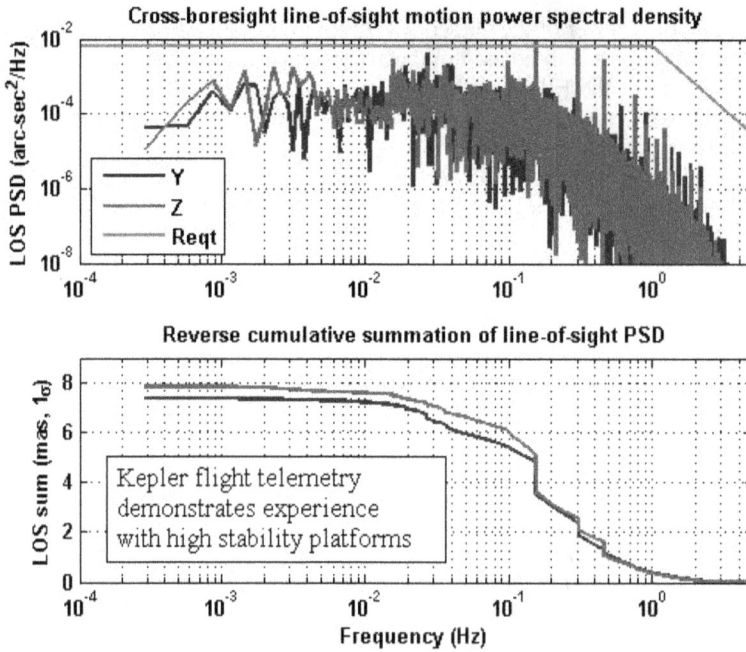

Figure 2.32. Flight telemetry was used to demonstrated consistent compliance with the stringent line-of-sight requirements necessary to support photometric performance. Figure courtesy of Ball Aerospace.

were brought down by firing the thrusters to apply an external torque in the opposite direction of the SRP torque once every three days. The reaction wheels were sized to carry up to six days' worth of SRP induced momentum, but they were despun after only three days to keep the rotor speeds below 3000 RPM (50 Hz), which was the frequency of the first rocking mode of the photometer's primary mirror.

Careful analysis of the long-term trends in the Kepler data reveals three timescales of dynamic behavior.

1. Reaction wheel speeds gradually increased over the three day desaturation interval. Friction between the rotors and their housings would increase with speed, causing an increase in heat dissipation and a change in the thermal profile of the vehicle.
2. The Kepler spacecraft science attitude was changed by 90° every three months to maintain illumination of the solar panels. This would cause a sudden shift in the vehicle's thermal profile.
3. The Kepler spacecraft was in a slightly elliptical, 372 day heliocentric orbit. Motion toward or away from the Sun would create shifts in the vehicle's thermal profile.

2.3.8 Integration and Launch

On 2008 May 19 the Kepler photometer and spacecraft were integrated. Figure 2.33 shows the Ball technicians carefully lowering the photometer onto the spacecraft

Figure 2.33. The photometer and spacecraft are connected to complete years of development and prepare for observatory level testing. Photo courtesy of Ball Aerospace.

structure. In early July the solar array was added and six months of testing began on the fully integrated spacecraft. This testing would expose the hardware and software to all the rigors of launch and mission operations.

The testing included:

1. Comprehensive functional test designed to systematically exercise each element of the spacecraft including the redundant elements. This test also established the baseline performance to be used as a reference following the upcoming environmental tests.
2. Acoustics, separation, and pyro-shock tests. The acoustic test exposes the spacecraft to the acoustic noise levels predicted inside the rocket fairing

during the launch. This test was performed using approximately 108 large speakers arranged in 9 by 30 foot tall columns surrounding the spacecraft. Next came the separation and pyro-shock test to verify the spacecraft can survive the shock from the explosive bolts used to separate the spacecraft from the launch vehicle during launch.

3. Thermal vacuum and thermal balance test. For the TV/TB test. Kepler was placed into a test fixture inside large vacuum chamber. The thermal test fixture completely enveloped the spacecraft with a number of large panels covered in electrical heaters. The vacuum chamber walls are lined with a cryogenic shroud which can be flooded with liquid nitrogen. This setup allowed for a realistic simulation of Kepler's space environment. It would be exposed to the vacuum of space while one side of the spacecraft received heat simulating the Sun and the other was exposed to the cold of deep space. By re-configuring the heaters we were able to mimic various solar orientations that Kepler would see over the course of a quarter. For more than two weeks Kepler was operated from the hottest to the coldest extremes expected during the mission. In addition this test included a thermal balance test to calibrate and correlate the analytical models used to predict thermal performance against the actual performance.

4. Electromagnetic interference and compatibility tests. The EMI/EMC test was performed in a special chamber completely shielded from external radio waves. The spacecraft is placed into the chamber and commanded through its operational modes while being bombarded by test radio waves to verify extraneous radio signals will not interfere with the science mission. Additional tests are done to measure the amount and frequency of radio emissions coming from the spacecraft to verify it will not interfere with the launch vehicle or itself.

5. Mission scenario tests. Throughout the environmental test sequence, eight mission tests were conducted to verify the operations team, operations procedures, and spacecraft could work together in realistic mission scenarios.

6. Post environmental alignment test. After subjecting Kepler to the rigors of space a final alignment check was performed to insure the spacecraft and photometer were still properly optically aligned.

Four years of planning and preparations for the final integration and test phase paid off and Kepler moved through each test with no significant issues. On 2008 December 17 the final major Kepler review at Ball was conducted. This pre-ship review covered the state of the design and verification of the spacecraft including all of the records covering the construction of the vehicle and the testing to verify it would work correctly. The team presented the status of all the requirements laid out during the preliminary design phase (Figure 2.34) which showed Kepler was ready for the journey to the launch site. In addition the review covered the details of the upcoming shipment and the activities to be performed at the launch site. The test team had been working non-stop for months and was given a well earned break over the holidays.

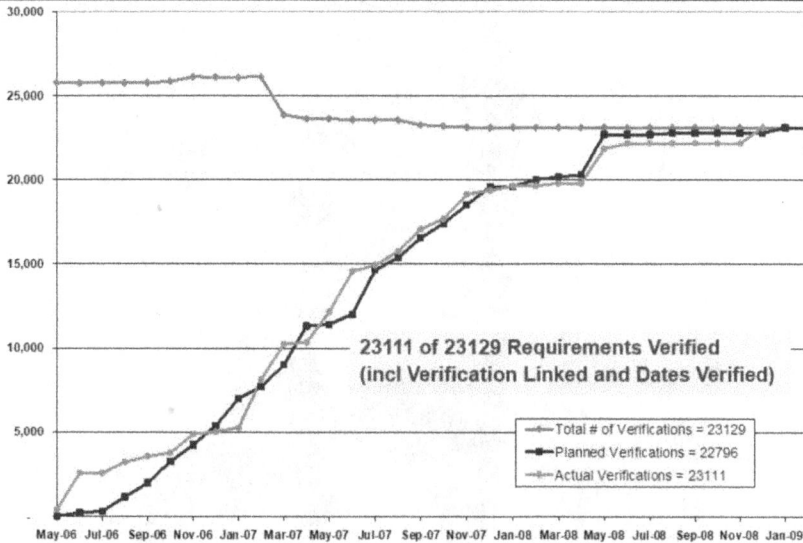

Figure 2.34. The spacecraft requirements status was presented during the review to authorize shipment of Kepler from Boulder, CO to Cape Canaveral, FL. Figure courtesy of Ball Aerospace.

Early on the morning of 2009 January 2 Kepler left the Ball facility on top of an oversized tractor trailer in a convoy headed for Florida. The trip to the launch site was relatively uneventful. Early on the first day the truck was stopped in Eastern Colorado due to a perceived paperwork issue at a weigh station. After a couple of hours and a call to the head of the Colorado Department of Transportation it was released and on its way again. Over the next four days the shipment would carefully navigate numerous bridges, underpasses, and a few highway detours. Mid-day on January 6th the truck rolled through the gate at the Astrotech facility in Titusville, FL where it was meet with a fresh team of technicians and a massive construction crane used to lift the container off the truck. A few hours later is was safely in a cleanroom and ready for the post shipment tests to verify nothing was damaged in shipment.

On February 3 Kepler was moved to the Hazardous Processing Facility at Astrotech to be fueled and integrated with the 3rd stage of the rocket. The hydrozine fuel used by Kepler is extremely toxic and corrosive. The fueling process requires the technicians to wear full hazardous materials suits with self-contained breathing apparatus. The decision to fuel a spacecraft is taken seriously since unloading the fuel at a later time would be difficult. Fueling involves carefully weighing the amount of fuel loaded. The lifetime of the mission is calculated based on the fuel load before launch. After fueling the spacecraft is lifted and lowered onto the third stage and a metal clamp band is installed to connect the two. This band is released after launch to allow the spacecraft and rocket to separate.

Figure 2.35. The Kepler space telescope and spacecraft fit inside the rocket faring high atop the Delta II launch vehicle. Photo courtesy of Ball Aerospace.

During the middle of the night on February 21 the spacecraft was transported 22 miles from Astrotech to the launch pad where it was place on top of the Delta 2 rocket. The rocket fairing was installed five days later and Kepler was ready to be launched. Figure 2.35 shows the top of the Delta II rocket containing Kepler sitting on the launch pad with half of the fairing installed. The figure shows only the very top of the launch stack with the rocket second stage, the third stage, and Kepler on top.

Kepler was launched on 2009 March 6 at 10:53 PM local time (Figure 2.36).

As the rocket cleared the launch tower all Kepler activities shifted to the Kepler Mission Operations Center at the Laboratory for Atmospheric and Space Physics

Figure 2.36. Launch from Cape Kennedy. The bulbous white top section of the rocket is the fairing inside of which rides the Kepler spacecraft as seen in Figure 2.35. Credit: United Launch Alliance.

at the University of Colorado. The spacecraft was successfully released from the launch vehicle and promptly entered safemode due to detection of a power charging error. The error was easily corrected and the ground operations team quickly determined Kepler was safely on orbit. Over the next 60 days the spacecraft went through a series of commissioning tests to incrementally check and initialize each part of the system. There were a typical number of issues discovered and each one had to be rigorously analyzed and adjustments and corrections implemented. One of the early photometer tests involved taking dark counts with the telescope cover still in place. We quickly discovered the telescope dust cover was not as light tight as desired so a work around was devised and the semi-dark data was acquired. A very subtle software bug was discovered which had required an extremely unlikely timing of events to occur but after a few days a fix was in place and progress continued. There were a couple of computer upsets caused by space radiation and the team revised the control modes of the spacecraft to minimize the effects. Overall the commissioning phase was challenging and consistent with the birth of a brand new spacecraft. At the end of commissioning Kepler was cleared to enter normal operations and the mission of exoplanet discovery was underway.

2.4 The Birth of the K2 Mission

Kepler performed well during its 3.5 year primary mission and meet the performance specifications as required enabling the many scientific discoveries described later in this book. The engineers and scientists on the operations team learned the idiosyncrasies and behavior of the spacecraft and systematically tuned it to provide improved performance.

In 2012 July, one of the four reaction wheels failed and in 2013 May a second wheel failed which ended Kepler's primary mission and set into motion one of the great space comeback stories in history.

K2 Mission Creation

Contributed by Kipp Larson, Kepler Mission Operations Manager, Ball Aerospace.

After the first reaction wheel failure (wheel number 2) in 2012 July, it was decided to leave that wheel turned off and proceed with science operations on just three wheels. While no attempts were made at that time to troubleshoot wheel 2, greater efforts were made to examine telemetry from the other three wheels to assess their health and attempt to anticipate any further performance degradation events. Because the design of the spacecraft's safemode required the use of reaction wheels to control vehicle attitude, plans were also begun to develop a new mode that would hold vehicle attitude using thrusters in the event of a safemode entry due to a new wheel failure.

This mode, called thruster-controlled safe mode or TCSM, was similar to the original safemode but used thrusters for attitude control rather than reaction wheels. It would allow the spacecraft to retain attitude control and stay power-positive in order to give the team time to assess any anomaly and to have a shot at some form of mitigation. While this would prevent the spacecraft from tumbling out of control, it would also burn fuel at a rate that would be unacceptable long term. It was recognized that a completely new way of maintaining attitude would be necessary in order to have a chance of saving the mission from a second wheel failure. A new attitude control mode, point rest state or PRS, was envisioned that would point the spacecraft at ecliptic north, allow solar pressure to tip the spacecraft backwards and then use a very small amount of fuel to push the top of the spacecraft back toward the Sun. In this way it would ping-pong back and forth, always staying power positive and in a good attitude for communication, but using only a tiny fraction of the fuel used for the original TCSM.

A schedule was put together to develop and test PRS over two months, but in early 2013 reaction wheel number 4 started to show signs of trouble. The telemetry indications of both short and long term friction increases that were also seen prior to wheel 2 failing indicated that wheel 4 failure was imminent, and the two month

---

development cycle was decreased to two weeks. The command sequences necessary to execute PRS were uploaded to the spacecraft on 2013 May 9. On May 14, reaction wheel 4 failed. In the following months tests were performed on wheel 2 in order to determine if it could be resurrected. While the wheel operated well for a few minutes, it ultimately slowed to a stop once again. As these tests were being planned and executed, a call went out to the greater aerospace community for ideas about possible missions that could be accomplished with only two wheels. The incredibly accurate pointing that Kepler had needed for planet detection that was easily achieved with three or four wheels was thought impossible with only two wheels. Thrusters alone would have caused too much movement for any useful photometry. At this point, most people thought that the planet finding mission was over, and that it was highly unlikely that a new two-wheel mission of any value could be created. Then, Doug Wiemer, an engineer at Ball Aerospace, had an idea.

In 2013 August, Doug proposed a possible mission concept in a short 5 slide PowerPoint presentation (see "Balancing a Spacecraft using Solar Pressure" below). It involved balancing the solar pressure in the roll axis along the symmetrical solar array panels that covered most of the length of the spacecraft. This would largely eliminate the need for constant control in that axis. The thrusters could, in theory, be used to help in this case. By firing opposing thrusters and making use of the small misalignments between them, the resultant force could make very small corrections to maintain the solar vector along the edge of the solar panels. None of this had ever been tried before, and it was not clear if the attitude control system could be reconfigured to operate with both the autonomy and precision that would be required. The presentation was given on a Friday afternoon, and Sue Ross and Katelynn McCalmont-Everton started working on the spacecraft simulator that evening to find out if it would work.

A new idea is generally presented in the form of a concept overview. Then, a set of requirements for the execution of the concept are put together and refined, the design based on those requirements was developed, reviewed and approved, and finally the design is tested in as many was as is necessary to ensure that it will work in the end. This lengthy and time consuming systems engineering process is typically necessary to ensure that very expensive spacecraft do not fail after they reach orbit. Under normal circumstances, developing an entirely new mission concept in this manner, reevaluating all aspects of spacecraft system performance and getting reviews and buy in from all stakeholders would have taken more than a year. In this case, there were only three months before NASA would decide, as part of its senior review, whether or not the program would continue to be funded.

Early tests on the spacecraft simulator showed promise, and development of the new mission paradigm began. Even as NASA was still evaluating other ideas from the community, the team at Ball started putting together a schedule of flight tests to incrementally demonstrate and evaluate the new approach. Each test included as many objectives as possible to make the most of deep space network (DSN) contact opportunities. Within the four person team at Ball, the goals for each test were discussed, debated, and settled on in one day. The next two days were initial flight product creation and testing on the simulator, followed by a day or two of final product generation at the Laboratory for Atmospheric and Space Physics (LASP) at the University of Colorado where the mission operations center was located. One more day was typically set aside for final review and briefing before the contact, if possible. The results of the flight test would be analyzed in one day, and the whole process would begin again. This weekly flight test cadence went on for approximately ten weeks starting in 2013 October.

The results of these tests lead to an accurate mapping of the solar array ridgeline that the solar vector would balance on, performance assessment of the thruster combinations and optimization of the new control algorithm, reconfiguration of the spacecraft's finepoint capability to hold cross boresight pointing to better than 0.05 arcsec 1 sigma, ground tools for creating and verifying the over 28,000 time-tagged commands that would be running on board for 88 days at a time, re-evaluation of the entire spacecraft's subsystems and fault protection and a reconfiguration of the algorithm for holding the Ka antenna pointed to Earth along with a new link budget to manage the spacecraft recorder volume and downlink times correctly.

This development was done with only four people working full-time at Ball and two more at LASP, with several others working part-time. The budget was the same as during the nominal mission, without extra resources and without anyone working overtime. This was accomplished by starting with the tried and true systems engineering principles and distilling them down to only what was absolutely necessary. Many corners were cut, but any team member was free to throw a red flag if any aspect of a test was not sufficiently thought out or adequately tested on the simulator. Multiple sets of eyes on the flight ops team reviewed each flight product, but external product reviews were seldom done. The NASA scientists at Ames were largely focused on making the science case for the mission and selecting possible areas of the sky to look at, and they gave the flight operations team a great deal of freedom to pursue the testing as they saw fit. The team at Ball spent their entire workdays together in the same room with the vehicle simulator, eliminating the need for meetings or even email communication. Any question or concern was immediately heard by all and dispositioned quickly. Crazy ideas were welcomed and were not eliminated until they could be proven to be functionally impossible. Without all of these pieces falling into place, it is unlikely that the new mission would have been successful.

As it turned out, a week-long test in 2013 December collected data showing the telltale transit of an exoplanet, demonstrating that the new mission was not only feasible in theory, but that it could, in fact, still find planets. This was sufficient cause for NASA to continue funding the program, renamed K2, until the spacecraft ran out of fuel at the end of 2018. By that time, it had completed 19 campaigns leading to the discovery of 1266 exoplanet candidates and counting. That's 1266 planets that would never have been discovered if the Kepler mission operations team did not ignore the crowd and take it upon themselves to define what was possible. A more in-depth discussion of the technical details related to the K2 mission creation can be found in Larson et al. (2014).

2.4.1 K2 Mission Creation and Implementation

When the second reaction wheel on the Kepler spacecraft failed it was a sad day in the project office at NASA Ames Research Center. As the project scientist of the Kepler mission, I received a number of phone calls and emails from around the world with condolences and best wishes. Some of the science team members begin to announce to the world the terrible news that Kepler was dead. However, as you read in this chapter, the spacecraft engineers had yet to throw in the towel. I also pursued looking into alternative science possibilities that could be done with the spacecraft and photometer, albeit in light of almost complete uncontrolled drift or at best a back and forth rocking motion.

Working with the spacecraft team, it became clear what few possible modes the spacecraft might be able to achieve. The idea of ecliptic pointing, using solar pressure to balance the spacecraft, a never before tried technique, rose to the top as the best chance. I sent an open letter to the science community given them some best guess possibilities including pointing at the original Kepler field of view, but with very poor pointing accuracy and large fuel usage, pointing at other new fields of view, with the same restrictions, or performing a set of observing campaigns, each of approximately 80 days as the telescope moved in its orbit about the Sun. The fields accessible in this mode, looking from the spacecraft orbital plane, were those of the zodiac in the sky, we'd have an ecliptic survey observing around the entire sky about once each year.

However, given the new and unproven technique and the necessary funding from NASA to pursue such a plan, great science had to be possible, this could not just be a spacecraft testbed. Taking the community replies to my letter, 45 lengthy, multiple author whitepapers, I collated the science cases and noted that much of the ideas could be accomplished by observing fields along the ecliptic as it passed though the galactic plane as well as the galactic halo. The Kepler project was given a few months to test the techniques described in this chapter and they worked. Not perfectly at first, but better each time, from a day or two of stability eventually up to the 80 day campaigns.

The next step was to sell the K2 mission and its science goals to the NASA senior review. This peer panel, organized every two years, reviews all operating spacecraft and their requested budgets and gives rankings and recommendations to NASA. A

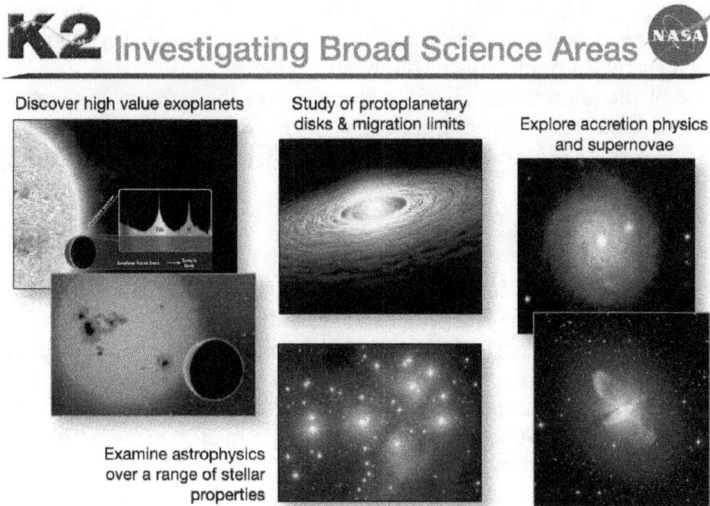

Figure 2.37. The expected broad science to be obtained at the start of the K2 mission. Credit: Howell/Stencil/NASA.

proposal and oral presentation for K2 was presented to the senior review panel[5] and in the end, K2 was ranked number one and fully funded for the next two years. This was an amazing accomplishment for a crippled and assumed dead spacecraft. K2 was a mission that was yet to be fully proven, was yet to achieve its full 80 day campaigns, and had no specific level one science requirements, all items usually needed and shown to be well-honed in order to pass the senior review. But the scientific promise of K2, using initial engineering and test campaign observations and the great science resulting from them, was clear (see Figure 2.37). We showed a path forward and the ability to make K2 succeed. Two years later, we proposed to the next senior review panel and were ranked number one again. K2 would operate for another two years.

2.4.2 Balancing a Spacecraft using Solar Pressure

Contributed by Doug Wiemer, Attitude Control Staff Consultant, Ball Aerospace.

The attitude system for the Kepler spacecraft was precise. Science attitude sensing was performed with selected starlight through the telescope itself, to ensure the very high accuracy and stability required to detect the subtle light variation transiting planets were making on their stars. The actuators, reaction wheels, were very low disturbance and three-axis control was accomplished with any three or all four wheels provided, when kept in a proper speed range.

The attitude system lived with a spacecraft design that was practical, but not ideally balanced, which allowed for significant and variable solar torques during normal operations. The heavy bus mass at one end and large hollow telescope tube at the other end meant that the spacecraft center-of-mass and center-of-solar-pressure were far apart. Worse yet one side of the spacecraft was Sun-warm and the other side space-cold amplifying the issue. The attitude system readily rejected these near constant torques. The reaction wheels had sufficient capacity to absorb several days worth of solar torque effects, between brief thruster momentum dumps.

The value of the science was exciting, so even after the primary mission was over, operations and data collection continued, with more exoplanet signatures, refining planet orbits, narrowing planet sizes, and determining possible habitability. And then a second reaction wheel failed.

K2

When the second wheel on Kepler failed my mind raced for a solution to keep it working. With just two wheels left, only two axes of the spacecraft can be controlled. A selected spacecraft-fixed vector can be inertially pointed but then the attitude about that vector cannot be controlled. And that vector must be well away from the momentum plane of the remaining wheels for it to work. That's not very science practical for a telescope that wants sub arcsec stability in all three axes for hours if not days. But there was a solution to do science again.

---

[5] A slightly modified version of the science sections of this proposal is available in Howell et al. (2014).

Figure 2.38 presents a motor vehicle analogy. With four or three wheels arbitrary pointing of the vehicle is achieved. With only two wheels system momentum becomes a key player in pointing.

If you inject the right momentum into the spacecraft with thrusters, and control pointing of the two axes of the telescope boresight at a desired inertial target, you get two axes of control (Y and Z) and one axis of stability (X), enough to make planet finding work. The injected momentum is initially absorbed by the two wheels and the spacecraft body about the telescope boresight. The spacecraft rotates about the telescope boresight, like it's on an invisible pole, until the rotating momentum plane of the two remaining wheels encompasses the inertial system momentum. Total system momentum goes completely into the two wheels. Body momentum about the space-craft boresight goes to zero and motion about the boresight comes to a gentle stop. Figure 2.39 shows momentum exchange between the two remaining wheels and the boresight axis of the spacecraft. It becomes obvious that the spacecraft will gently rotate about the telescope boresight to a stop at a specific angle given a specific system momentum.

By adjusting the system momentum with small thruster bursts, the attitude about the boresight is tuned so that a preselected set of fine guidance stars can be used. Small adjustments in momentum and attitude are accomplished by firing opposing thrusters at their minimum on time and using the minute resulting difference. The system is again stable and precise enough to find planets. And the versatile and configurable attitude software just needed new table values, no new software.

Now comes that nasty solar torque. For an arbitrary stellar target, one with reasonable solar orientation for operations, the solar torque changes the system momentum and causes the spacecraft to rotate slowly about the telescope boresight while the telescope boresight remains inertially fixed. This is not good enough. Science stars move across the photometer reducing and completely ruining their planet finding usefulness. At some point the fine guidance stars will fall out of their boxes and control is lost. But again there was a solution to do science again.

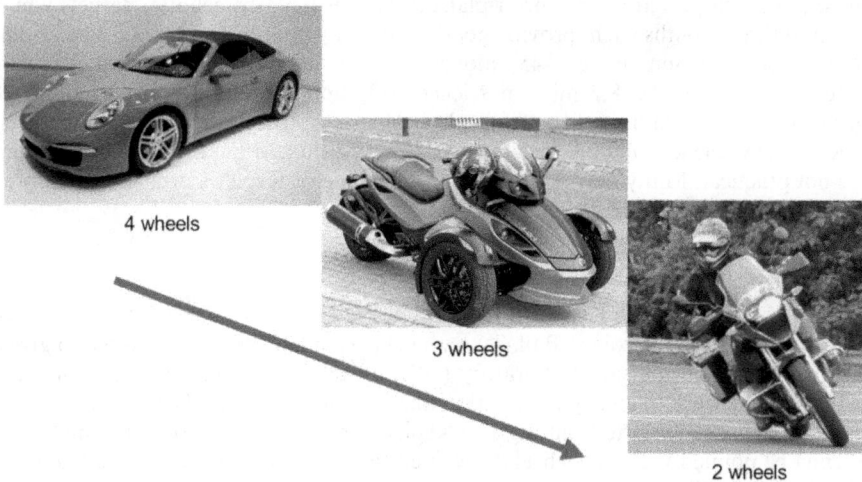

4 wheels

3 wheels

2 wheels

Figure 2.38. Riding on two wheels takes momentum to stay stable. With three or more wheels you can point in any direction at any speed but with two wheels you need momentum to stay pointed or you lose control. Figure courtesy of Ball Aerospace.

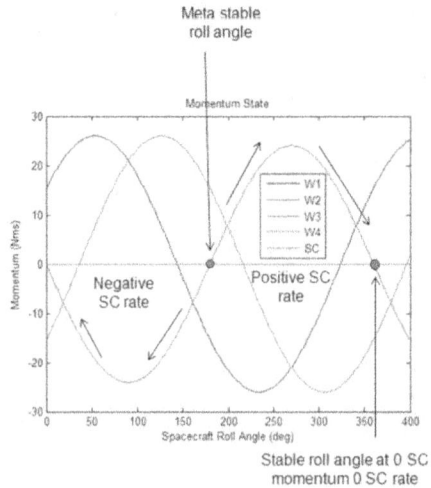

Figure 2.39. A slide from the 2013 July 2 Wiemer presentation explaining how Kepler can operate on two wheels. Figure courtesy of Ball Aerospace.

If viable science could be performed where the solar torque and momentum absorption is kept within the momentum plane of the two wheels, the spacecraft would again point inertially fixed for days between momentum dumps. Stellar targets were found which put the solar torque into the two-wheel momentum "null space." These science targets are in the orbit plane of Kepler (i.e., the ecliptic), targets which for up to three months each, provide good solar array pointing and a null space solar torque while the Sun moves ±45° along the solar array spine. Calibrations were performed prior to the K2 mission science collection to determine true null space torque attitudes given imperfections in physical spacecraft symmetry.

K2, a new science mission, became possible even though the original field of view was not practical. Many new sky regions in the plane of Kepler's orbit were explored during the four years of K2's extended operation.

Downlink

I had fun figuring out the Kepler to K2 attitude puzzle in 2013 but had to get back to many other tasks in my role at Ball. Two years later, in late 2015, I was asked to give a paper at the Jet Propulsion Laboratory (JPL) on the K2 attitude control system. It made me think back to a nagging question that I hadn't solved, and hadn't had time to think about, until I started putting the Kepler paper together. Could the high gain antenna be pointed with two wheels? I wanted to solve that problem before the paper that November. One more time there was a solution.

The high gain antenna boresight is in the plane of wheel momentum so two axis pointing control, which would have been good enough, is degenerate. Only a single axis of a pointing control is nominally possible, not two. That's not at all good enough.

The solution used very nonstandard control path manipulation. Control of the two telescope boresight axes was retained. Once prepositioned for downlink, but used without modification, the antenna drifts off the Earth rapidly given the non-null space solar torques existing in the downlink attitude. Cross coupling and splitting of one control axis was the solution.

The first of the two control axes (about Y) was left untouched. It could be used to control one axis of the telescope or to control one axis of the high gain antenna, which was 90 degrees away from the telescope boresight. Its error signals were defined for the first antenna axis instead of the first telescope boresight axis.

The second telescope boresight control axis was aligned with the antenna boresight (about Z) and had to be modified to be useful. The proportional and derivative error control components of the second telescope pointing axis (Z) were retained for stability and to prevent control degeneracy. The integral component was rerouted to the second antenna pointing axis (X) error. This would effectively become a slowly changing command to the Z-axis to get desired X-axis pointing.

Coupling between the second antenna pointing axis (about X) and the second telescope boresight pointing axis (about Z) had to be created. This was accomplished through momentum coupling; by injecting momentum to produce different magnitude wheel speeds. If the spacecraft rotated about the second axis of telescope pointing (about the boresight of the antenna Z) the spacecraft would twist and produce predictable motion in the second antenna control axis (X). Antenna pointing would be near perfect at the cost of drifting some about its boresight over time.

Over two days of wheel-only downlink the spacecraft would slowly rotate 20° about antenna boresight maintaining solar array pointing and telescope safety. Antenna pointing at the Earth was accurate to 0.003° but limited by star tracker performance. Because of the better pointing, downlink data rates were increased significantly over thruster control pointing, so downlink times could be reduced. Fuel was also saved to perform more science campaigns.

Safe Mode

A year or so later I figured out how to make a safe mode operate indefinitely, using the Sun to dump momentum. This isn't a spin the spacecraft to average the solar torque to zero; it's a mode that automatically and actively rotates the spacecraft to use the solar torque to control onboard momentum. This again needed no new code, just new values.

Control is similar to telescope boresight pointing since the solar array normal is well away from the wheel momentum plane so two axis Sun pointing control is readily achieved. Pointing the array accurately at the Sun is not critical, but it is different than telescope pointing in that wheel speeds increase as solar momentum is accumulated instead of decrease, due to wheel and target geometry. If left alone, wheel speeds would reach their limits and control would be lost.

Moving the desired Sun vector from 90° to 135° away from the telescope boresight solves the problem given the wheel geometry. This swaps the stable and meta-stable locations, which are 180° apart about the sunline, while still retaining 71% solar array input. Now wheel speeds decrease with solar momentum accumulation at the stable attitude, until they go through zero speed, which instantly puts the system at the meta stable attitude. After a while at the meta stable attitude, the spacecraft rotates

±180° about the sunline to arrive at the new stable attitude. The process repeats indefinitely.

To visualize the dynamics of the spacecraft, think of a repeating series of hills and valleys 180° apart with a ball settled stable in one valley. The ball's location defines attitude about the sunline—the stable attitude. On top of a hill would be the meta stable attitude. The height of the valley floor at the ball represents the system's momentum. The valley floor rises where the ball is located (while the nearby hills drop at the same time). This represents the solar torque effect on system momentum—the wheel speeds getting smaller. When the wheels go through zero speed is when the landscape becomes flat—the system has no momentum. As momentum continues to accumulate the hills become valleys and valleys become hills. The ball on top of the new hill is now meta stable. It will stay there with no disturbances but there are always small disturbances. After a while, the ball rolls either left or right to an adjacent valley. This is the same as the spacecraft rotating either left or right to a new stable attitude—the same attitude is achieved by rotating either clockwise or counterclockwise 180°. The process starts again.

If some arbitrary about sunline momentum exists in the system, the spacecraft won't just flip back and forth as described, it will rotate only in one direction faster at times and slower at others as the wheels are able to absorb less and more of that momentum based on attitude. This pointing could continue indefinitely, for as long as the hardware survives. It may still be working now.

Finding ways to make Kepler do things with just two wheels was a real joy for me. There were so many capabilities in the dynamics and software of Kepler to use it was just like finding the right pieces of the puzzle and putting them together in the right way, which was never obvious.

Optimizing K2 Pointing

Contributed by Katelynn McCalmont-Everton, Kepler Attitude Control Engineer, Ball Aerospace.

The K2 mission utilized the same concept of coarse and fine pointing as Kepler. The fine guidance sensors (FGS) located on the four corner CCDs centroided on specified guide stars in their fields of view. The star trackers were used to determine the vehicle attitude in coarse point to great enough precision that the fine guidance stars were reliably in their pixel of interest blocks (POI). Once the FGS POIs showed enough star intensity representing the expected guide stars, attitude determination was handed over from coarse point to fine point. This was a capability that was very much in doubt during the development of K2 since fine pointing relied on precise control ability from the reaction wheels. K2 first demonstrated the ability to transition from coarse point into fine point using only two wheels during the 2013 December test leading up to the NASA Senior Review submission in 2014. While in fine point, K2 proved capable of <1 arcsec precision in the cross boresight axis leading to confidence that this mission would be capable of finding planets using only two reaction wheels.

K2 was designed such that the solar pressure could remained balanced on the vehicle for up to 80 day science campaigns. On-orbit solar calibration tests were performed to map out the exact balance "ridge" along the solar-panel apex of the vehicle. However, any slight deviation from this ridge as the Sun traversed the solar array over 80 days causes slight imbalances in torque. This torque causes attitude error to slowly build up in the uncontrolled axis of the vehicle. By keeping the system momentum-biased with the two remaining reaction wheels, the effect of the imbalance solar torque was minimized. However, a way of correcting the about-boresight error was needed to keep targets within their science apertures on the focal plane. The operations team devised a scheme to use thrusters to carefully correct about-boresight errors. By using specific pairs of thrusters, >90% of the thruster force of each thruster was canceled out by the opposing thruster in the pair. The remaining <10% of the thruster force applied a small, precise torque on the spacecraft that could be used to trim the about-boresight error of the vehicle to within 15 arcsec. Specific pairs of thrusters used for coarse and fine adjustments were identified and shown in Figure 2.40. Testing after the second wheel failure and before the start of the K2 mission allowed the operations team to calibrate the thruster control system to precisely tune the about-boresight errors. Figure 2.41 shows thrusters 3 and 6 firing for 60 ms each to cause a larger, coarse correction of 700 arcsec. These larger "mamabear" tweaks, as they were named on board the vehicle, were used to clean up larger attitude errors that resulted from wheel resaturation events. Smaller fine tweaks were used to keep the system within the ±15 arcsec deadband during science collection. This allowed the science targets to remain in their prescribed apertures during the full 80 day science campaigns. Figure 2.42 illustrates the effects on the X attitude error of various thruster pairs.

During the Kepler prime mission, the solar pressure force caused a momentum build-up on the vehicle that caused the reaction wheels to increase in speed during science collection. Every three days, the thrusters would fire and desaturate that momentum. During K2, the solar pressure management was designed to operate in the opposite fashion. The thrusters would fire to spin up the reaction wheels, injecting momentum into the system. Over the course of two days, the solar pressure would cause the reaction wheels to lose speed before the thrusters were used to resaturate the wheels again.

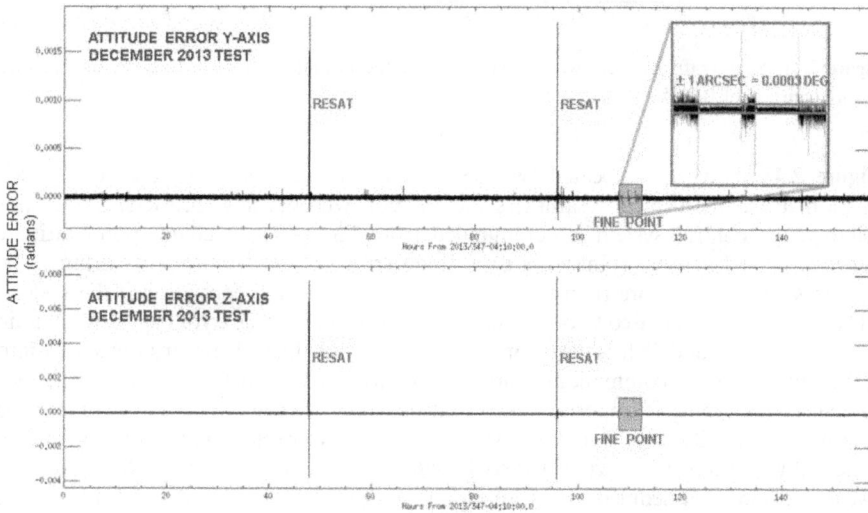

Figure 2.40. Pointing accuracy and attitude error. Figure courtesy of Ball Aerospace.

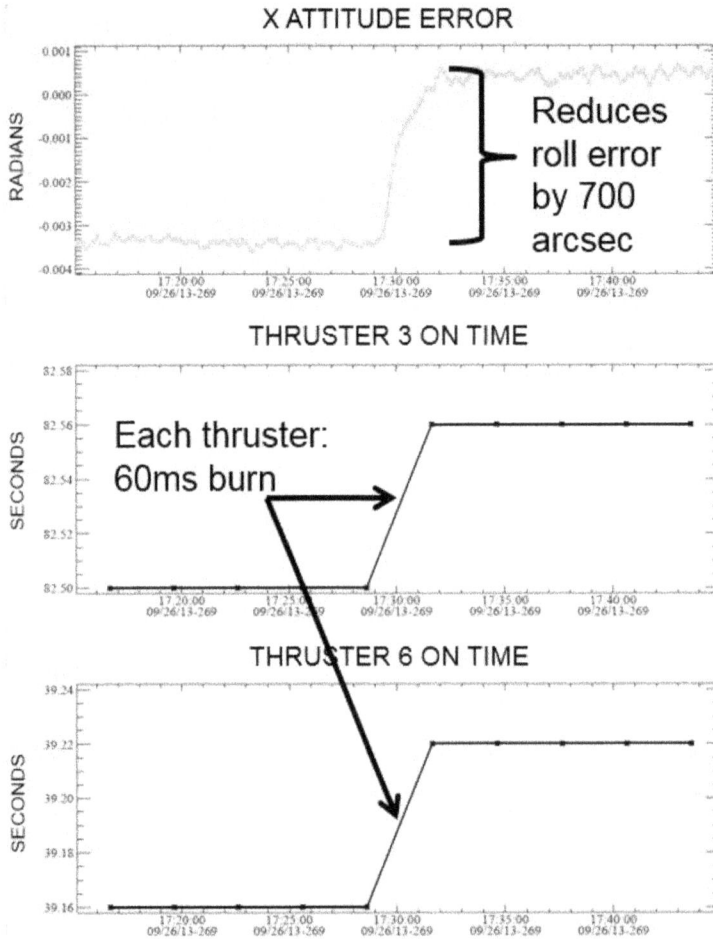

Figure 2.41. Spacecraft thrusters were used to tweak the attitude and provide precision pointing. Figure courtesy of Ball Aerospace.

Figure 2.43 shows the repeated two day cadence of science operations for the K2 mission. First, the vehicle resaturates the reaction wheels using thrusters. To do this reliably, the control system is commanded from fine point to coarse point attitude determination for more robustness to attitude errors induced by thruster firings. The thursters fire for no more than three minutes to re-inject momentum into the system. Thruster pairs are then used to tweak the about-boresight attitude error back down within the control deadband. When the error is acceptable, the vehicle then transitions back into fine point. All of this commanding happens autonomously within one 30 minute science cadence. This timing is important such that only one science integration is lost to the resaturation activity every two days. After the resat is complete, uninterrupted science is collected with no thruster firings for six hours. Every six hours for the next two days, a tweak session is scheduled to clean up any about-boresight drift that may have accumulated during science collection. During the course of the campaign, the momentum state of the vehicle stays within 12–20 N m s using the resaturation concept of operations.

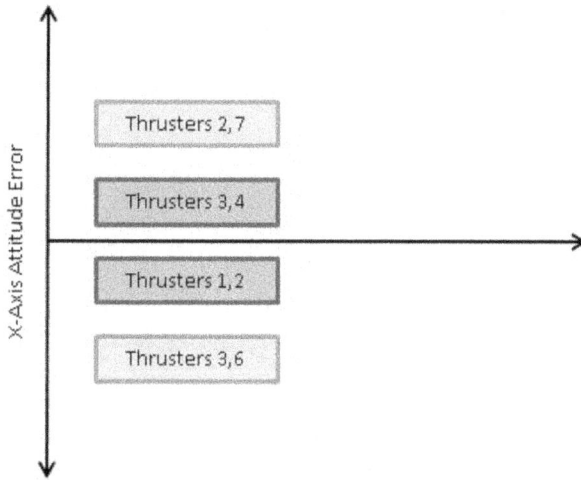

Figure 2.42. Specific thruster pairs were chosen for strength and affect. Figure courtesy of Ball Aerospace.

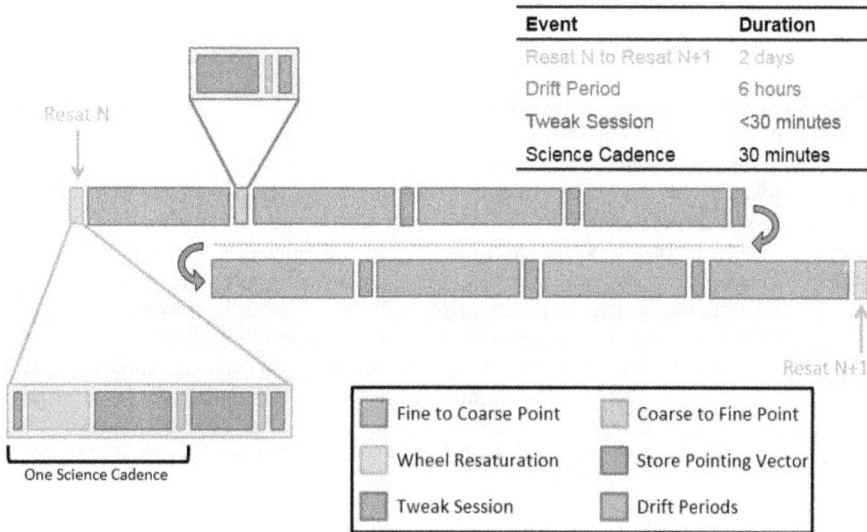

Event	Duration
Resat N to Resat N+1	2 days
Drift Period	6 hours
Tweak Session	<30 minutes
Science Cadence	30 minutes

Figure 2.43. Periods of reaction wheel resaturation, tweaks, and periods of drift established the science cadence. Figure courtesy of Ball Aerospace.

Developing thruster-controlled safe mode (TCSM) was essential to the advent of the K2 mission. It allowed for a safe, reliable place for the vehicle to regress in the event of an unforeseen fault. However, TCSM burned fuel at a high rate during dwell times and could use up to 40 science days' worth of fuel just on entry to the mode alone. The Kepler mission required storing momentum in the two remaining reaction wheels to keep the boresight of the telescope steady. By design, this momentum was dumped on entry to TCSM causing large amounts of fuel to be used. In 2017, more than three

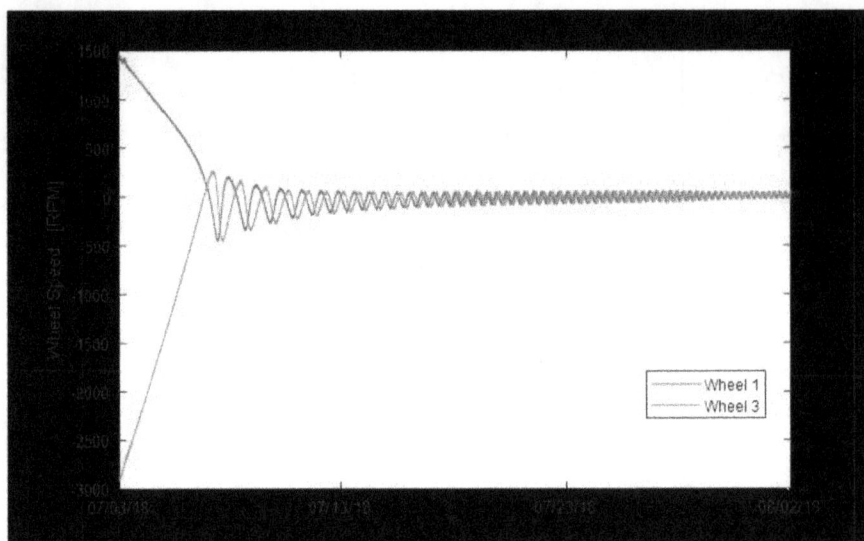

Figure 2.44. Development of a two wheel thruster safe mode was instrumental in minimizing fuel use during anomalies. Figure courtesy of Ball Aerospace.

years after the start of K2, the operations team deployed two wheel safe mode (TWSM) to the vehicle on orbit. This was preceded by months of development and testing at Ball Aerospace to ensure that the new safe mode was a safe regression state for all entry scenarios. The new two wheel safe mode did not use any fuel upon entry or dwell. This was a vast improvement over TCSM knowing the vehicle could respond to faults and use zero fuel indefinitely. The novel mode allowed for any momentum build-up in the system to prescribe the rotation rate of the vehicle. Over time in Kepler's heliocentric orbit, the momentum level in the system would fluctuate, but never build up to greater than the storage capacity of the wheels. Figure 2.44 shows on-orbit data for the TWSM entry after Campaign 18 was aborted due to low fuel pressure readings. TWSM was a key part of the end of mission scenario for K2. When the pressure readings indicated that the fuel was running low mid-way through C18, the operations team had a safe place to send the vehicle to immediately cease fuel usage and conserve any fuel remaining to downlink the science data.

The K2 mission was approved by the NASA senior review and began officially collecting science data in 2014. After the approval of the mission, the operations team began prioritizing ways to conserve life-limiting fuel. Two modes of operation required 100% thruster-based solutions at the start of the K2 mission: safe mode and downlink mode. In order to downlink the data stored during the 80 day science campaigns, the vehicle was commanded to maneuver to point the body-fixed high gain antenna within the 0.5° bandwidth to close the telecommunications link with the Earth. For the first 8 science campaigns, this was achieved successfully using thruster-only control at a high hydrazine expenditure rate. Starting with Campaign 9 in 2016, the operations team deployed two wheel stored data downlink (TWSDDL) mode on board the spacecraft. This mode used thrusters to spin up the two remaining reaction wheels once in the downlink attitude. After the wheels were spun up to pre-determined speeds, the momentum in the system was used to keep the antenna-boresight axis of the vehicle steady and pointed at Earth. A novel control architecture impacting the proportional, derivative and integral gains of the control system allowed the spacecraft to keep the antenna pointed at Earth while the vehicle slowly drifted about that axis. This still allowed the communications link to be kept with Earth while expending no fuel. Kepler could remain in this TWSDDL mode for greater than 65 hr without expending any fuel. Figure 2.45 shows the about-antenna-boresight drift rate of TWSDDL using telemetry data for Campaigns 9–15. TWSDDL could be reliably tracked and expected to last at least 60 hr before reaching the Sun-keep-out fault limit, shown by the red dashed line. In addition, TWSDDL provided >20 times better pointing precision than thruster-based downlink modes, allowing the data rate to be increased to complete the downlinks quicker. Over the remainder of the K2 mission, through the downlink of Campaign 19 data, TWSDDL was estimated to save roughly 1.5 campaigns' worth of fuel.

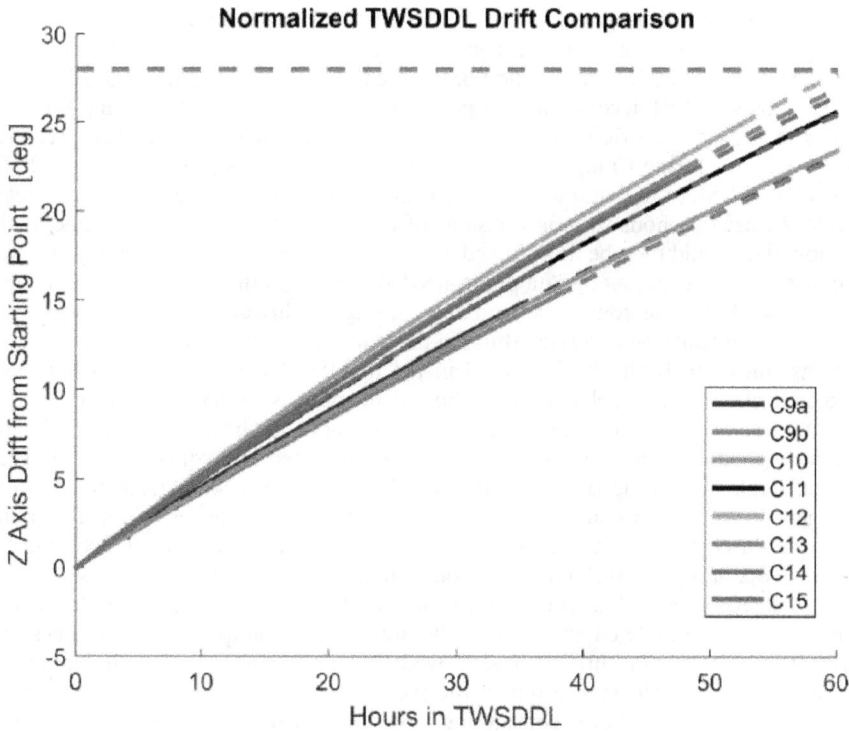

Figure 2.45. Two wheel science data downlink mode enabled transmission of final data from the K2 mission. Figure courtesy of Ball Aerospace.

Maximizing Science with Remaining Fuel

Contributed by Colin Peterson, Aerospace Operations Engineer, Ball Aerospace.

The Kepler mission operations team was prepared for the impending failure of the second reaction wheel in 2013 May. Two new thrusters-only modes of operation were developed and flight ready to give the Kepler team options following the reaction wheel failure. The first mode called thruster-controlled safe mode (TCSM) would be entered autonomously if triggered by the fault management software to enter safe mode on thrusters to ensure positive power and attitude control that would not be possible on only two reaction wheels. This mode would consume hydrazine fuel at a rapid rate of 7 thruster on-time seconds per hour, which was equivalent to about 0.8 kg of fuel every 10 days. For comparison, the Kepler mission from 2009–2013 used fuel at about 0.8 kg yr^{-1}. The initial launch load of about 12 kg of fuel had been reduced to 7 kg of fuel remaining in the tank at the time of the second reaction wheel failure. TCSM would burn through that quickly and only give about 90 days of additional mission life. The second thrusters-only mode was a fuel-efficient standby state called point rest state (PRS). This quiescent mode burned fuel at about 0.2 thruster seconds per hour. This gave the Kepler team time to explore future science missions. By the time the K2 mission started in early 2014, significant fuel had been used during on-orbit development and characterization of the K2 Mission concept. There was about 5 kg of fuel remaining and early estimates predicted three years of mission life, or about 12 K2 science campaigns to last into early 2017. The fuel budget was refined by incorporating early K2 actual fuel usage data, which showed fuel consumption at a rate of about 1.2 kg yr^{-1}. Accounting for the implementation of some operational fuel efficiencies provided a conservative extrapolation of the prediction to about Campaign 16 in early 2018. A detailed fuel accounting was performed after each campaign going forward, and by the time of the original estimated end of fuel at Campaign 12, the new predicted end of K2 was during Campaign 18 in the middle of 2018, as shown in Figure 2.46.

As the K2 Mission continued toward Campaign 18, the mission operations team developed new methods to look for signs of running out of fuel. By this time in the mission, data could only be downlinked at a rate of 10 bps. The team innovated a way to collect just the exact set of data points needed to get a picture of thruster usage, tank pressure, and attitude control performance during all thruster events. The team was looking for any out-of-family signatures in this data that could be indicative of the tank running out of fuel. On 2018 June 22 in the middle of Campaign 18, as had been predicted, the signs of fuel starvation showed themselves. A significant drop in tank pressure was observed as a "knee" in the pressure curve, as shown in Figure 2.47. This provided good evidence that the tank had been depleted of hydrazine fuel, and the internal tank membrane that separated the hydrazine propellant from the gaseous nitrogen pressurant was now pushing against the tank fuel outlet. From this point forward, thruster performance would be uncharacterized and unpredictable as the result of operating on fuel remaining only in the thruster lines. Campaign 18 was aborted so the science data could be downlinked before all thruster capabilities were gone. This was completed successfully. Before starting Campaign 19, fuel lines and thruster heater temperature set points were increased outside operating limits to squeeze as much performance out of the fuel in the lines as possible. Campaign 19 was started as scheduled even considering it was unknown operating territory. Despite an initial failure to start Campaign 19, the fuel performance still suggested quality science could be obtained. Campaign 19 was restarted and two weeks of good science

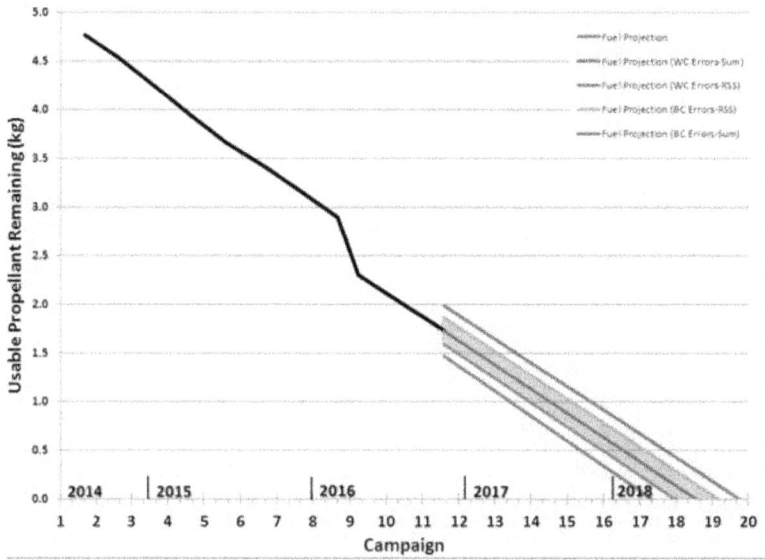

Figure 2.46. Modeling allowed use fuel prediction during the mission. Figure courtesy of Ball Aerospace.

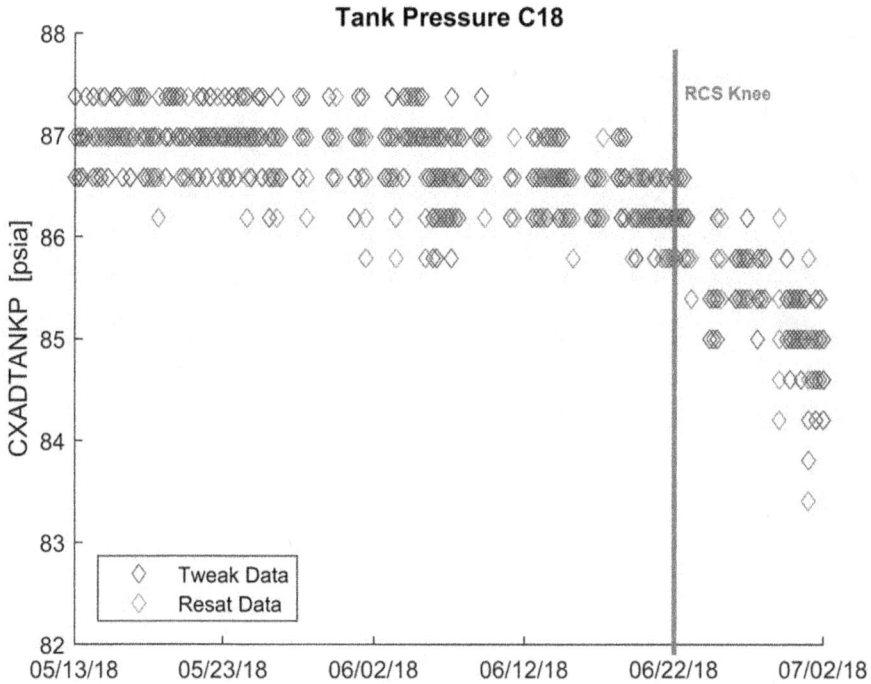

Figure 2.47. The pressure sensor in the fuel tank provided an indication of remaining fuel. Figure courtesy of Ball Aerospace.

was completed before the spacecraft pointing performance deteriorated, and thrusters could not keep target stars in their science apertures. The campaign was aborted on 2019 September 25. All Campaign 19 science data was successfully downlinked, and with the slimmest glimmer of hope, Campaign 20 was started. The spacecraft was not able to achieve science pointing mode, and entered Safe Mode in two days with the tell-tale sign that fuel was indeed exhausted. Figure 2.47 shows the tank pressure in the final months of K2, with a precipitous pressure drop upon starting Campaign 20. The end had arrived, welcomingly almost two additional years later than originally anticipated. Being able to still operate and collect science with an empty fuel tank, operating on only fuel in the lines for three months, seemed miraculous.

2.4.3 The End of the Mission

On 2018 November 15 a small operations team met in the Kepler Mission Operations Center in Boulder, CO and a group of engineers and scientists joined in via telecon from the NASA's Ames Research Center. Our purpose was to send the final commands to the spacecraft and bid farewell to Kepler. At 9:42 Mountain Standard Time we verified the radio carrier signal from the spacecraft had been disabled. Over the previous two decades countless scientists and engineers had toiled through seemingly insurmountable engineering and physics challenges to imagine, build, test, and fly the Kepler mission. I offer a thanks and acknowledgment to each and every person that made Kepler and its discoveries possible. Goodnight Kepler.

References

Borucki, W. J. 2016, RPPh, 79, 036901

Howell, S. B., Sobeck, C., Haas, M., et al. 2014, PASP, 126, 398

Larson, K., McCalmont, K., Peterson, C., & Ross, S. 2014, in SpaceOps 2014 Conf., Kepler Mission Operations Response to Wheel Anomalies, AAIA 2014-1882, OCMSA–Operations Experience, https://doi.org/10.2514/6.2014-1881

The NASA Kepler Mission

Steve B Howell

Chapter 3

Exoplanets

Jessie L Christiansen, Stephen R Kane, Susan E Mullally, Steve B Howell, David R Ciardi, Calen B Henderson, Andrew M Vanderburg, Fergal Mullally and Rachel Street

3.1 Designing a Mission to Find Other Earths

Stars are faint, but planets are much fainter. The light reflected into space by the Earth is 10 billion times fainter that the light from the Sun. To find planets, astronomers must typically use indirect techniques to infer the presence of a planet too dark to be seen. By the time Kepler launched in 2009, most planets had been found by the radial velocity or "wobble" technique. The radial velocity method uses careful observations to measure changes in the speed at which a star is moving toward or away from us due to the gravitational tug of a nearby planet. But this technique is limited to bright, nearby stars and relatively large planets (typically more massive than Neptune).

Kepler used a different technique, called the transit method. Instead of searching for the faint glimmer from the planet, the transit method waits for the orbiting planet to pass between the telescope and the star, blocking a small portion of the star light and making the star appear ever so slightly fainter for a few hours (see Figure 3.1). The amount of light blocked is tiny, 1% for a big planet like Jupiter around the Sun, and less than 0.01% for a rocky planet the size of the Earth. But crucially, these drops in brightness are stronger than the typical flickering in stellar brightness, and therefore detectable.

There is a catch. Most planets—over 99.5% of them—can never be detected by this method because their orbits pass above or below the line of sight to the star as viewed from the Earth. To find transiting planets, you would have to look at about 200 stars with planets to detect just one transit. Worse, while it takes the Earth 365 days to complete one orbit of the Sun, for any given observer it only transits the disk of the Sun for 12 hours in that orbit, and only once a year. Blink, and you'll miss it.

These two challenges drove Kepler's design. By launching into space, and into an Earth trailing orbit, the telescope could stare at the same patch of sky almost

doi:10.1088/2514-3433/ab9823ch3

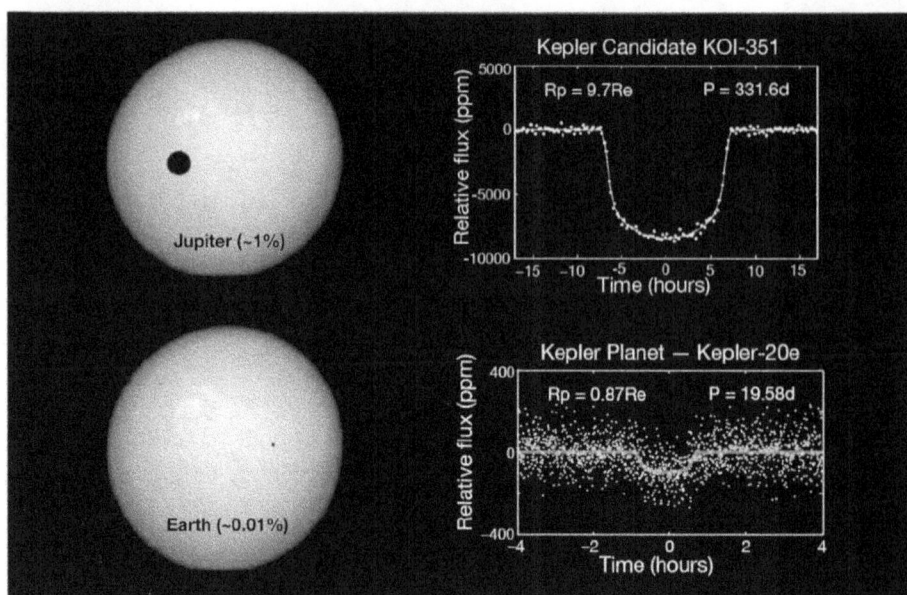

Figure 3.1. Left: Relative sizes of Jupiter (top) and Earth (bottom) as they transit the Sun. Right: Real transits, as seen by Kepler. A Jupiter-sized planet (top) blocks out about 1% of the star's light. An Earth-sized planet produces a much smaller dip, so small it no longer stands out against the noise in the data. Credit: NASA/Ames/Stencil.

uninterrupted for years at a time.[1] An optical telescope on the Earth can only observe during the night, and even the Hubble, in low Earth orbit, can typically only observe for 45 minutes at a time before either the Sun, or the Earth, blocks its view. Kepler had no such constraints.

To look at enough stars to be sure to detect enough transits, the telescope had a large field of view and was equipped with an enormous CCD camera. While the sensor in a typical smart phone is smaller than a grain of kosher salt, Kepler's sensor was as big as a dinner plate. The area of the sky it covered was similarly huge, over 100 deg^2 at a time or as much sky as you cover with your entire hand held at arms length. A single snapshot would capture literally millions of stars.

These two requirements posed a problem. Kepler took a full image continuously every 30 minutes. But the spacecraft was so far away that it would take longer than 30 minutes to download each picture in its entirety and on-board memory storage could not keep up with this large data collection. Worse, the only radio dishes big enough to detect the weak signals from Kepler were shared between all other NASA deep space missions, thus they could only spend a few days a month pointed at Kepler.

The solution was to define small regions within each image around the brightest stars as useful and worth saving. On-board software would clip out and save these

---

[1] The telescope would stop observing for a day each month to download the data it had collected and to upload software updates.

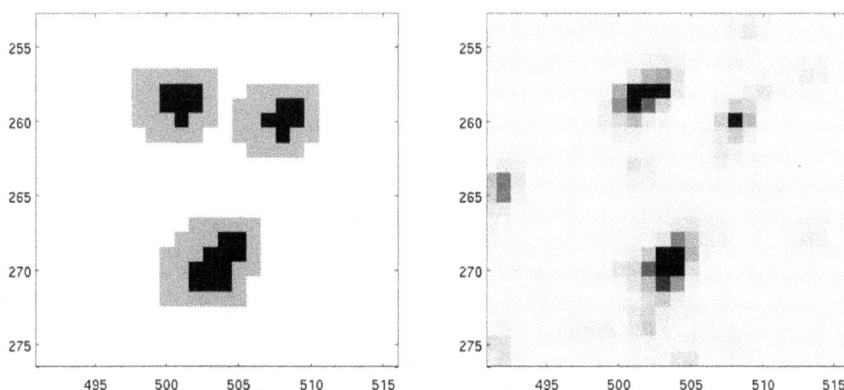

Figure 3.2. The limited bandwidth for download meant that most of Kepler's data had to be discarded before it was downloaded to the Earth. Careful analysis was necessary before launch to ensure the most important data was selected to be saved. The black and gray pixels in the left panel here show the pixels selected for download from a small patch of the field of view. The right panel shows the location of the stars in that patch. (From Bryson et al. 2010.)

"postage stamps" around each interesting star, discarding the rest of the data (Figure 3.2). The resulting data set was small enough to be saved on-board, and downloaded once a month. Kepler could save data on over 150,000 stars at any one time,[2] of which about 20,000 were bright enough that we could hope to detect the tiny dip from a transiting, small Earth-sized planet.

Of course at the time of launch, the Kepler science team had no idea how many planets they would find. Based on radial velocity surveys, they knew that between one in every 10–20 Sun-like stars hosted a massive, Jupiter-sized planet in a short period orbit. There were arguments that these Jupiters had formed much further away from their parent stars and migrated inwards for a variety of reasons, destroying any inner planets in the process. From the 20,000 brightest stars in the survey, in the worst case scenario the mission might only find 10–20 planets. "When I joined a year after launch, the team was confident we'd find lots of planets, but it was clear that lots meant tens, not thousands," recalls Susan Mullally, who joined in 2010. "We really weren't ready for what we found."

3.2 Sipping from the Fire Hose: The Deluge of Exoplanet Discoveries

3.2.1 Behold! New Worlds

The Kepler data, with its rapid cadence, long baseline, and super precision photometry revealed the variety of exoplanets that exist in our Galaxy. Kepler found small, lava worlds because of its high photometric precision, worlds in one year orbits because of the long baseline, and worlds around binary stars because of the

---

[2] 512 of the postage stamps could be readout and saved every minute, the rest, the standard long cadence observations, were readout and saved every 30 minutes.

uninterrupted coverage. Here we outline some of the more notable discoveries made possible for the first time by the Kepler and K2 missions.

Naming Exoplanets

Contributed by Stephen Kane, University of California, Riverside.

One important aspect of exoplanet discovery to keep in mind is that its roots lie firmly within the realm of stellar astrophysics. In particular, observations of binary stars had been occurring for many decades before the first exoplanet discovery and many lessons were learned from the radial velocity and eclipse (transit) observations of eclipsing binary stars. All exoplanet hunters needed to do was to push these techniques into a small mass/size regime, that is, find companions that are "sub-stellar" in nature.

For many years, observations of stars had led to discoveries of new stellar companions, previously unknown due to their small size and/or small angular separation from the primary star. These findings morphed into a naming convention such that the primary (brightest) star was called "A," the next brightest was called "B," and so on. For example, the brightest star in the night sky, Sirius, was found in 1862 to have a white dwarf companion. As such, the primary is called Sirius A and the white dwarf was named Sirius B.

Since many of the exoplanet observational techniques were borrowed from binary stars, the exoplanet naming convention also borrowed from the same arena. However, to make it clear that the discovered companions are sub-stellar rather than stellar, it was decided to use a lower case letter rather than upper case. Therefore, the first planet discovered within a system is denoted "b" (with the host star being "a" in this case, but that designation is not used), the second planet "c," etc. As for the stellar binary case, the letter "A" is reserved for the host star, being the primary object within the system. For example, the Sun-like star 51 Pegasi was one of the first to have a planet discovered, which was then named 51 Pegasi b.

As can be seen, the name of the star is usually used as the prefix for the exoplanet name. However, there are numerous occasions when the host star is sufficiently faint that it does not have an easily recognizable name, and is often recorded in terms of its numerical sky coordinates.[3] In those cases, it is common for the star to receive a new designation that adopts the name of the survey that discovered the planet, for example WASP-10 b or Kepler-11 d. In the latter case, the name may be interpreted as the third planet found to orbit Kepler-11, which in turn was the eleventh planetary system discovered by the Kepler mission.

Like any sufficiently complex book-keeping exercise, the exoplanet naming convention is not without its complications. As noted above, planets are designated in order of discovery, not distance from the host star, and so the names of planets within a system may not be in alphabetical order with increasing distance from their host star. Furthermore, it is occasionally found that a previously reported planet is a false-alarm and so some systems may have letters missing from their exoplanet inventory. Planets that orbit more than one star can be complicated, but such cases are a solvable problem by merging both binary star and exoplanet naming schemes. For example, a planet

---

[3] For example, the star 2MASS J19092683+3842505 is also known as Gaia DR2 2099606483621385216 but is better known in the exoplanet community as Kepler 21. The numeric names are often referred to as the star's "phone number" and usually derive from large star catalogues.

orbiting one component of a binary would use the name of the star including either the "A" or "B" component designation followed by the letter of the planet (for example, 16 Cygni B b). A planet orbiting both stars would use both stellar component designations followed by the planet letter (for example, Kepler-16(AB) b). For all of its complications, it can definitely be said that the current exoplanet naming system is information rich as it contains a partial record of the pathway to discovery for each planet.

3.2.2 Terrestrial Worlds

Given that the primary aim of the Kepler mission was to detect terrestrial planets, the first such rocky planet detection by Kepler was highly anticipated. The first nine transiting systems observed by Kepler were a mixture of previously discovered and newly identified systems consisting of gas giant planets Borucki et al. (2011a). The exquisite photometry provided by Kepler for these systems generated considerable excitement, as the data not only provided significantly updated parameters for known systems, but demonstrated the quality of the Kepler data and its capability in the detection of new worlds.

The discovery of the first rocky planet occurred mere months into the observations made by the Kepler spacecraft. The new planet, dubbed Kepler-10b, was measured to have a radius of 1.47 Earth radii and a mass of 3.72 Earth masses (Weiss et al. 2016). This placed the planet firmly within the terrestrial regime and was thus hailed as the first rocky planet discovered by Kepler, achieving an important milestone for the planet-hunting mission (Batalha et al. 2011). The top panel of Figure 3.3 shows the discovery light-curve for the planet, published by Batalha et al. (2011). Due to the relatively small signal of the planet crossing the star, the discovery included eight months of data to fully tease out the planetary signature. However, a remarkable aspect of the planet was its close proximity to the host star. With an orbital period of 0.84 days, the planet lies at a distance from the host star that is less than 2% of the Earth–Sun distance. The planet is thus likely tidally locked to the host star, such that the same side always faces the star. Furthermore, the planet has an extreme temperature difference between the day and night sides, as the day side is highly irradiated by the star (see bottom panel of Figure 3.3). Further examination of the Kepler data led to the conclusion that the day side may in fact be covered with lava, explaining the phase variations observed in the Kepler data (Rouan et al. 2011). This first terrestrial planet discovery by Kepler became the fascinating subject of both exoplanet characterization and formation scenarios.

3.2.3 Habitable Zone Worlds

The primary aim of the Kepler mission extended beyond the discovery of rocky worlds. A fundamental purpose was to measure how common rocky planets are that might have conditions suitable for life (Borucki et al. 2010). This is generally represented by rocky planets that lie within the habitable zone (HZ) of their host star (Kasting et al. 1993; Kopparapu et al. 2013, 2014), defined as the region around a

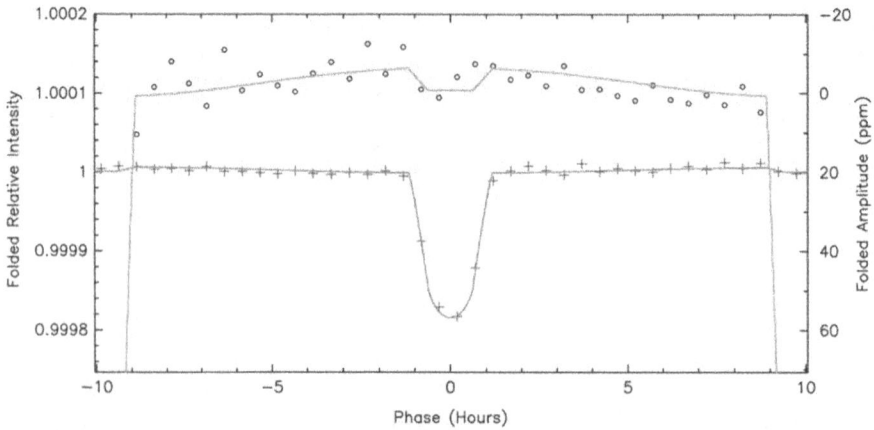

Figure 3.3. Top: Discovery data from the Kepler mission of the planet Kepler-10b (Batalha et al. 2011), the first rocky planet discovered by Kepler. Bottom: Artists depiction of Kepler-10b, a rocky planet orbiting so close to its star that the surface is blasted with radiation, resulting in an extremely hot day side of the planet. Credit: NASA/Kepler Mission/Dana Berry.

star where water may exist in a liquid state on the surface of a planet if there is sufficient atmospheric pressure. Finding such HZ planets is not trivial given the combination of their small size and relatively large distance (i.e., long orbital periods) from their respective host stars, and so required a minimum time baseline of Kepler photometry before such detections could occur.

The first significant finding in the context of Kepler HZ planets occurred with the discovery of Kepler-22b, a 2.4 Earth radius planet in the HZ of a star similar to the Sun (Borucki et al. 2012). Although the planet is unlikely to be rocky in nature due to its relatively large size, the discovery demonstrated that Kepler photometry was sufficient to probe into the HZ of the host stars being monitored. A further success arrived with the discovery of the five-planet system, Kepler-62 (Borucki et al. 2013).

The outermost planets, e and f, have radii of 1.6 and 1.4 Earth radii respectively, and so are likely to be super-Earth planets with a rocky composition. The e and f planets also lie within the HZ of the host star and provided the first Kepler planet discoveries that aligned with the mission objectives of detecting terrestrial planets in the HZ. Another milestone was achieved when the planet Kepler-186f was discovered using Kepler observations (Quintana et al. 2014). The particular significance of the discovery was due to the planet having a size of only 10% larger than the Earth, and orbiting a star much smaller than the Sun. The similarity in size between Kepler-186f and Earth created substantial confidence that the surface conditions could be inferred as being likely temperate, the extent of which generated an enormous amount of publicity. Shown in Figure 3.4 is an artist depiction of the Kepler-186 system in direct comparison to the solar system, demonstrating both the similar sizes of Kepler-186f and Earth and also the compact architecture of the Kepler-186 system. Since Kepler-186f is located toward the outer edge of the HZ, the planet is depicted as having liquid surface water but with relatively large regions of polar ice.

Even with the numerous individual HZ planet discoveries, most of those found were in the HZ regions of low-mass stars since the relatively short orbital periods and larger transit depths of such HZ planets made them significantly easier to find. Thus, the discovery of a possible terrestrial planet in the HZ of a star almost identical to the Sun represented a further milestone achievement of the Kepler mission. The planet, named Kepler-452b, has a radius of 1.6 Earth radii and an orbital period of 384 days, bearing remarkable similarity to the orbit, if not size, of the Earth (Jenkins et al. 2015). It is unknown if the planet in this case is truly of rocky composition, but the detection of the planet is a legacy of the power of Kepler data to discover planets in even Earth-like orbits. The success of Kepler in finding

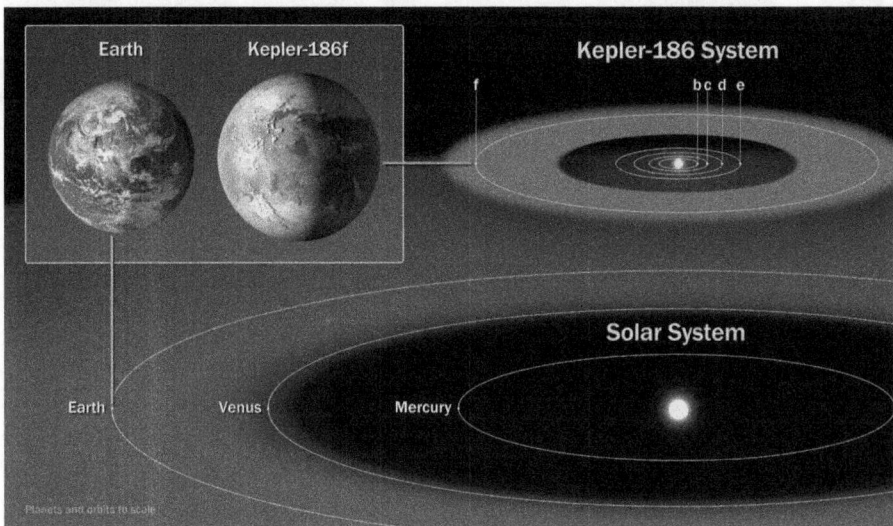

Figure 3.4. An artist depiction of the Kepler-186 system, showing a comparison of the relative sizes of Kepler-186f and Earth, and a size comparison of the Kepler-186 system to the solar system. The habitable zone region around both Kepler-186 and the Sun is shown in green. Credit: NASA Ames/SETI Institute/JPL-Caltech.

HZ planets orbiting a variety of star types continued throughout the remainder of the mission. Toward the end of the Kepler mission, a catalog of HZ planets was published to both encapsulate the HZ planet findings of Kepler and to analyze the statistical properties of the sample (Kane et al. 2016). This work identified a total of 29 potentially terrestrial planets with radii less than two Earth radii within the broadest definition of the HZ (the so-called "optimistic" HZ) and further found that planets are equally common inside the HZ compared with the region interior to the HZ where conditions are likely too hot for surface liquid water. Figure 3.5 shows a representation of the HZ as a function of stellar temperature and the energy received by the planet compared with the Earth. The color-shaded region, bounded by the "Recent Venus" and "Early Mars" lines, indicates the extent of the optimistic HZ and the 29 planets identified by Kane et al. (2016) are shown with their Kepler designations. By providing the first large sample of terrestrial HZ planets, the Kepler mission enabled the first robust calculations on the occurrence rate of planets that may be similar to Earth.

3.2.4 The First Circumbinary Planets

With the vast amount of precision photometry acquired on many thousands of stars, it was anticipated that many unusual planetary systems outside of the primary mission goals would be discovered. One of the most exciting of these discoveries came in 2011 when Kepler data was used to detect a planet orbiting around the center of mass of two stars, otherwise known as a circumbinary planet (Doyle et al.

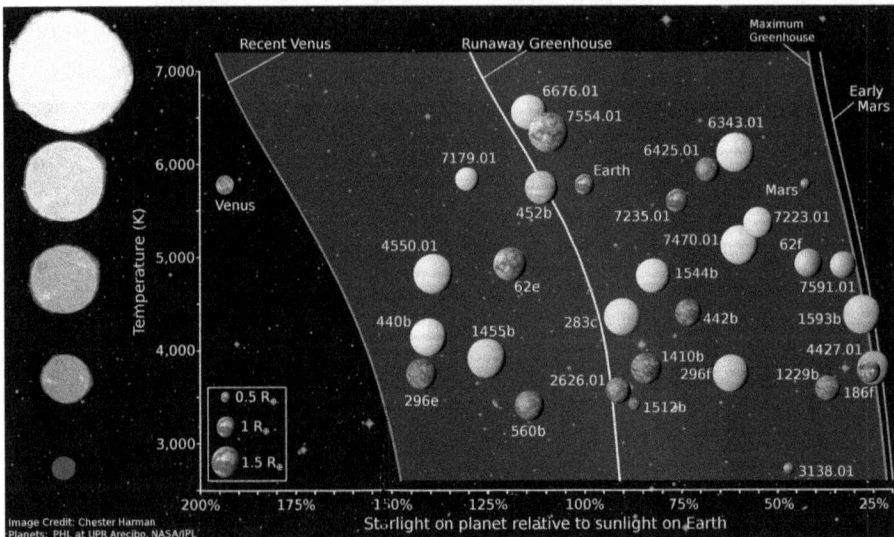

Figure 3.5. A representation of the HZ, showing the dependence on stellar temperature and the energy received by the planet from the star relative to the energy received by the Earth from the Sun. The color-shaded region, bounded by the red and orange lines, indicates the extent of the HZ. The location of Kepler planets are shown with their Kepler designations and the solar system planets of Venus, Earth, and Mars are shown for comparison. Credit: Planets:PHL@UPRArecibo/NASA/JPL/Chester Harmon.

2011). In such cases, the modeling of the data is understandably complex as one needs to account for the motion of the stellar binary in conjunction with the orbiting planet. The new planet was named Kepler-16(AB)b, where the "AB" represents the A and B components of the binary star. Shown in the top panel of Figure 3.6 are the Kepler photometry for the system. The data clearly show the eclipses of the stars on each other, shown in blue and yellow. Careful analysis of the data revealed the presence of the planet as it transited both the A (green) and B (red) components of the binary. These data were first presented at an exoplanet conference held in Jackson Hole, Wyoming, in 2011. The fantastic nature of the system combined with the impressive analysis by the discovery teams prompted a standing ovation from the audience, something quite rare for science talks. The media hailed the discovery as a real-life "Tatooine" from the movie series Star Wars, although the planet is neither terrestrial nor habitable (Kane & Hinkel 2013). Numerous artists depictions

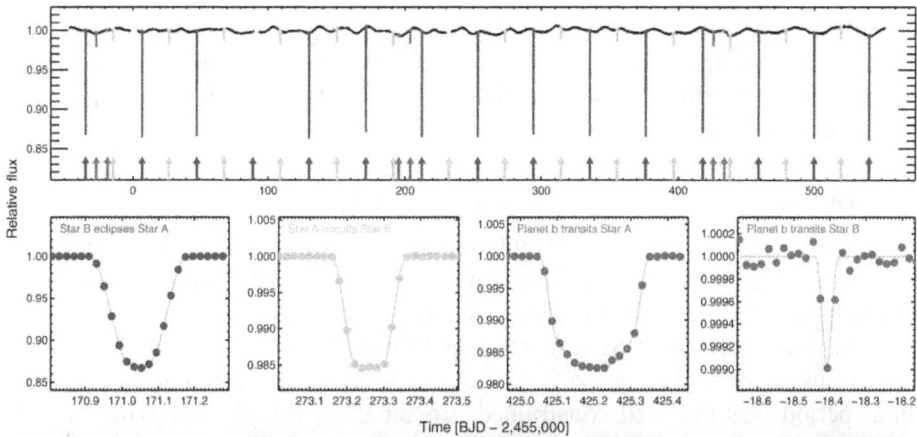

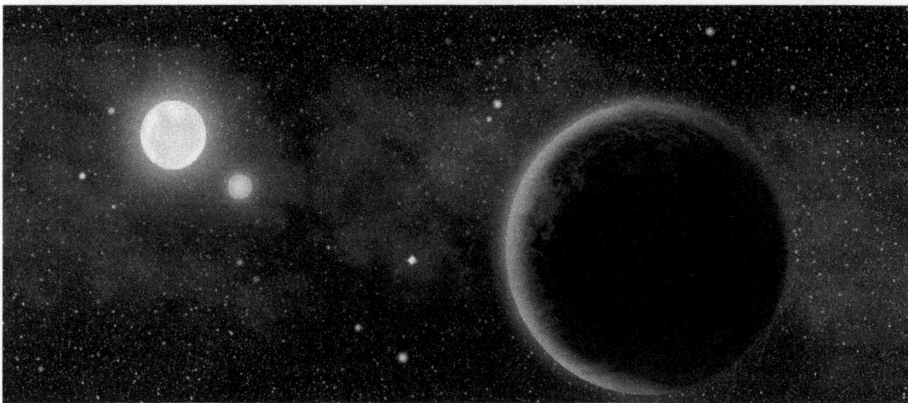

Figure 3.6. Top: Kepler photometry of the binary star designated Kepler-16. The two components of the binary star regularly eclipse each other, seen in the blue and yellow transits. The presence of a planet orbits the binary was revealed via the transits shown in green and red (From Doyle et al. 2011. Reprinted with permission from AAAS.). Bottom: An artists impression of the planet Kepler-16(AB)b orbiting the two stars of the Kepler-16 system. Credit: Ara Syndicate.

of this planet with two "suns" were created, including the example shown in the bottom panel of Figure 3.6.

The discovery of Kepler-16(AB)b triggered a gold-rush of circumbinary planets in the Kepler data, especially as there was already known to be a strong prevalence of eclipsing binaries observed by Kepler (Prša et al. 2011). New circumbinary system discoveries included Kepler 34 and Kepler 35 (Welsh et al. 2012), Kepler-47 (Orosz et al. 2012, 2019), Kepler-413 (Kostov et al. 2014), Kepler-453 (Welsh et al. 2015), and Kepler-1647 (Kostov et al. 2016). The circumbinary planets discovered are all gas giants with a variety of orbits around a range of stellar types. These circumbinary planet discoveries from Kepler observations have led to a new field of exoplanet exploration, and have shed light on the variety of planetary system configurations that are possible.

3.3 Exoplanets Discovered with K2

Due to K2's larger combined field of view and greater number of stars observed, its planet discoveries are in many cases more diverse and extreme than the original Kepler mission. In this section, we give a short overview of some of K2's most interesting discoveries.

3.3.1 Benchmark Systems Around Nearby, Bright Stars

The very first K2 planet discoveries fulfilled the mission's promise to discover small planets around bright stars. The first K2 planet, a super-Earth/mini-Neptune around a bright K-dwarf called HIP 116454, was detected in data from a short engineering test of the spacecraft's new pointing strategy (Vanderburg et al. 2015b). HIP 116454 b only transited once during the 9 days of engineering data (see Figure 3.7), so its orbital period was not well constrained. Fortunately, the planet's host star, HIP 116454 (also known as K2-2), was bright enough that radial velocity observations could both determine the planet's precise orbital period and measure its mass.

The discovery of a small planet around a star bright enough for detailed follow-up observations in just nine days of engineering data was a sign of future successes to come. Shortly after the release of the K2's first full-length observational campaign, Crossfield et al. (2015) reported a system of three planets around a nearby M-dwarf called K2-3. Particularly exciting was a small (1.5 $R_\oplus$) outer planet called K2-3 d orbiting within the circumstellar habitable zone. Planets the size of K2-3 d often lack thick gaseous envelopes (Rogers 2015), generating excitement that K2-3 d might be a rocky, potentially-habitable planet, but follow-up radial velocity observations have shown that this is likely not the case (Damasso et al. 2018; Kosiarek et al. 2019). In that same Campaign 1 data set, another nearby M-dwarf was found to host a small (2.4 $R_\oplus$) planet in its habitable zone (Montet et al. 2015). Radial velocities (Cloutier et al. 2019) and Hubble Space Telescope transit spectroscopy observations (Tsiaras et al. 2019a; Benneke et al. 2019) showed that while the planet is not Earth-like and has a thick hydrogen/helium envelope, its atmosphere contains water vapor.

As the K2 mission progressed and more of the sky was observed, it revealed planetary systems with increasingly complex architectures and around ever brighter

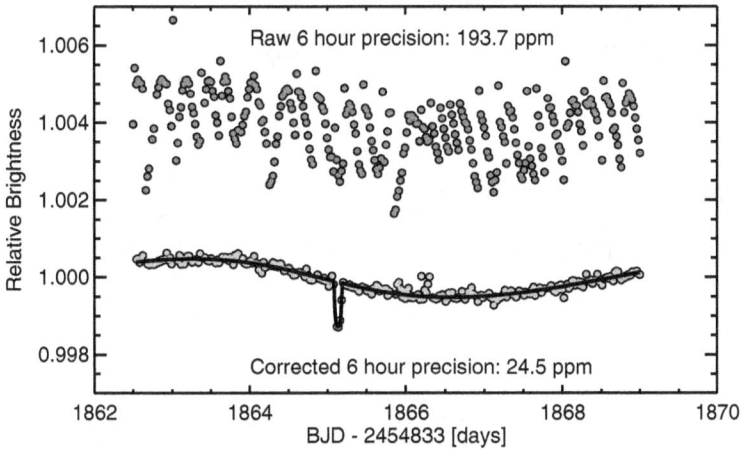

Figure 3.7. HIP 116454 discovery light-curve. Blue points are raw aperture photometry, and orange points are the systematics-corrected light-curve. The black line is a model of the transit and low-frequency light-curve variability. Only one transit of HIP 116454 b was observed during the nine days of K2 engineering data, but follow-up radial velocity observations were able to measure the planet's mass and orbital period and confirm the mission's first planet discovery. Reproduced from Vanderburg et al. (2015b). © 2015. The American Astronomical Society. All rights reserved.

stars. The brightest host star for any confirmed or validated K2 planet is HD 212657 (also known as K2-167 b; Mayo et al. 2018), with a V-band magnitude of 8.24. An even brighter planet candidate around the $V = 6.9$ star HD 73344 has been reported by Yu et al. (2018) and awaits confirmation or validation. Multi-planet systems discovered around slightly fainter host stars like HD 3167, HD 106315, and GJ 9827 (Vanderburg et al. 2016c; Crossfield et al. 2017; Rodriguez et al. 2017; Niraula et al. 2017; Rodriguez et al. 2018a) are providing rich prospects for follow-up observations. Planets in all three systems have radial velocity mass measurements (Christiansen et al. 2017; Gandolfi et al. 2017; Barros et al. 2017; Teske et al. 2018; Prieto-Arranz et al. 2018; Rice et al. 2019), one planet in each system is being observed by the Hubble Space Telescope (Crossfield 2017), and two of the planets (HD 3167 c and HD 106315 c) have measured spin/orbit angles (Zhou et al. 2018; Dalal et al. 2019).

A particularly interesting and challenging multi-planetary system orbits the bright ($V = 8.9$) star HIP 41378. Initial K2 observations in Campaign 5 detected two super-Earths with periods of 15.6 and 31.7 days, and three larger planets which each only transited once during the ≈75 days of data (Vanderburg et al. 2016a). The system's high multiplicity, apparently long orbital periods, and very bright host star made HIP 41378 an attractive target for future observations, but without precisely measured orbital periods for the outer planets, follow-up opportunities were limited. Fortunately, with its fuel reserves dwindling toward the end of the mission, K2 re-observed HIP 41378 in Campaign 18 and re-detected the transits of two of the three outer planets detected in Campaign 5. These new data yielded asteroseismic measurements of the host star's parameters (Lund et al. 2019) and a crucial set of precise possible orbital periods for the two planets (Becker et al. 2019; Berardo et al. 2019). Armed with this new information, Santerne et al. (2019) used hundreds of radial velocity observations from spectrographs around the globe to uniquely determine the period and mass of the outer planet HIP 41378 f. The planets in the HIP 41378 system have surprisingly low densities: HIP 41378 f is nearly as large as Saturn, but has a mass smaller than that of Neptune. The bright host star and extremely low surface gravity makes HIP 41378 a very appealing target for transit spectroscopy observations with either the Hubble Space Telescope or James Webb Space Telescope.

3.3.2 Unusual Architectures

The original Kepler mission discovered systems with spectacular architectures by continuously observing stars for nearly four years. Though K2 did not have the luxury of such long observational baselines on individual systems, it still discovered systems with new and unusual planetary architectures by virtue of sheer numbers. Early on in the mission, the planetary system around K2-19 garnered interest due to a nearly exact 3:2 period ratio between two sub-Saturn sized planets and the detection of transit timing variations in ground-based follow-up observations (Armstrong et al. 2015). Later analyses detected (Vanderburg et al. 2016b) and validated (Sinukoff et al. 2016) a third, interior Earth-sized planet in the system. Follow up transit and radial velocity observations (Barros et al. 2015; Dai et al. 2016; Narita et al. 2015; Petigura et al. 2020) yielded precise orbital parameters and mass and radius measurements for the two sub-Saturns and showed that despite their close proximity to the 3:2 period ratio, the two planets are not in mean motion resonance.

For two decades after the discovery of the first hot Jupiter, no close planetary companions had ever been discovered in a hot Jupiter system despite numerous attempts (Miller-Ricci et al. 2008; Steffen et al. 2012a), so the K2 discovery of two small planets orbiting near the hot Jupiter WASP-47 b was a surprise (Becker et al. 2015). WASP-47 is known to host at least four planets: a transiting hot Jupiter (discovered from ground-based observations by Hellier et al. 2012), two small transiting planets flanking the hot Jupiter (see Figure 3.8), and an outer Jovian companion detected in radial velocities (Neveu-VanMalle et al. 2016). Follow-up observations quickly measured the planet masses (Dai et al. 2015; Almenara et al. 2016; Sinukoff et al. 2017; Weiss et al. 2017; Vanderburg et al. 2017), the hot Jupiter's spin/orbit angle (Sanchis-Ojeda et al. 2015a), and dynamical calculations constrained the system architecture to be nearly flat (Becker & Adams 2017; Vanderburg et al. 2017; Becker et al. 2017). The discovery of the WASP-47 planetary system has raised more questions than it has answered. How did the system form, and did it form in the same way as most other hot Jupiters (Huang et al. 2016)? The presence of small, close-in planetary companions and the system's flat architecture suggest that WASP-47 cannot have formed by high eccentricity tidal migration (Rasio & Ford 1996), but did the planets all form far from the host star and migrate in (Lin et al. 1996), or did they form in situ (Batygin et al. 2016)?

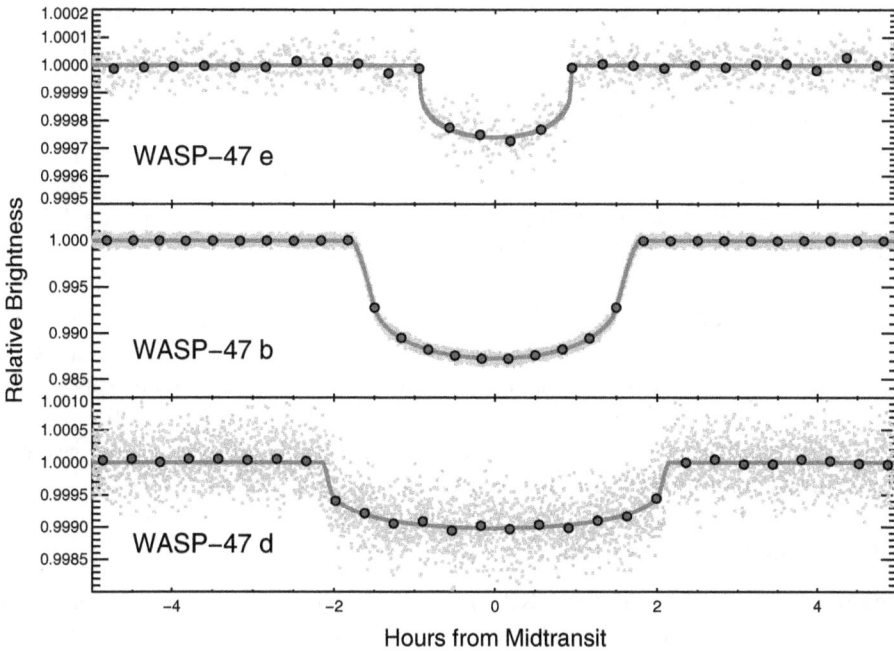

Figure 3.8. K2 light-curves of the hot Jupiter WASP-47 b (middle panel), and the two smaller companions: the ultra-short period super-Earth WASP-47 e (top panel) and the Neptune-sized WASP-47 d (bottom panel). Reproduced from Becker et al. (2015). © 2015. The American Astronomical Society. All rights reserved.

The planetary system orbiting the K-dwarf star K2-266 appears ordinary at first glance, but has a surprising architecture on closer inspection. Rodriguez et al. (2018b) validated at least four planets around K2-266, and identified two other low signal-to-noise ratio planet candidates. The system contains an inner planet with a 16 hour orbital period, two super-Earths near or in a 4:3 mean motion resonance with periods of 14.7 and 19.5 days, a sub-Earth-sized planet in a 7.8 day orbit, and two candidate sub-Earths in 6 and 56 day periods. Detailed modeling of the transit light-curve revealed that while the outer system (3 planets and 2 candidates) all have inclination angles between 87° and 90°, and are likely close to co-planar, the inner 16 hour planet has an inclination angle of 75° and therefore must be misaligned from the rest of the system by at least 12°. Forming a compact, near-resonant multi-planet system with a single inner planet with a significant misalignment is a dynamical challenge and will require future study to explain.

3.3.3 Ultra-short Period Planets

K2 excelled at discovering ultra-short period (USP) planets, or planets with orbital periods less than one day. USP planets are intrinsically rare and almost always smaller than 2 $R_{\oplus}$ (Sanchis-Ojeda et al. 2014), so they were well-matched to K2's ability to cover large areas of sky with high photometric precision. Many of the well-characterized USP planets discovered by K2 are in multi-transiting systems (Becker et al. 2015; Vanderburg et al. 2016c; Rodriguez et al. 2018b; Adams et al. 2017; Malavolta et al. 2018), but the USP planet with the shortest period observed by K2, K2-137 b (Smith et al. 2018), appears to orbit alone. This planet orbits on the brink of tidal disruption and therefore likely has an elevated iron/silicate ratio (Price & Rogers 2020). The brightest stellar hosts to USP planets permit the detection of thermal emission or reflected light from the daysides of these planets. Malavolta et al. (2018) reported the K2 detection of the optical phase curve on the 6.8 hour period planet K2-141 b, which combined with observations with Spitzer should help constrain the planet's albedo and temperature (Kreidberg et al. 2018). The overall sample of USP planets discovered by K2 and characterized by other observations has significantly improved our understanding of the composition and architectures of USP systems. K2 discoveries have revealed that most USP planets appear to be rocky with iron/silicate ratios consistent with that of Earth (Dai et al. 2019) and has shown that significant misalignments (like that between K2-266 b and the rest of its system) may be common in the shortest period planets (Dai et al. 2018).

3.3.4 Exoplanets Through Time

A major scientific driver for the K2 mission was to observe young stars in open clusters and associations. The ecliptic plane happens to contain some of the most famous open clusters in the sky, including Hyades, Pleiades, M67 and Ruprecht 147, the Upper Scorpius association, and the Taurus star-forming region. So far, K2 data have revealed four planets (and one candidate) in the Hyades (Mann et al. 2016a, 2018; Ciardi et al. 2018; Livingston et al. 2018; Vanderburg et al. 2018), eight planets (and one candidate) in Praesepe (Obermeier et al. 2016; Mann et al. 2017; Pepper et al. 2017; Rizzuto et al. 2018; Livingston et al. 2019), one planet (and one brown dwarf) in Ruprecht 147 (Nowak et al. 2017; Curtis et al. 2018), a system of planets in Taurus (David et al. 2019b, 2019a), and a planet in Upper Scorpius (David et al. 2016; Mann et al. 2016b). An important result from these discoveries is that young planets may be puffy—the sample of young cluster planets around M-dwarfs have preferentially larger radii than planets around older field M-dwarfs (Mann et al. 2017). Future observations of these planets to determine masses (and therefore planet densities), eccentricity, and atmospheric characteristics will probe the time evolution of these properties in exoplanet systems (Thao et al. 2019).

3.3.5 Citizen Science

Citizen scientists have played a major role in discoveries from the K2 mission. Two major citizen science projects, Planet Hunters (Fischer et al. 2012) and Exoplanet Explorers (Christiansen et al. 2018), provided a web interface for citizen scientists to inspect and search K2 data online. These efforts resulted in the discovery of dozens of planet candidates and interesting individual systems (Schmitt et al. 2016; Christiansen et al. 2018; Feinstein et al. 2019). Sometimes citizen scientists work more independently and publish their own results. Two citizen scientists, LaCourse & Jacobs (2018), searched K2 light-curves for single transit events using the LCTools software (Schmitt et al. 2019), and compiled a list of 164 candidates, which they published as a Research Note of the AAS. In other cases, citizen scientists work closely with professional astronomers to contribute to discoveries. Some famous examples of planets and systems first identified by citizen scientists include: HIP 41378 b/c/d/e/f (Vanderburg et al. 2016a), WASP-47 d/e (Becker et al. 2015), and K2-141 c (Malavolta et al. 2018).

Citizen scientists specialize in discovering unusual signals which are not necessarily flagged by automatic transit search algorithms (Figure 3.9). These discoveries often include unusual transits or eclipses, with shapes indicating asymmetric occulting objects (Ansdell et al. 2019; Rappaport et al. 2019a). These transits may be caused by dusty debris clouds passing in front of their stars or a planet/star with an orbiting disk or ring system. K2 observations also yielded the Random Transiter (Rappaport et al. 2019b), which is perhaps the most perplexing citizen science discovery since Boyajian's star (Boyajian et al. 2016). The Random Transiter (also known as HD 139139) was observed by K2 to have 28 transit-like dips in its light-curve over the course of 78 days of observations, but the dips had no discernible

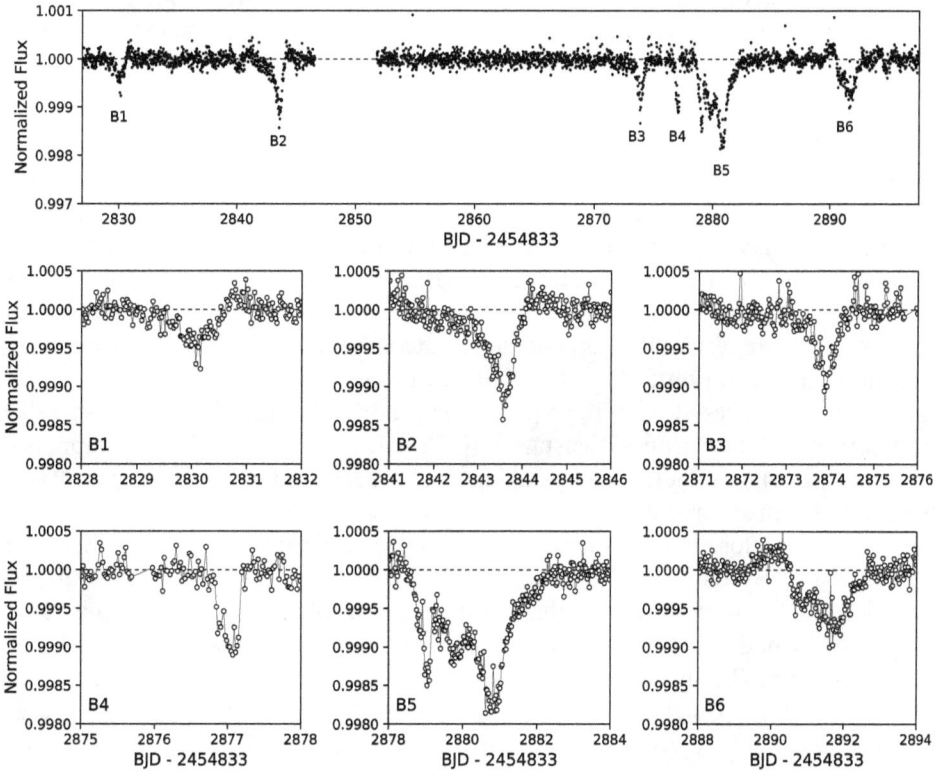

Figure 3.9. K2 light-curve of EPIC 235240266 (top row) and close-up views (middle and bottom rows) of the aperiodic and asymmetric dips identified by citizen scientists (Ansdell et al. 2019). Ansdell et al. (2019) refer to these transits as "little dippers" and speculate that they are caused by star-grazing exocomets. The little dippers demonstrate how citizen scientists can make important contributions to exoplanet research by detecting transits which might have been missed by automatic pipelines. Reproduced from Ansdell et al. (2019), by permission of Oxford University Press on behalf of the Royal Astronomical Society.

periodicities. Rappaport et al. (2019b) considered a slew of different scenarios to explain the dips, including transit timing variations, circumbinary planets, and dusty dipper activity, but found no satisfactory explanation. Theoretical and observational work by the wider community (Schneider 2019a, 2019b; Brzycki et al. 2019) have posed and ruled out some new scenarios, but the origin of the Random Transiter remains a mystery.

3.4 Planets in (Young) Star Clusters

Some ground-based observational surveys took on the task to search for exoplanets orbiting stars within star clusters. The importance of such work is discussed in Lada & Lada (2003). Notable surveys, all of which searched for transiting objects and all turned up empty, were performed by Gilliland et al. (2000) using the Hubble Space telescope to search for exoplanets in the globular cluster 47 Tucanae, Howell et al. (2005) who surveyed the open star cluster NGC 2301, and Pepper & Burke (2006) who looked for exoplanets in NGC 1245.

It was not until Kepler and K2 observed open star clusters that photometric transits were detected and exoplanets discovered. The resulting exoplanet discoveries from these two missions are discussed in detail in Chapter 4.

3.5 Disintegrating Planets

The interiors of planets are very much a mystery because of the difficulty in remote sensing observations to "dig" down below the surface into the interior. Even on our own Earth we have only sampled material some 12,262 m below the surface, about half way to the Earth's mantle. However, nature has provided a way to peer inside an exoplanet: by ripping it apart and sending that material into a tail of gas and dust. These systems are teaching us one possible outcome for the final stages of planetary evolution, disintegration followed by falling onto the star.

An unexpected result from the Kepler mission was the discovery of the so-called disintegrating planets around the stars KIC 12557548b and KOI 2700b (Rappaport et al. 2012, 2014), characterized by asymmetric transits and rapidly changing transit depths. The transits are believed to be due to dusty effluents from roughly Mercury-mass planets in close-in orbits that are undergoing rapid mass loss (Perez-Becker & Chiang 2013). By the end of the Kepler mission, only two such objects had been identified, so the discovery of a third disintegrating planet by K2 (Sanchis-Ojeda et al. 2015b) had the potential to significantly increase our knowledge of these unusual systems (Figure 3.10). Sanchis-Ojeda et al. (2015b) identified an unusual short-period transit signal around the M-dwarf K2-22. Like the disintegrating planets discovered by Kepler, the transits of K2-22b rapidly changed depth (on timescales of ~1 day), but unlike KIC 1255 and KOI 2700, the shape of K2-22's transits were not particularly asymmetric. Sanchis-Ojeda et al. (2015b) modeled the dust clouds around K2-22 and concluded that dissimilar to KIC 1255 and KOI 2700, the dust around K2-22 likely forms clouds that both lead and trail the source planet, causing the transits to appear more symmetric. K2-22 helped demonstrate the diversity in this still mysterious class of exoplanets. In addition to this unusual

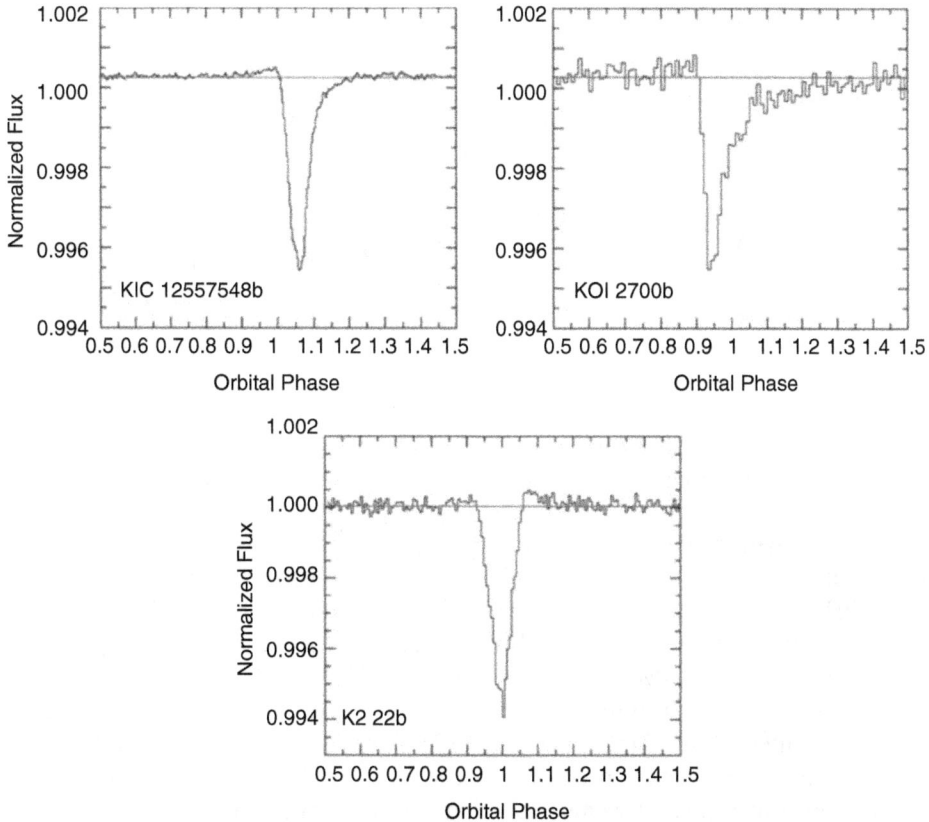

Figure 3.10. The average transit shapes of the transiting planets orbiting main sequence stars. Reprinted by permission from Springer Nature: Springer, van Lieshout & Rappaport (2018).

transit shape, all three orbit a main sequence star with an orbital period of less than a day and show variability in their transit depth. Each of these systems provide an opportunity to study the interiors of rocky planets.

The average shape of the observed transit is well modeled by a small, hot, rocky body (similar in size to Mercury), releasing gas and dust. This gas and dust form a tail which creates the slow rise in brightness after the transit. The increase in brightness just before the transit likely occurs because of star light traveling away from the Earth being scattered off of the dust tail and back toward the Earth. This scattering occurs throughout the event, but is only significant compared to the other effects immediately before the rocky body transits (van Lieshout & Rappaport 2018).

Given the estimated size of the bodies and the estimated amount of gas and dust required to create the transit, it is estimated that such disruptions have been ongoing for about 103 yr, 1 Myr, and 1 Gyr, respectively in the three systems discussed above (van Lieshout & Rappaport 2018). These relatively short timescales imply that such objects are rather rare, however it is possible that more will be found with future transit surveys. Future observations will continue to explore the size of the gas and dust spewing off of these systems, shedding light on our understanding of exoplanet (interior) composition and evolution.

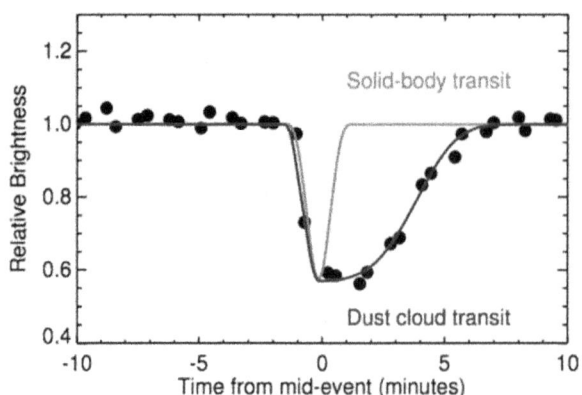

Figure 3.11. The shape of the averaged transit from WD1157+017 (black points). Two models show what a transit shape from a solid body and what a transit shape from a dust cloud would look like.

A similar phenomenon to the disintegrating planets was discovered to be taking place around the white dwarf WD 1145+017 (Vanderburg et al. 2015a). Vanderburg et al. (2015a) identified a periodic transit-like signal around WD 1145+017 that repeated every 4.5 hr. Ground-based photometry and a more careful inspection of the K2 light-curve revealed that the transits of WD 1145+017 changed depth and were asymmetric (see Figure 3.11). However, unlike the transits of the main sequence disintegrating planets like KIC 1255, the light-curves of WD 1145+017 seemed to show evidence for multiple objects in close orbits. Given these differences and the white dwarf host, it seems likely that a different mechanism is at play around WD 1145+017 than around the main-sequence disintegrating planets. Over the past couple of decades, a scientific consensus has emerged that many white dwarf stars show evidence of accreted planetary materials in their spectra (Zuckerman et al. 2010). Because the surface gravity is so high on a white dwarf, it is not possible for such materials to exist for long at the top of the atmosphere where we detect see it. These materials are believed to be debris from rocky bodies scattered inwards by gravitational interactions and tidally disrupted by the white dwarf's extreme gravity (Debes & Sigurdsson 2002). The transits of WD 1145+017, therefore appear to be caused by this disrupted planetary debris. For the first time, WD 1145+017 has given astronomers the chance to study planetary disruption in real time.

3.6 Planets for Atmospheric Characterization

Most of Kepler's exoplanets are orbiting stars that are too dim to allow us to learn much more than the exoplanet size and the exoplanet's distance from the star. However, for the more massive planets orbiting some of the brighter stars in reasonably short orbital periods, it is possible to measure the mass of the planet and in some cases begin to constrain the composition of the planet's atmosphere. Since their discovery, the Hubble Space Telescope and the Spitzer Space Telescope have been able to gather observations that teach us about the atmosphere's of some of the Kepler and K2 discovered exoplanets.

One of Kepler's first confirmations was Kepler-7b Latham et al. (2010), a planet about half the mass of Jupiter but about 1.5 times larger in radius. Using Spitzer, Demory et al. (2013) was able to create a map of the atmospheric surface of this hot Jupiter planet. The detected high reflectivity of Kepler-7b, measured just before the planet passes behind the star, reveals that there are clouds in its upper atmosphere.

Because K2 was able to observe more of the sky, it was able to measure more bright and nearby stars. Planets discovered around these stars also have atmospheres that can be studied. K2-18 b is an example of one of these gems. Using the Hubble Space Telescope, Benneke et al. (2019) found evidence of water in the atmosphere of this temperate, mini-Neptune-sized planet. Tsiaras et al. (2019b) show how the transit depth changes with wavelength from 1.1 to 1.6 microns using the Hubble Space Telescope (Figure 3.12). The increase in the observed transit depth around 1.4 microns is consistent with water vapor in the atmosphere.

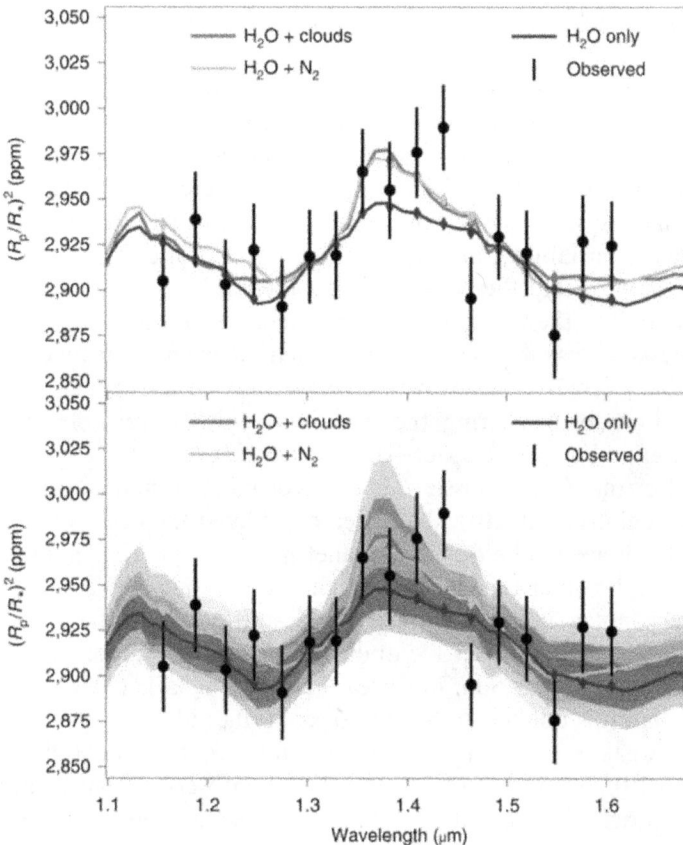

Figure 3.12. Transmission spectra of K2-18 b as published by Tsiaras et al. (2019b). The black points show the measurements. Four models containing water are over-plotted. The lower plot shows the 1 and 2 sigma uncertainty ranges for each model. Reprinted by permission from Springer Nature: Macmillan Publisher Ltd, Nature, © 2019.

3.7 Did Kepler Detect Earth Analogs?

The Kepler mission was the first scientific experiment capable of detecting the transit signal from an Earth-sized planet in a one year orbit around a Sun-like star. So, it is a natural question to ask, does the Kepler data set contain any Earth-analogs? In truth, only time will give us the answer. Currently there is no confirmed planet from Kepler that can be truly called an Earth twin with high confidence. A few exoplanets were detected that got close to being an Earth analog and it is worth discussing what we know of these discoveries.

To start, we emphasize that with transits alone, it is impossible to determine if a planet is truly Earth-like. Transits only reveal certain properties of a planet: the planet radius, the orbital period, and the distance from the star. Other information is needed to determine what the planet is made of and if it has liquid water or an atmosphere. For most of Kepler's planets, these types of observations are impossible with current facilities and technology. As a result, when discussing similarities to Earth, the comparison stops at the planet size and whether it is the right distance away from its star to be in the habitable zone, sometimes known as the Goldilocks zone.

Even if you define Earth analogs to simply mean that the planet is the right size and the right distance from the right kind of star, Kepler data alone are not enough to be convincing that you have a signal from a transiting exoplanet. When considering a signal consistent with an Earth twin, there are several other scenarios that could make the same signal, for example, an eclipsing binary blended with the target star, stellar variability, or various instrumental noise. Transits are frequently validated by showing that the possibility that these false positive scenarios are very unlikely compared to the possibility that the signal comes from a planet. However, some of the tests we use to rule-out these scenarios are not effective for very weak signals containing few transits, as would be the case for a true Earth analog signal. Full confirmation by measuring the mass will only be possible with improved instrumentation and/or future space-based telescopes.

At the end of the Kepler mission, both Kepler-186f (Quintana et al. 2014) and Kepler-452b (Jenkins et al. 2015) had been considered the best examples of Earth-like planets. Each are unlike our own planet in key ways. Kepler-186f is certainly rocky and sits in the outer edge of the star's habitable zone, however it orbits a cool, small star very much unlike our Sun. Such stars are known to produce large, powerful flares that may impact the ability of planets to support life. Kepler-452b orbits a star similar to our Sun, however the planet is larger than our Earth. It is large enough that it is unlikely to have a rocky surface like our Earth. Additionally, because of the weakness of the transit signal, some recent work (Mullally et al. 2018; Burke et al. 2019) calls into question whether the observed transits are significant enough to confirm this world as a validated planet without obtaining additional observations.

Current estimates of the occurrence of Earth analog planets, based on extrapolating Kepler's measurement of the frequency of super-Earths, indicate that the signal of most transiting Earths is buried in the light-curves detected by Kepler.

Eventually someone might find a way to reliably tease the evidence of these worlds from Kepler's measurements, either by applying better data analysis methods that can account for known sources of noise in the data or by first detecting the presence of an Earth-like planet using other observations of these stars. Kepler may have detected a true Earth analog, but without more information we have not been able to determine which of the small dips in the light-curves are the needle and which are just another piece of hay.

3.8 Overall Demographic Trends

Even before attention moved to calculating "Eta-Earth" ($\eta_\oplus$), the frequency of rocky planets orbiting in the habitable zone of Sun-like stars (see Section 3.12), we noticed ever more interesting trends emerging as the exoplanet radius-period diagram became filled in. Figure 3.13 shows the new parameter space probed by the Kepler discoveries, compared to the previously known planets, extending out to nearly 4 yr orbital periods and down to 0.24 $R_\oplus$. Underlying these discoveries is a rich landscape of new information about the population of planets observed by Kepler, a selection of which is discussed here.

3.8.1 Small Exoplanets Are Common

The first feature of Figure 3.13 which stands out is that planets smaller than $\sim 4\ R_\oplus$ ($<0.5\ R_{\text{Jupiter}}$) vastly outnumber larger planets. Given that smaller planets are harder to detect via the transit method than larger planets, any detection bias in this image only serves to exacerbate the difference even further—there are likely many more missing small planets than large planets. This was noted in the earliest planet candidate catalogs (Borucki et al. 2011a), and examined in more detail in Howard et al. (2012). Figure 3.14 shows the number of planets per G+K spectral type star observed by Kepler; a rapid rise is evident in the number of planets at smaller radii. An occurrence rate of 13 ± 0.8% for planets with radii 2–4 $R_\oplus$, orbiting G+K stars with orbital periods shorter than 50 days has been determined. Later studies confirmed this result; although caution should be used in separating the planet radius distribution from the orbital period distribution, they have largely been fit independently and in each case confirm that planet frequency increases with decreasing planet radii. The putative turnover or plateau in the occurrence rate for planets smaller than $\sim 2\ R_\oplus$ in Figure 3.14 was also borne out by later studies; see Section 3.8.3 for more details.

The trend of small planets outnumbering large planets is perhaps unsurprising,[4] given previous examples in astronomy such as the stellar initial mass function or galaxy mass function where the largest objects are the rarest, however their location at short orbital periods was initially a challenge to core-accretion formation theories (see Howard et al. 2012 and discussion therein). Having a significant sampling of the

---

[4] Unsurprising indeed as smaller size objects always outnumber larger size objects. Take, for example, microbes versus insects versus humans, or the size distribution of stones in a stream bed or asteroids in the main belt.

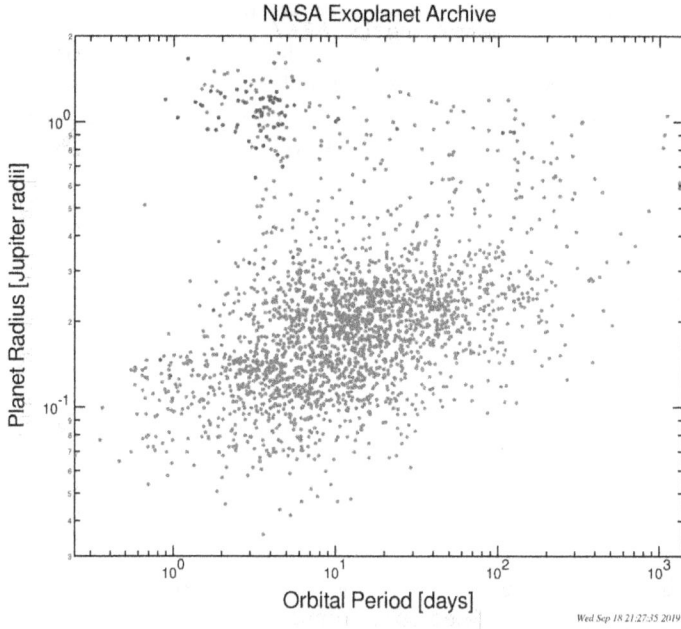

Figure 3.13. The confirmed planets discovered prior to 2010 with measured radii and periods are shown in blue, and the confirmed planets discovered by the Kepler mission are shown in red.

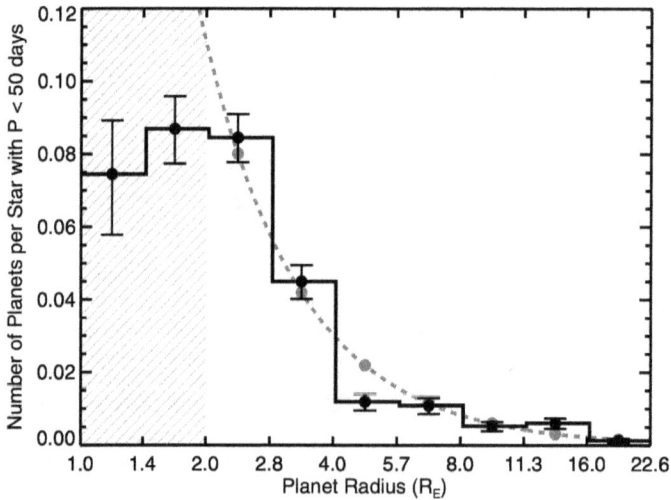

Figure 3.14. An estimate of the underlying distribution of planet sizes for planet candidates orbiting GK stars in the Kepler sample with periods shorter than 50 days. Adapted from Figure 5 of Howard et al. (2012). © 2012. The American Astronomical Society. All rights reserved.

parameter space from large to small planets over a range of orbital periods is crucial for providing input into the ongoing refinement of planet formation and migration theories, both of which must sculpt the distribution of planets ultimately observed by Kepler.

3.8.2 There Are More Planets than Stars

One of the most exciting discoveries out of the Kepler mission is the fact that planets outnumber stars in our Galaxy. There are a number of studies that constrain exoplanet demographics across the main sequence studied by Kepler, from F stars (1–4 $M_\odot$, 6000–7000 K) to M dwarfs (0.1–0.6 $M_\odot$, 2300–3900 K). All indicate that the total frequency of planets increases with decreasing stellar mass, i.e., smaller stars are orbited by more planets. Figure 3.15 shows one example of this increase for small (0.5–2.5 $R_\oplus$) planets with short orbital periods (<10 days), based on results from Hardegree-Ullman et al. (2019), Mulders et al. (2015), and Dressing & Charbonneau (2015). Extrapolating to longer periods, the number of planets per star increases. For instance Dressing & Charbonneau (2015) find that there are 2.5 ± 0.2 small planets from 1–4 $R_\oplus$ per M dwarf with periods shorter than 200 days. Since M-dwarfs comprise ~75% of the stars in the Milky Way Galaxy, it follows that there are, on average, more planets than stars. And that is for the small, short-period planets!

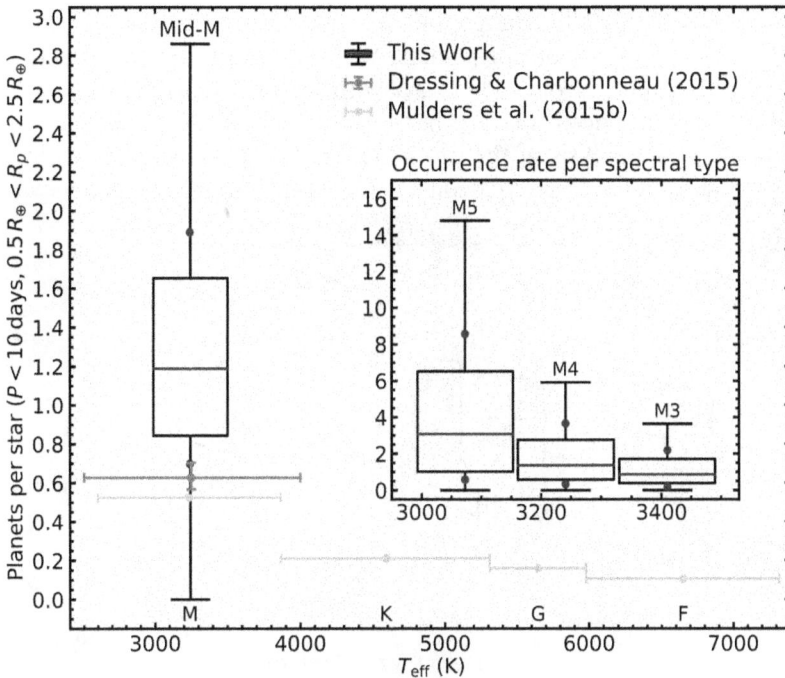

Figure 3.15. The frequency of planets per star with periods <10 days and radii from 0.5–2.5 $R_\oplus$, show an increasing frequency with decreasing stellar effective temperature. The inset shows the frequency of the same planets broken out by spectral type for the mid-M dwarfs. Reproduced from Figure 16 of Hardegree-Ullman et al. (2019). © 2019. The American Astronomical Society. All rights reserved.

If you integrate out to wider orbits and larger planets, the number of planets per star increases even further. Planets are everywhere, and we have the Kepler mission to thank for this knowledge.

One might be tempted to probe the trend described above in more detail. Intuitively, you could assume that the mass of material in a protoplanetary disk was proportional to the mass of the central star, and you would be correct (see, for example Andrews et al. 2013). But then, wouldn't it follow that increasingly massive stars, having more and more material available with which to form planets, would have an increasing frequency of planets? However this is the opposite of what we observe (see Mulders et al. 2015), and remains an open question of research.

3.8.3 Short Period Radius Gap

Early analyses of the Kepler planet candidate sample were limited in the precision with which the stellar radii, and therefore planet radii, could be measured. These uncertainties were, in fact, obscuring a very interesting underlying trend, shown in Figure 3.16. This figure, reproduced from Fulton et al. (2017), shows the radius distribution of the Kepler planet candidates orbiting stars with stellar properties measured by the California Kepler Survey, using high-resolution spectra obtained with Keck/HIRES. Reducing the uncertainties in the stellar radii from 25% to 11% on average allowed the bi-modality of the underlying planet radius distribution at short periods (<100 days) to emerge, the first clear demarcation of the super-Earth and mini-Neptune populations of planets. We note that a further reduction of the stellar radii uncertainties from 11% to 3% with the incorporation of data from the ESA Gaia mission did not reveal further substructure (Fulton & Petigura 2018).

There is ongoing analysis of the properties of this radius gap, which inform the processes by which it could be created. The location of the radius gap appears to be

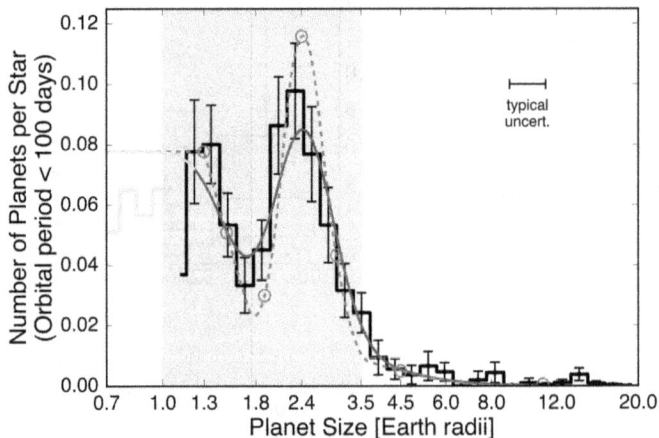

Figure 3.16. The radius distribution of planet candidates with periods <100 days, showing a clear bi-modality in size distribution between planets with radii 1–1.8 $R_\oplus$ (super-Earths) and 1.8–4 $R_\oplus$ (mini-Neptunes). Reproduced from Figure 7 of Fulton et al. (2017). © 2017. The American Astronomical Society. All rights reserved.

related to the mass of the stellar host star, i.e., for higher mass stars, the gap moves to larger radii (see, e.g., Figure 8 of Fulton & Petigura 2018); and to be related to the period of the planet, i.e., for longer period planets, the gap moves to smaller radii (see, e.g., Figure 2 of Van Eylen et al. 2018). As you will see later in this chapter, the radius determination of an exoplanet based solely on its transit depth can have some unforeseen complications. A major one occurs if the exoplanet host star is a binary, causing the transit depth to provide an incorrect planet radius. Teske et al. (2018) studied how this consequence, when applied to Kepler planets, would effect the observed planet radius gap.

There are two competing theories attempting to explain the radius gap. One is that the gap is sculpted by photoevaporation of the planets' primordial atmospheres by stellar insolation (see, e.g., Owen & Wu 2013; Lopez & Fortney 2014). In this scenario, planets above some critical mass cut-off are able to retain their atmospheres, and those below have their atmospheres stripped away, producing very few planets with intermediate-range radii. A second is core-powered mass loss, where radiation pressure from the cooling planetary core blows away light atmospheres, while allowing larger atmospheres to be retained (see, e.g., Ginzburg et al. 2018; Gupta & Schlichting 2019). This too is still an active area of exoplanet research.

3.8.4 Trends Observed in Multi-planetary Systems

There is no better indication that we were unprepared for the number and variety of planetary systems that the Kepler mission would uncover than that the website[5] built to coordinate the follow-up and confirmation of the planet candidates was designed to accommodate only one kind of system: a star and a single planet. After the first year of observations the entire website and underlying database had to be redesigned on the fly to cope with the undeniable truth: Kepler was finding hundreds of multi-planet systems. At current count,[6] Kepler has discovered 567 planets that are in two-planet systems, 305 planets in three-planet systems, 158 planets in four-planet systems, 79 planets in five-planet systems, 17 planets in three separate six-planet systems, and finally a single eight-planet system, KOI-351 (Shallue & Vanderburg 2018). Two notes about the previous statement: (i) additional planets in these systems were discovered by non-Kepler facilities, and (ii) these are the minimum numbers of planets in those systems—additional undiscovered planets would push these numbers even higher.

Attempts have been made to tease out the trends in multi-planet systems. Early analyses indicated that, for planets smaller than Neptune, inner planets were typically smaller than outer planets (Ciardi et al. 2013). Further study of a larger sample of planet candidates found that planets in the same system typically had similar radii—systems consisting of all small or all large planets (Weiss et al. 2018). This "peas in a pod" scenario, shown in Figure 3.17 has subsequently been refuted

---

[5] Originally KFOP (the Kepler Follow-up Observing Program), then CFOP (the Community Follow-up Observing Program) and now ExoFOP (the Exoplanet Follow-up Observing Program), found at http://exofop. ipac.caltech.edu.
[6] On the NASA Exoplanet Archive as of 2019 September 19.

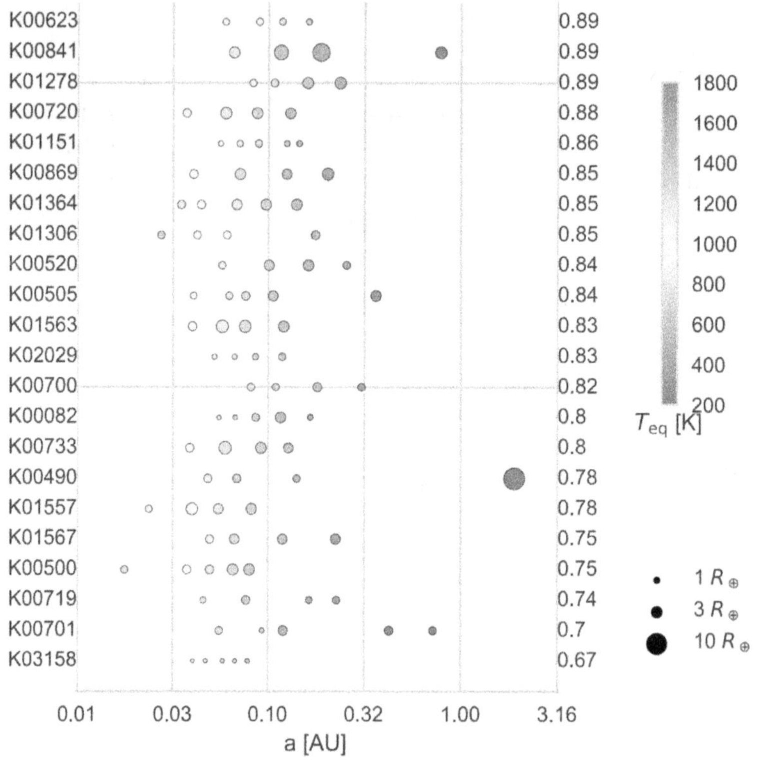

Figure 3.17. A sample of multi-planet systems from Kepler, showing the relative sizing and spacing. Most systems contain planets with similar radii, as compared to radii randomly drawn from the full distribution of radii in the Kepler planet sample. Modified from Figure 1 of Weiss et al. (2018). © 2018. The American Astronomical Society. All rights reserved.

by Zhu (2020; see e.g., Weiss and Petigura 2020) as due to detection biases; further analysis is needed to confirm or refute the hypothesis. If it is indeed true that planetary systems typically form planetary systems of similar sizes, we wish to discover why.[7]

Another aspect of multi-planet systems to examine is the degree to which they are dynamically packed. The planets in our solar system are fairly far flung, exerting little dynamical influence on each other at this mature stage of planetary evolution. Outside our solar system, hot Jupiters are the most readily detected transiting planets, and it is somewhat unfortunate that they turn out to be "lonely," typically having no additional detectable planets, compared to warm Jupiters and Neptunes, which in contrast are often found to have companion planets (Steffen et al. 2012b). Analysis of the Kepler multi-planet system spacing

---

[7] Note however, that this scenario is not true in our solar system. We have small terrestrial planets, ice giants, and larger gas giants but no super-Earths. Additionally, while our solar system planets span orbital periods from 88 days to over 200 yr, most of the Kepler discovered solar systems are compact having planets with very short orbital periods, most far less than that of Venus.

finds that the typical planet gap is ~20 mutual Hill radii (Weiss et al. 2018), and that the period ratios tend to pile up just wide of the mean motion resonances (Lissauer et al. 2011; Fabrycky et al. 2014).

One question that remains unanswered by the Kepler mission is whether our solar system is typical. There is no single upcoming mission or project that will be sensitive to solar system analogs, but by piecing together the different parts of parameter space probed by the upcoming surveys (e.g., TESS, PLATO, WFIRST), we may be able to discern whether other systems are configured similarly to our own.

3.9 Advances in the Hunt for Exoplanets

3.9.1 Introduction to Identifying Transits

The Kepler mission observed approximately 170,000 stars for 4 years at a 30 minute cadence. A transiting Earth-size planet would only produce 3, 12 hr long dips in the light-curve during this time period. As a result, out of the approximately 70,000 Kepler observations per star, only ~70 of those observations would be dimmer (by only 84 parts per million—about 0.01%) because of the transiting Earth-size planet. Complicating the problem, the star, the instrument, and the telescope all cause changes in brightness measurements. The data set is large enough and the signal-of-interest weak enough, that looking at all the data by only a few experts would not likely be efficient nor repeatable.

The Kepler missions solution to how to search the light-curves for transits was to create software called the transiting planet search module (TPS; Jenkins 2017; Jenkins et al. 2010). This set of algorithms basically looked for periodic transits in the data by matching the data to U-shaped transit trains of various periods and durations. Once a signal was found, it was validated by the data validation module of the pipeline where several tests were applied in order to determine if the signal was likely the result of a planet transit, or some other (perhaps completely false) source. TPS was not the only algorithm developed to search the Kepler data for transits. But each had the same problem. The software search produced a complete, but fairly unreliable list of transit candidates that still needed vetting by a human expert in order to identify those dips that were caused by true planet transits. In this section, we explore some of the methods that different groups attempted to tackle this complex data question.

3.9.2 Kepler Objects of Interest

Kepler discovers exoplanets by looking for U-shaped transit signals in the time series data that it collects. The original intent of the Kepler object of interest (KOI) catalog was to identify and record transit event in the Kepler data. However, not all transit-looking signals are exoplanets. Dips in the brightness of a star could also be caused by some other astrophysical source (usually an eclipsing binary), or from unaccounted for noise in the instrument. The best way to determine if the signal is indeed from a planet is to obtain additional observations (see Section 3.10.1) that confirm the exoplanet's existence or rule-out some of the known false positives. As a result, the KOI list became, at first, a list of targets that needed follow-up observations. Since this was the first time we looked at the Universe with such high photometric precision, there were loads of interesting phenomena that people wanted to investigate further. So, planet transits, unusual pulsators, and eclipsing binaries were added to the KOI list, with a note indicating whether the signal was likely caused by an exoplanet or not. One of the most notable, non-exoplanets added to the KOI table, was KOI-54, an eccentric binary system with large *increases* in brightness. KOI-54 is now known to be the prototype of a class of binary stars called the heartbeat stars (Welsh et al. 2010; Thompson et al. 2012). By the end of the Kepler mission, the KOI table shifted from only a to-do list for future observations and evolved into a catalog of transiting exoplanet candidates that would enable exoplanet occurrence rates out to 1 year orbital periods.

Sculpting a KOI List
"Every block of stone has a statue inside of it and it is the task of the sculptor to discover it," is a quote attributed to Michaelangelo. It is a good analog to the process of creating Kepler objects of interest. The process of creating a KOI catalog starts with the transit planet search module of the Kepler pipeline (Jenkins 2017). This software module searched the 180,000 light-curves for potential periodic dips and returned the time when each dip occurred. The signals that reached a certain detection threshold (i.e., a detectable dip) were called threshold crossing events, or TCEs. The TCE list is very complete, in that it finds most transit-like events with a period longer than half a day, but it is also exceptionally unreliable in that it finds all sorts of other signals that are not transiting exoplanets. The job of creating a KOI table was to go through the tens of thousands of TCEs to find the few thousand likely exoplanet transit signals. Unlike popular belief, creating an exoplanet catalog does not involve examining light-curves by hand to identify transits. It is the opposite, the job is to identify and eliminate those detections (TCEs) that are not caused by transiting exoplanets, a task called vetting. The scientist's job is to determine which TCEs are simply unwanted stone, and which make up the statue of exoplanets hiding beneath.

To see the size of the problem, see Figure 3.18. It shows the period distribution of the Kepler TCEs, the KOIs, and the planet candidates. Only about 10% of signals found by the Kepler pipeline ultimately became planet candidates. By allowing the detection algorithm to return so many false positives, it was able to search deeper and therefore find the signals of the smallest planets in the widest orbits.

How to identify and remove the "false positives," those TCEs that were not KOIs, was a technological and scientific achievement in itself. Kepler scientists found clever ways to look at the data alone to rule-out the possibility of a transiting exoplanet.

Many false positives are identified by simply looking at the shape of the phase folded light-curve. The TCE list is littered with stars that change brightness because of stellar pulsations, stellar spots, or binary systems. By looking for a second, smaller, brightness change similar to that of the reported transit indicates that we actually found an eclipsing binary. By looking to see if the transit shape is more like

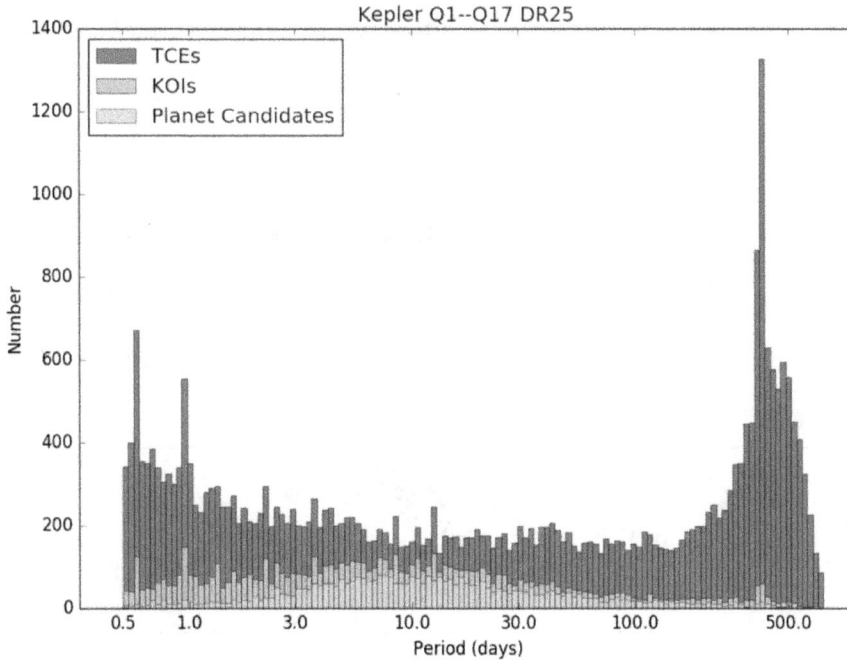

Figure 3.18. A histogram of the TCEs, KOIs, and planet candidates identified in the Kepler data. The large number of TCEs found around 372 days is caused by the orbital period of the spacecraft around the Sun. Credit: Jeffrey Coughlin.

a sine wave than a U-shape, then likely the source is stellar spots or pulsations. See Figure 3.19 for some of the false positives seen in the Kepler TCE list.

Eclipsing binaries hiding behind a closer, brighter star is another type of false positive. An eclipsing binary shows deep V-shaped eclipses, but if that light is diluted by the light of another star the signal can mimic a transiting exoplanet. Since the Kepler pixels cover a small, but significant, patch on the sky, it is possible for a second stellar system to appear in the same pixel. Bryson et al. (2013) showed that the time series images could be used to find evidence of these background eclipsing binaries. Essentially if you measure the location of the light both in and out of transit, then it is possible to identify these background binaries with the Kepler data alone.

Another successful way of eliminating false positives was to look for events from different stars that occurred at the same time. Between instrumental electronic cross-talk and reflected light, it is possible for the light of one star to contaminate another (Coughlin et al. 2014; Van Cleve & Caldwell 2009). By matching the transits times across all the CCDs it is possible to find these instances and remove them from the final KOI list.

At long periods, one of the common sources of false positives was an electronic artifact known as a "rolling band." These artifacts are caused by a spatial pattern in the detectors background level that moves across the chip in response to changes in

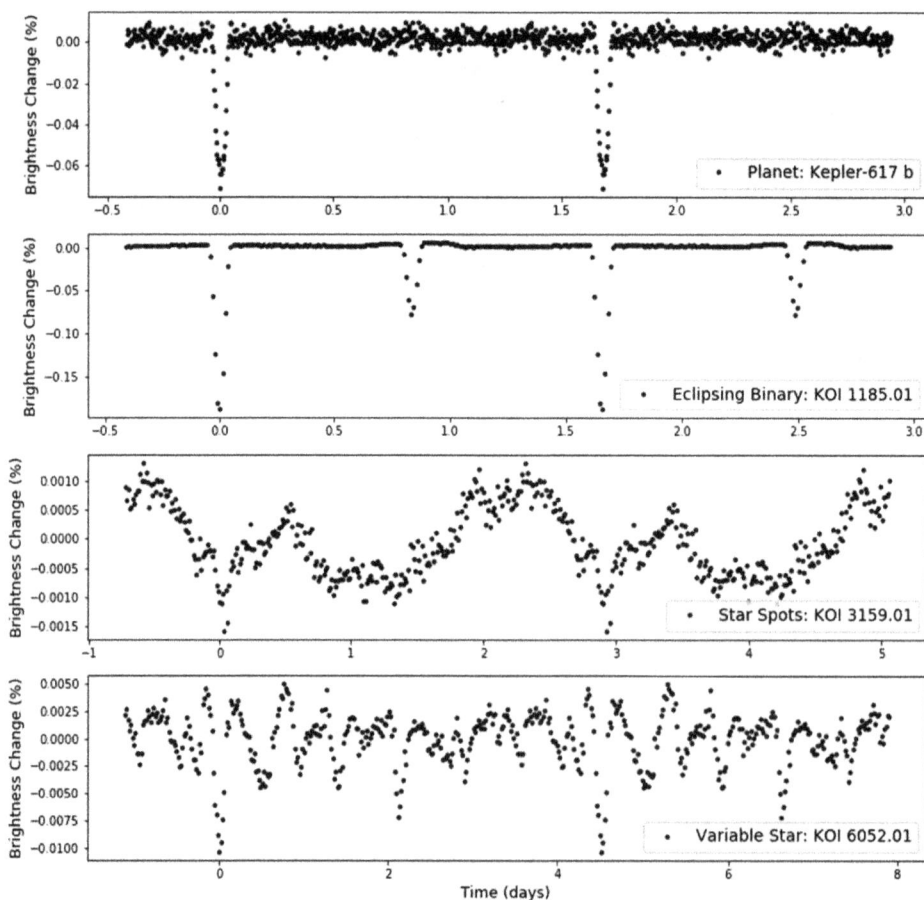

Figure 3.19. Example light-curves (folded on the detected period and binned) of TCEs found by the Kepler pipeline. The first is a transiting planet and the other three are examples of false positives, stars which show brightness variations that can be confused with planet transits. Credit: Susan Mullally.

the temperature of the detector and its electronics (for more details, see Van Cleve & Caldwell 2009). Rolling bands show up as rapid, chaotic changes in the light-curves during part of the quarter but only occur for stars that are imaged on certain CCD detectors. Since the Kepler spacecraft was rotated by 90° every quarter year (to keep sunlight on the solar panel arrays), stars would only show these variations once, and sometimes twice, per Kepler year as the star rotated onto one of the affected detectors. As a result, many of the TCEs with a period near 372 days (the length of a Kepler year) are false positives caused by rolling bands (see the large number of TCEs at 372 days in Figure 3.18).

The Evolution from Humans to Robots
Measuring whether one of these false positives is the reason for the TCE was initially done by a group of Kepler and community scientists looking, by eye, at a set of

graphs which highlight various features of the data. This group was known as TCERT, the Threshold Crossing Event Review Team. As they scrutinized those plots they would continue to find other sources of false positives and learned a lot about the Kepler data and sources of false positives in the process.

Unfortunately there were a few downsides to vetting in this way. First, it took a lot of time. For each planet search, the dozen TCERT scientists had to go through more than 20,000 TCEs; the last list of TCEs was allowed to be very complete and produced 32,532 TCEs. Second, sometimes new ways to identify false positives were discovered after already looking through many of the TCEs, causing those TCEs to be revisited. Third, scientists are not always consistent, the evaluation of a TCE may depend on whether a scientist had a particularly strong cup of coffee that morning! Fourth, in order to measure how well the vetting activity was doing it also needed to vet known signals, those simulated to be true transit candidates or true false positives. This last piece is required in order to accurately measure the occurrence rates; otherwise one might think that the scientists preferentially included true small planet candidates if one was biased toward finding planet signals.

Because of these downsides to manual vetting, it was decided to fully automate the KOI catalog. The first attempt to do this was by McCauliff et al. (2015) who used a random forest algorithm on the many measurements of the data produced by the Kepler Data Validation module. While somewhat successful, it required extensive training based on previously vetted TCEs and did not give easy to identify reasons for its decisions.

Instead of a complex machine learning approach, the TCERT team decided to determine algorithms that mimic the manual activities they were already doing. In this way the scientists could more easily understand why TCEs became false positives but still allowed them to automatically vet the TCEs. This effort started in the sixth exoplanet catalog by determining an algorithm to detect the evidence for background eclipsing binaries (Mullally et al. 2015). Next came a classifier (using dimensionality reduction and nearest neighbors) that looked at the shape of the phase folded light-curves of each TCE and determined if it looked like one of the already identified transit signals (Thompson et al. 2015). This algorithm removed much of the rolling band signal and many of the over contact and deep eclipsing binaries. Other algorithmic measures of the data followed that automatically detected evidence that the TCE was caused by an eclipsing binary, by a rolling band, or by other instrumental effects. This software became known as the Robovetter.

The Robovetter not only operated on the TCEs produced by the data, but also looked at simulated transits and simulated false positives so that it could be tuned to find only the most reliable transiting exoplanet candidates while eliminating most true false positives. The final catalog correctly identified 85% of the transiting exoplanets with periods less than 100 days with only 2% of them being due to false positives. At longer periods, where there were only a few transit signals to work with, the catalog includes 75% of the transits in the TCE population but is contaminated so that about half of the catalog is actual false positives. While this is a disappointment for those wanting to find individual long period planets, the fact that the biases of the catalog were measured means it is possible to use the KOI catalog as a whole to determine how common long period exoplanets are in our Galaxy.

KOI catalogs went from highly complete, manually-vetted catalogs which took thousands of person hours to produce (Borucki et al. 2011a; Batalha et al. 2013; Burke et al. 2014; Rowe et al. 2014) to a catalog that could be vetted in less than 5 minutes (Mullally et al. 2015; Coughlin et al. 2016; Thompson et al. 2018). No catalog is perfect, but many of the imperfections of this last catalog are measured and so it is more ideal for occurrence rate studies.

The Final KOI Catalog

If you compare the early KOI catalogs to the later ones, you will notice that as the data set got longer, the mission started finding longer period planets, and also smaller planets. The last few catalogs worked on removing known sources of false positives and automating the effort. In the final catalog, see Figure 3.20, you can see the large abundance of planets smaller than Neptune and the lack of Jupiter-sized planets in short period orbits. Also you can see at long periods how the detection efficiency drops off, so that only a few possible worlds were found with a radii similar to that of the Earth.

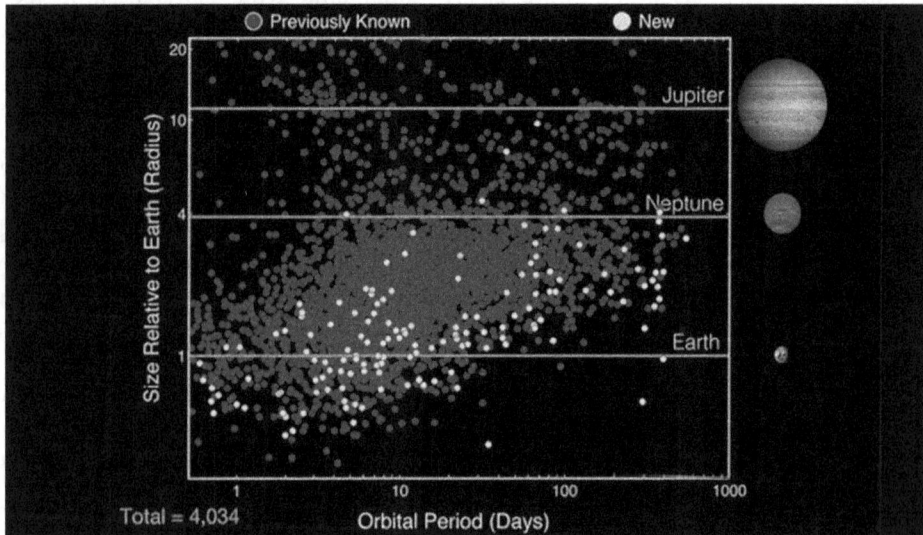

Figure 3.20. The planet candidates at the end of the Kepler mission (Thompson et al. 2018). The newly identified candidates are marked in yellow. As the length of the data increased, longer period planets were found. While many of these candidates cannot be confirmed, odds are most of these are indeed truly exoplanets. Credit: Wendy Stenzel/NASA/Ames Research Center.

3.9.3 Citizen Science Projects

There are interesting transit signals for which the planet-hunting software algorithms perform poorly. These include: very long period planet signals, with only one or maybe two transits observed by Kepler or K2 (falling short of the three-transit minimum required by the Kepler mission software); non-periodic transit signals, such as those of planets orbiting eclipsing binary stars, or exhibiting large transit timing variations, where the software cannot fold the transit signals on top of each other with a single period; and transit signals in light-curves with large amounts of correlated stellar activity on similar timescales (such as pulsating stars). However, the human brain is exceptionally good at pattern recognition, and is able to identify transits in all of the above cases.

Planet Hunters

The first and most extensive citizen science project to use Kepler and K2 data was Planet Hunters.[8] Volunteers were shown 30 day sections of detrended Kepler and then K2 data, and asked to mark any transit-like events they could see. Planet Hunters was very successful: PH-1 b is a circumbinary planet found in a quadruple star system (Schwamb et al. 2013); PH-2 b is a Jupiter-sized planet in the habitable zone of its star (Wang et al. 2013); PH-3 d is a mini-Neptune in a system with two previously identified planets that was originally missed due to very large TTVs (Schmitt et al. 2014b); many new planet candidates (Schmitt et al. 2014a; Wang et al. 2015); and perhaps most famously, Tabby's Star (Boyajian et al. 2016). Figure 3.21 shows the full Kepler light-curve of Tabby's Star, with 10 separate dimmings spread throughout the 4 year observing baseline. Current theories to explain the light-curve are largely focused on extinction by circumstellar material orbiting the central star, for instance clouds of dust, asteroids, or comets. The Planet Hunters project is continuing to serve the citizen science community with data from the NASA K2 and TESS mission.

Exoplanet Explorers

Another significant citizen science project hunting for exoplanets was Exoplanet Explorers.[9] In this project, volunteers were asked to assess potential planet candidates already identified by automated algorithms in the K2 data, in order to discard false positives and speed up the process of identifying the most likely planet candidates. This was critical for the K2 phase of the mission where a new campaign was being observed every ~80 days, leaving very little time for data reduction, signal detection and vetting, and following up good candidates before the next set of data appeared. Volunteers on the Exoplanet Explorers project identified the six-planet K2-138 system (Christiansen et al. 2018), and K2-288B b, a temperate super-Earth in a binary system (Feinstein et al. 2019). The Exoplanet Explorers project is currently retired.

---

[8] http://planethunters.org.
[9] http://exoplanetexplorers.org.

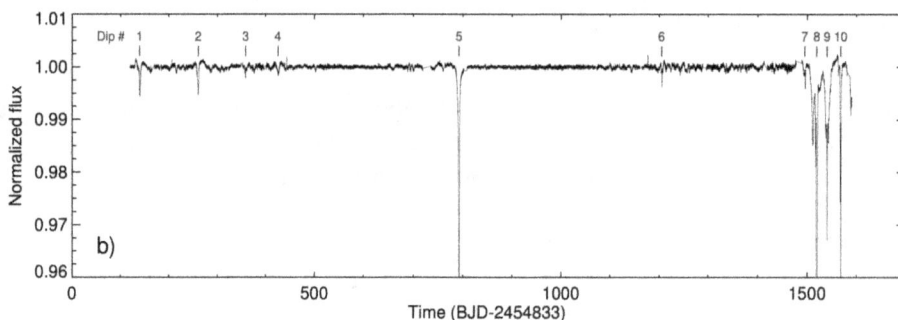

Figure 3.21. The Kepler light-curve of KIC 8462852, showing multiple deep, aperiodic dips. Modified from Figure 1 of Boyajian et al. (2016), by permission of Oxford University Press on behalf of the Royal Astronomical Society.

Musicians and Music Influenced by the NASA Kepler Mission Exoplanet Discoveries

Contributed by Henri Scars Struck and Steve B. Howell.

- *The Kepler Mission* (Band)
  https://music.thekeplermission.com/music.
  This band is not "the satellite," just a band of five people who play reverb-rock and perform shows around Los Angeles, California. Guitar/Keys: Kyle Biane; Bass/Vocals: Nick della Cioppa; Guitar/Vocals: Toby Mason; Drums: Hyke Shirinian; Guitar/Vocals: John Theodore.
- *In Isolation*—Trappist-1 (A Space Anthem)
  https://www.youtube.com/watch?v=9_5Yt1dp9yI.
  Not your usual love song; *In Isolation's* new single "TRAPPIST-1" presents the desire to leave Earth and relocate 40 lt-yr away on a potentially-habitable planet in the Aquarius constellation.
- *Gary Blackwood*—Exoplaneteers
  from the 2017 March 13 Astronomy on Tap, der Wolfskopf in Pasadena
  https://youtu.be/CwDEAnV-uHc.
  Filmed during an "Astronomy on Tap" event, NASA's Exoplanet Program Office program manager, Gary Blackwood, regales us with his original "Exoplaneteer" lyrics sung to the MIT fight song.
- *Henri Scars Struck and Steve B. Howell*—Beyond Me
  A Grammy Award winning musician and the Kepler Mission project scientist team up to use real Kepler light measurements and create "Beyond Me"—a sound installation.
- *Coma Niddy*—The Exoplanet Song
  https://www.youtube.com/watch?v=Fgm4Uvwz_FQ.
  An original science inspired music video performed on the YouTube channel "Mike Likes Science."
- *Eric Kauffmann & Nrgmind*—Exoplanet
  https://www.youtube.com/watch?v=eHJ78dI0uYs.

- *Jim Ocean*—SuperEarth
    An original from California Bay Area singers/songwriters/music producers Jim and Kathy Ocean. A Kepler mission inspired song about life on a Super Earth. Part of the "Astronaut Lullabies" collection
    https://jimoceanmusic.com/track/693543/super-earth.

3.10 Exoplanet Confirmation

3.10.1 Observational Follow-up Efforts

Finding the transit events in the Kepler and K2 data, identifying candidate systems and producing the list of KOIs is but the first stage of planetary discovery. These discoveries begin as candidates, but the path to confirmed planet requires the elimination of many astrophysical false positive scenarios. And even the planet candidates that have been validated as a true planetary body require additional analysis to derive reliable system parameters, including the masses and radii of the planet, the planet orbital properties, the characteristics of the host star, and the presence of other stars or planets in the system. All of this additional information requires a substantial observational follow-up effort primarily using ground-based telescopes. The Kepler data by themselves are not sufficient to accomplish the scientific goals of planetary discovery and characterization.

Kepler was the first exoplanet space mission to employ a ground-based follow-up observation program dedicated to the validation of the planetary candidates identified by the project—and the Kepler Follow-up Observation Program (i.e. the KFOP) was performed as a service to the community as part of the general effort for the primary scientific goal of the mission which was the determination of $\eta_\oplus$, the frequency of rocky planets. The follow-up observations fell into primary categories of effort: the elimination of astrophysical false positives and the determination of accurate planetary parameters.

Fundamentally, the Kepler imaging instrument only measures the change in the apparent brightness of the target star as the planet passes in front of the star (Figure 3.22). Thus, one of the most fundamental measurements that must be made is the determination of the stellar radius. If the star is not the size that you think it is then neither is the planet, as the transit depth is simply the ratio of the two. A primary objective of the KFOP was obtain spectroscopy of as many KOIs as possible. Over a 6 year period (2009–2015), the KFOP spectroscopically observed 2667 KOIs (and 614 standard stars) with the sole purpose of determining the stellar temperatures, metallicities, and surface gravities of the candidate host stars (Furlan et al. 2018). These properties where then compared to stellar models to derive stellar radii ($R_{t\star}$) from which the planetary radii (R_p) could be determined (Mathur et al. 2017). For the completeness and occurrence rate determinations, the Kepler project needed to determine the stellar properties for all of the targets—and the KFOP yielded the stellar properties for the stars that were thought to host planetary candidates (Figure 3.23). Precise and accurate stellar radii are critical but not sufficient to the

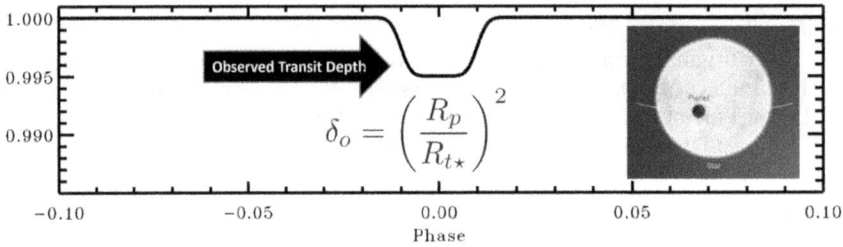

Figure 3.22. A simple planetary detection where the depth (δ_o) of the observed transit is proportional to the ratio of the planetary radius (R_p) to the radius of the star being orbited ($R_{t\star}$). Based on work in Ciardi et al. (2015).

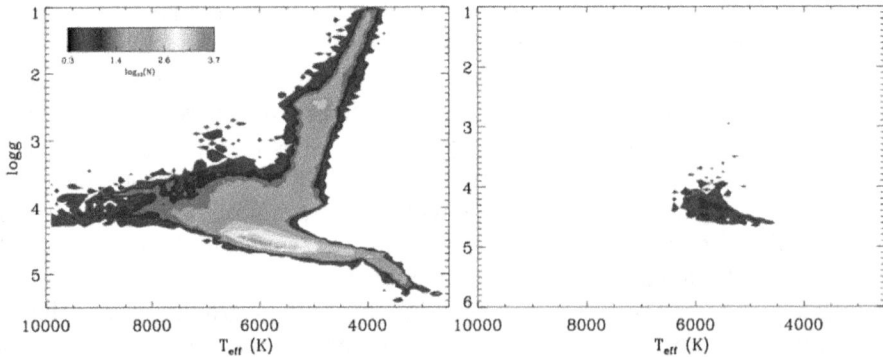

Figure 3.23. H–R diagram of the stars in the entire Kepler target list (left) and the KOIs that were observed by the KFOP (right). The measured stellar parameters are used to derive the stellar radius, which in combination with the observed transit depths, are used to determine the planetary radii (see Figure 3.22). Reproduced from Mathur et al. 2017. © 2017. The American Astronomical Society. All rights reserved.

determination of the planetary radii. For example, spectroscopic observations were crucial to the discovery of the short planet radius gap (see Section 3.8.3).

However, it is not always that simple. Other astrophysically occurring phenomena can masquerade as a planet transiting a star. The most insidious arrangement is the chance alignment of a previously unknown eclipsing binary that is blended with the target star as viewed along the same line of sight by the Kepler spacecraft. As a result, the light of the brighter target star fills in the eclipse depth of the fainter eclipsing binary making the eclipse appear as if it is a transit around the brighter star (Figure 3.24). The transit depth of an Earth-sized planet crossing a solar-like star is approximately 0.01%; a background eclipsing binary that has an eclipse depth of only 10% but is 500 times fainter ($\sim$ 7 magnitudes) than the target star can make the target star appear to have an Earth-sized planetary transit (Figure 3.25). Thus, a search for possible background stars that may be blended with the planetary candidate host star had to be made, and high resolution imaging was the tool of choice.

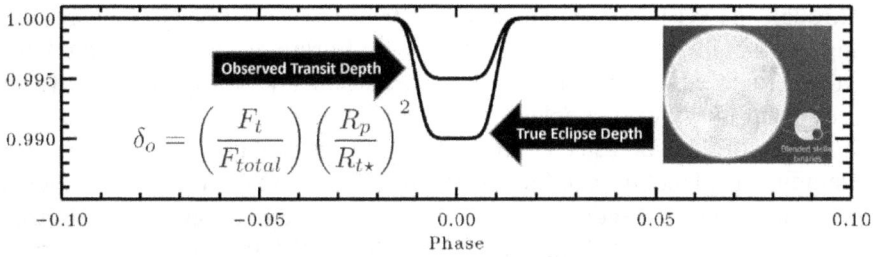

Figure 3.24. A fainter eclipsing binary blended with a brighter target star (as seen by Kepler) can appear to be planetary transit around the brighter star because the stellar eclipse depth is diluted by the brighter star. The diluted eclipsing binary thus appears to be a transiting planet around the brighter star. Based on work in Ciardi et al. (2015).

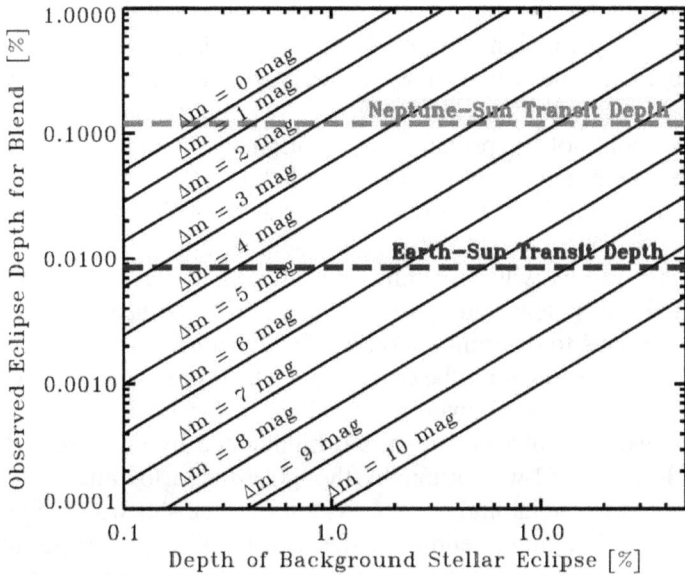

Figure 3.25. Fainter eclipsing binaries blended with a brighter target star (as seen by Kepler) can appear to produce a much more shallow planetary transit when in fact the true event is deeper and produced by a stellar eclipse. Based on work in Ciardi et al. (2015).

Toward the goal of identifying blended eclipsing binaries that may be masquerading as planetary candidates, the KFOP undertook a high resolution imaging survey of the KOIs—the second primary objective of the follow-up observation program. In this context, high resolution imaging refers to "better-than-seeing-limited" imaging and the KFOP utilized two primary techniques to achieve imaging observations near the diffraction limit of the telescopes. The first technique was optical speckle interferometry (e.g., Howell et al. 2011). This technique utilizes the interference pattern inherent within individual rapid exposures of stars, each image appearing to be full of speckles that are produced by the atmosphere. The second technique was near-infrared adaptive optics imaging (e.g., Adams et al. 2010). In the end, through the KFOP and contributions from the community, high resolution imaging was obtained for all of the KOIs (e.g., Furlan et al. 2017; and references therein) where it was found that 30% of the KOIs had a nearby companion within 4″ or closer—that is, within 1 Kepler pixel.

While most of these detected companions were not background eclipsing binaries masquerading as planetary candidates, the blended companion stars, whether truly bound companions or background stars, all made the detected planetary transits appear more shallow and thus, for 30% of the KOIs, the initial planetary radii estimates were underestimated. Much like the case of a blended eclipsing binary (Figure 3.24), if another star is unknowingly blended with the target, as viewed by Kepler, the light from the additional star will fill in the transit and make the transit appear shallower than it is in reality (Figure 3.26).

In such a scenario, if the additional star is equal in brightness to the primary target star that hosts the planet, the planet radius will be underestimated by $\sqrt{2} \approx 1.4$. As the companion star gets fainter and fainter, then the dilution caused by the companion, and hence, the planetary radius correction, asymptotically approaches zero (Ciardi et al. 2015). However, if the planet actually transits the companion star and not the primary star as originally assumed, then the planetary radius can be grossly underestimated (Figure 3.27).

In this configuration, because the true host star is much smaller and fainter than the primary star, that is, the star originally assumed to host the planet, the estimate of the planetary radius will be significantly too small—by factors of 2 or more (Figure 3.28). So, if a stellar companion is detected to be near a Kepler host star, care must be taken as to determine whether the planet orbits the brighter star or the fainter. In many cases, it cannot be determined which star the planet actually orbits and this fact dominates (see however, Howell et al. 2019).

An exciting example of blended stars is highlighted by the case of Kepler-1652 (KOI-2626). Kepler-1652 was originally thought to be a low-mass star hosting an Earth-sized planet in the habitable zone. As a result, the star was a high priority for the high resolution imaging to eliminate the eclipsing binary false positive scenario. While the eclipsing binary scenario was removed, the system was found to be a

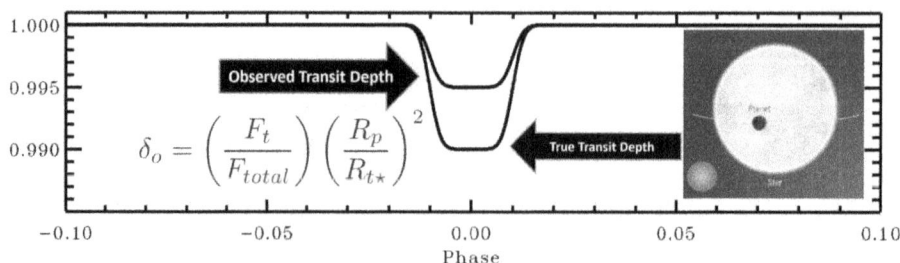

Figure 3.26. A fainter star blended with a brighter target star (as seen by Kepler) can make a planetary transit around the brighter star appear to be more shallow than it is in reality. As a result, the planetary radius, in the absence of knowing about the additional star, will appear to be smaller than it actually is. Based on work in Ciardi et al. (2015).

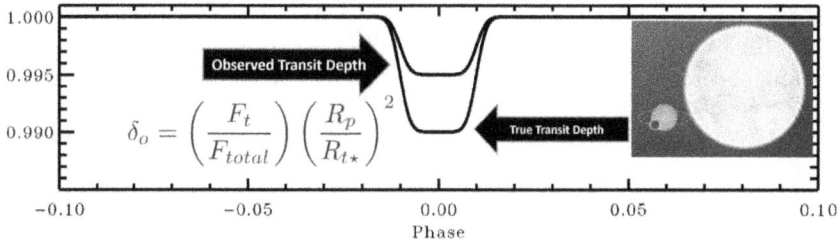

Figure 3.27. A fainter star, hosting a planetary companion, blended with the brighter target star (as seen by Kepler) can make a planetary transit around the brighter star appear to be more shallow than it is in reality. As a result, the planetary radius, in the absence of knowing about the additional star, will appear to be smaller than it actually is (Ciardi et al. 2015).

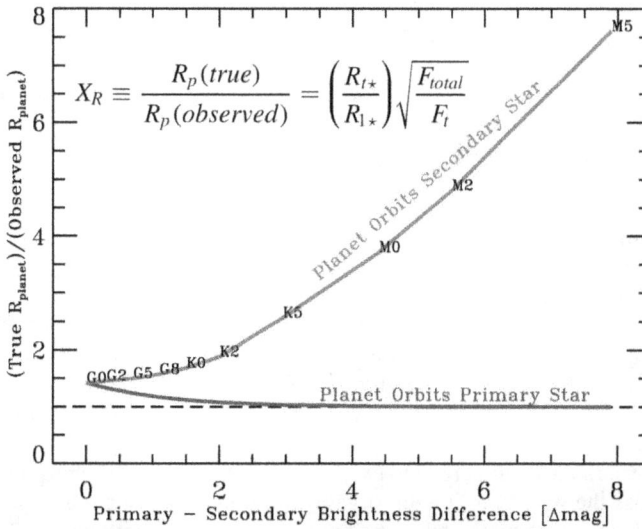

Figure 3.28. Planetary radii correction factors for an example Kepler candidate host star that is of similar mass to the Sun. If the solar-type host star has a fainter (lower mass) companion that stellar companion. Based on work in Ciardi et al. (2015).

Kepler-1652	Stellar Temperature and Radius	Planet Radius
Assume Single Star	3480 K 0.35 R_{Sun}	1.12+/-0.16 Re
Component A	3650 K 0.48 R_{Sun}	2.04+/-0.33 Re
Component B	3520 K 0.42 R_{Sun}	2.37+/-0.44 Re
Component C	3400 K 0.32 R_{Sun}	2.58+/-0.62 Re

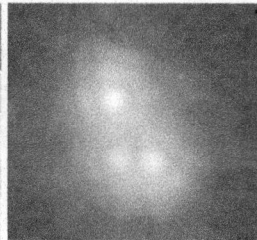

Figure 3.29. Kepler-1652 (KOI-2626) was originally thought to be a single low-mass star hosting an Earth-sized planet in the habitable zone. High resolution imaging revealed it to be a triple star system which turned the Earth-sized radius estimates into super-Earth sized radius estimates. Reproduced from Cartier et al. (2015). © 2015. The American Astronomical Society. All rights reserved. Torres et al. (2017).

multiple star system with three low-mass stars (Cartier et al. 2015). In the end, it was found that the planet was real (Torres et al. 2017), but it was ambiguous as to which star the planet really orbited and what the real size of the planet could be (Figure 3.29).

Follow-up Observation Coordination Stories

Contributed by the KFOP.

The Kepler photometric data were so good—like nothing we had ever seen before for so many stars at once—that when the first data were shared with the KFOP in 2009, many of the KFOP members thought they were looking at simulation data and not real data from the spacecraft. When the KFOP was informed that the data were indeed real and that the first Jupiter-sized planets were detected with a significance never before achieved for a transit survey, one of the KFOP members remarked "Um … wow … my guess is that we are going to have our work cut out for us 'cause these data mean we are going to discover things we never thought we could."

And in fact, from 2009–2014, the KFOP team collectively observed for more than 3000 nights at some of the worlds largest and most powerful telescopes. That average of 500 nights of observing time per year meant that on any given night during the northern summer months when the Kepler field was available to be observed, two or more astronomers were at two or more telescopes collecting data on Kepler targets. On average, each member of the small group of KFOP astronomers observed more than 200 nights each over that 6 year period. During that time, some of the KFOP observers had observing time on two telescopes simultaneously. One KFOP member was on Kitt Peak observing at the Mayall 4 m and the WIYN 3.5 m telescopes on the same nights, driving back and forth between the two to start exposures on one and move to the next target on the other.

Everyone was so dedicated to, and excited by the science and the project that no one wanted to miss the weekly follow-up organization telecons—including over vacation. One KFOP member joined the telecon while at a water park with his family. When the KFOP member joined the telecon, he tried to find a quiet place away from all of the people at the park but did not realize that he was standing next to a loud speaker that was piping "bird sounds" into the park. On the telecon, one member of the KFOP remarked, "Hey, whoever is in the tropical rain forest, could you please mute your phone … it's hard to hear about the science above the macaws?"

The contributions of the follow-up program became so important that the world-wide community joined in—from professional astronomers to citizen scientists. One of the true legacies of the Kepler project was the worldwide collaborative effort that was formed to characterize and follow-up the Kepler candidates. The coalitions that formed for the sake of the science and the community have carried forward in ways that are hard to enumerate. This was especially true for the K2 mission which had an open data policy from the start. But one concrete example is the TESS follow-up observation program (TFOP). TESS is the planetary transit hunting mission that NASA launched in 2018 to find planets around nearby stars and, like Kepler and K2, the planetary candidates require the same type of dedicated follow-up observing. But now, instead of a few dedicated KFOP observers, the TFOP is made up of hundreds of astronomers, both professional and citizen scientists, from around the world, who observe TESS targets every single night and share those data and results with the world through the fully open access NASA ExoFOP[10] archive portal.

3.10.2 Statistical Confirmation

Early on in the Kepler mission it became clear that it was going to be impossible with available resources to measure a mass for every planetary candidate found in the Kepler data. Seeing the star wobble (in a radial velocity sense) at the same period as the transit is the gold standard for confirmation that the planet is really there and not caused by another source. However, with current instrumentation and technology, radial velocity confirmation is not possible for most of the small, rocky exoplanets. Statistical confirmation, or validation, is a process to weigh the odds that the observed signal is caused by a transiting exoplanet against the odds that it is caused by one of many other astrophysical scenarios, such as a background eclipsing binary. If those other scenarios are rare in comparison, then it is still possible to call the planet a bona-fide planet discovery. The majority of Kepler's planets have been statistically validated in this way. Because only limited follow-up observations are required, many can be validated and published all at once, making it possible to add hundreds of exoplanets to the confirmed planets table in one day.

Validation was first introduced using a technique called BLENDER (Torres et al. 2011; Borucki et al. 2012). This method generates synthetic light-curves for a large number of possible false positive scenarios and compares them to the transit found in the Kepler photometry. This was followed shortly by another method called PATIS. Several handfuls of planets statistically validated in this way, including the first habitable-zone planet orbiting a Sun-like star, Kepler-22b (Borucki et al. 2012). Comparing fully modeled light-curve shapes was expensive, and so Morton et al. (2016) introduced the idea of using simpler, trapezoid-type, model for the transit shape. In this way scientists were able to speed up the processing and validated over a thousand of Kepler's exoplanets (Morton et al. 2016; Tsarias 2019a, 2019b).

Another method to statistically validate planets in multi-planet systems was introduced by Lissauer et al. (2012). Here the authors argued that if there were two sets of transit trains (or more) at different periods, it is unlikely that these can result from multiple false-positive signals, and thus multi-candidate planet systems are true planets. With these arguments they were able to validate over 800 multi-planet systems (Rowe et al. 2014).

When Kepler launched one of the largest worries was in how to distinguish between the true exoplanet transit signals and the plethora of background eclipsing binary stars as a source of false positives. From the large effort of ground-based telescopic follow-up and statistical validation efforts, it became clear that Kepler's planet candidates, as a population, were not largely polluted by eclipsing binaries and that, in fact, the vast majority of the discoveries are real exoplanets.

---

[10] https://exofop.ipac.caltech.edu.

3.11 Using K2 to Observe Microlensing Events

3.11.1 A New Era of Ground-based Survey Facilities

The years leading up to the launch of the Kepler spacecraft and encompassing its operation witnessed an evolution in ground-based microlensing searches for exoplanets. In 2006, the Microlensing Observations in Astrophysics (MOA) collaboration began the MOA-II microlensing survey on the newly constructed 1.8 m MOA-II telescope located at Mt. John University Observatory in New Zealand. This telescope also featured an upgrade to a camera with a 2.2 deg^2 field of view (FoV) (Sako et al. 2008). Several years later, in 2011, the Optical Gravitational Lensing Experiment (OGLE) entered its fourth phase of operation, OGLE-IV, on its 1.3 m telescope at Las Campanas Observatory in Chile after a new camera with a 1.4 deg^2 FoV was installed (Udalski et al. 2015a). Most recently, the Korea Microlensing Telescope Network (KMTNet; Kim et al. 2016) began science observations in 2015. KMTNet consists of three 1.6 m microlensing survey telescopes each equipped with a 4.0 deg^2 FoV, with one node located at Cerro Tololo Inter-American Observatory (CTIO) in Chile, one at South African Astronomical Observatory (SAAO) in South Africa, and one at Siding Spring Observatory (SSO) in Australia. These survey telescopes began observing several tens of square degrees with cadences as frequent as 4 hr^{-1}, allowing them to detect ~2000 unique microlensing events each year and also densely cover the perturbations in the magnification structure arising from the presence of possible planets.

This new regime of exoplanetary microlensing led to a corresponding increase in the planet detection rate and also facilitated improved characterization of those planetary systems. On the date of the launch of the Kepler spacecraft, a total of eight microlensing planets had been discovered, including the first two-planet system discovered via gravitational microlensing (Gaudi et al. 2008; Bennett et al. 2010). This number more than doubled in the next 4 years, up to 20 by the end of 2012, and it nearly doubled again in the subsequent 4 years.[11]

3.11.2 Lens Characterization

The higher cadences and improved photometric precisions of the new array of survey telescopes furthermore opened up multiple avenues for characterizing the lens systems giving rise the observed microlensing events. There are three observables that provide a mass–distance relation for a given lens systems. Two of them, the Einstein radius θ_E and the microlens parallax π_E, are related to the physical properties of the lens and source by:

$$\theta_E \equiv \sqrt{\kappa M_\ell \pi_{rel}}, \quad \pi_{rel} = \pi_E \theta_E = \mathrm{AU}(D_\ell^{-1} - D_s^{-1}), \qquad (3.1)$$

where M_ℓ is the mass of the lensing object, D_ℓ and D_s are the distances to the lens and source, respectively, π_{rel} is the relative lens–source parallax, and $\kappa \equiv 4G/(c^2\mathrm{AU}) = 8.144$ mas/$M_\odot$. Measuring either θ_E or π_E (and assuming the distance to the source is

---

[11] Data taken from the Confirmed Planets table, NASA Exoplanet Archive, doi: 10.26133/NEA1.

approximately known, which is generally the case for microlensing surveys toward the Galactic center) yields one relation between M_ℓ and D_ℓ. Determining the flux of the planet-hosting lens star, F_ℓ, measuring the extinction toward the lens system, and applying a mass–luminosity relation provides a third mass–distance relation for the lens system. Any two of θ_E, π_E, and F_ℓ can give a unique solution for M_ℓ and D_ℓ, and the various methodologies can be cross-checked in the rare instances in which all three can be measured.

The Einstein Radius θ_E

The high, multi-band cadences of OGLE-IV, MOA-II, and KMTNet, most commonly in *V*- and *I*-band, lead to routine measurements of the Einstein radius θ_E, providing one channel for relating M_ℓ and D_ℓ. Specifically, the flux observed at a given time t during a microlensing event, F_{obs}, is the sum of the intrinsic flux of the background source star F_s multiplied by the time-dependent magnification A and the blend flux F_b from any other stars in the seeing disc of the observing facility:

$$F_{obs}(t) = A(t) \cdot F_s + F_b. \tag{3.2}$$

Thus, by observing the event in multiple filters at multiple times t with different magnifications $A(t)$, it is possible to determine the apparent color of the source. Then θ_E can be derived with the aid of color-surface brightness relations (see, e.g., Yoo et al. 2004; Kervella et al. 2004) via

$$\theta_E = \theta_*/\rho, \tag{3.3}$$

where θ_* is the angular radius of the source and ρ is the normalized angular source size. The parameter ρ can be measured if an observed light-curve includes the passage of the source over or near a caustic structure. Since the perturbations induced by caustics are what typically lead to the inference of a possible planet in the lens system, ρ, and by extension, θ_E, can be measured.

For completeness we also note that it is possible to measure θ_E via the entirely separate channels of multi-epoch high-precision astrometry to measure the shift in the centroid of the source over the course of the microlensing event (see, e.g., Hog et al. 1995; Miyamoto & Yoshii 1995; Walker 1995; Lu et al. 2016) or interferometry to resolve the images of the lens source during the event (see, e.g., Delplancke et al. 2001; Dalal & Lane 2003; Rattenbury & Mao 2006; Dong et al. 2019). However, these methodologies are employed far less frequently, largely due to a variety of additional observational constraints.

The Microlens Parallax π_E

A second avenue for obtaining a mass–distance relation for the lens can be achieved by measuring the microlens parallax π_E. One way this can be done is by observing the asymmetries that result from the acceleration of the Earth's orbit around the Sun (see, e.g., Refsdal 1966; Gould 1992) through what is known as the orbital parallax. While this generally is feasible only for microlensing events with longer timescales, such that the timescale of the ongoing microlensing event spans a larger fraction of the Earth's orbital phase, increasing the probability of distinguishing the acceleration-induced asymmetries from an otherwise symmetric light-curve, measuring the orbital parallax has become more common, in part due to the improved photometric precision and higher cadences of OGLE-IV, MOA-II, and KMTNet.

Another channel involves taking simultaneous observations from (at least) two widely separated observing locations and is referred to as the satellite parallax effect. Each observer views the same microlensing event but with slightly different geometries, which manifest themselves via shifts in the time and magnitude of the peak magnification. This was first employed through a target of opportunity (ToO) program with the Spitzer space telescope in 2005, leading to the discovery of a binary lens most likely residing in the galactic halo, in the direction of the Small Magellanic Cloud (Dong et al. 2007). Nearly a decade later, the first sustained campaign to use a space-based telescope as a platform from which to systematically obtain satellite parallax measurements of π_E began with a 100 hr pilot program on Spitzer. Among other discoveries, this led to the first satellite parallax measurement for an isolated star (Yee et al. 2015b) as well as the first such measurement for a planetary event (Udalski et al. 2015b), the light-curves of which are shown in Figure 3.30.

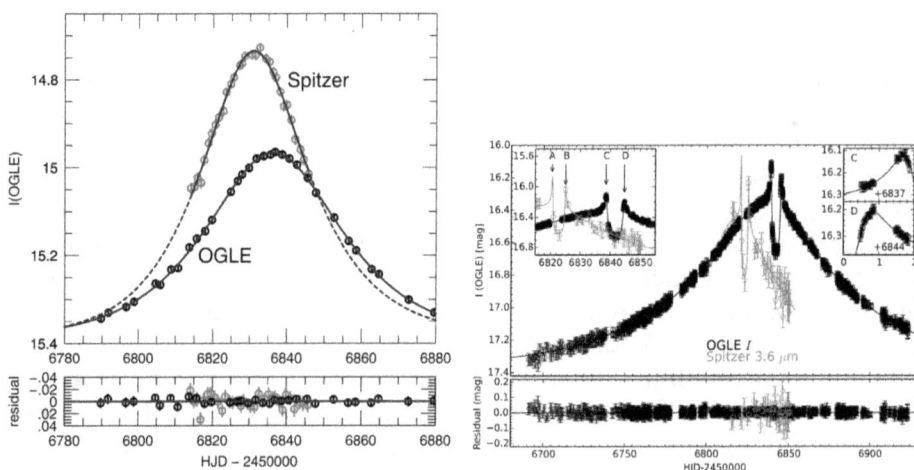

Figure 3.30. The light-curves of single-lens microlensing event OGLE-2014-BLG-0939 (OB140939; left panel), reproduced from Yee et al. (2015b; © 2015. The American Astronomical Society. All rights reserved), and planetary microlensing event OGLE-2014-BLG-0124 (OB140124; right panel), from Udalski et al. (2015b; © 2015. The American Astronomical Society. All rights reserved), as seen in I-band by OGLE-IV from the Earth (black points) and by Spitzer in its 3.6 micron band (red points). The different viewing geometries between Earth and Spitzer, in its Earth-trailing solar orbit, lead to differences in the magnification structure. For OB140124, the capital letters denote the four caustic approaches as seen from the Earth and from Spitzer. By utilizing the satellite parallax effect, the mass of the isolated lens OB140939L was measured to be $M_\ell = 0.23 \pm 0.7 M_\odot$ at a distance of $D_\ell = 3.1 \pm 0.4$ kpc, while the planet OB140124L b was determined to have a mass of $\sim 0.5 M_{Jup}$.

This ultimately led to a 6 year campaign with Spitzer to conduct targeted follow-up of microlensing events first detected by the ground-based optical surveys in order to first measure the relative abundance of planets in the galactic disk and the galactic bulge to help determine the galactic distribution of planets (see, e.g., Yee et al. 2015a).

The Timescale Distribution
In addition to facilitating the measurement of the mass of and distance to individual lens systems, the advent of the new networks of microlensing survey telescopes also gave rise to new statistical explorations of the frequency and mass function of solivagant planetary-mass objects (SPlaMOs), which are not gravitationally tethered to any host object, stellar or otherwise. The Einstein crossing time t_E is defined as:

$$t_E \equiv \frac{\theta_E}{\mu_{rel}}, \tag{3.4}$$

where μ_{rel} is the relative lens–source proper motion. By combining Equations (3.1) and (3.4), we see that $t_E \propto M_\ell^{1/2}$, such that a microlensing event with a shorter timescale arises from a lensing object with a correspondingly lower mass (marginalizing over the other physical properties of the lens and source). Specifically, an event for which $t_E \approx 1$ day (~ 1 hr) may be caused by a lensing object that is a Jupiter-mass (Earth-mass) SPlaMO. From Equation (3.4), however, the measurement of t_E for an individual microlensing event provides little information about the physical properties of the lens system, since the timescale itself is a degenerate combination of M_ℓ, D_ℓ, and D_s as well as the relative transverse motion of the lens–source system. Nevertheless, by examining an ensemble of events for which the selection effects are well understood and that have been corrected for the survey detection efficiency, it is possible to make statistical statements about the aggregate sample.

Sumi et al. (2011) did precisely this by examining 474 events observed in the first 2 years of operation of the MOA-II survey, from 2006–2007, and that they determined to be well characterized, following a series of quality cuts. The left panel of Figure 3.31 shows the observed timescale distribution, along with expectations from multiple theoretical models. By comparing the observed timescales with detection efficiency-corrected models, which include stellar, stellar remnant, and brown dwarf populations, they found an excess of short-timescale events, with a peak at $t_E \sim 1$ day, above the theoretical expectations from the models. Given that their sample excludes a wide variety of possible astrophysical contaminants, including cosmic ray hits and lensing events due to high-velocity stars, they concluded that the most likely cause of these observed short-timescale events was a population of solivagant or widely separated Jupiter-mass objects. They went on to estimate that these objects are 1.8 times as common as main sequence stars in the Galaxy, and that constraints from direct imaging imply that these objects must be solivagant.

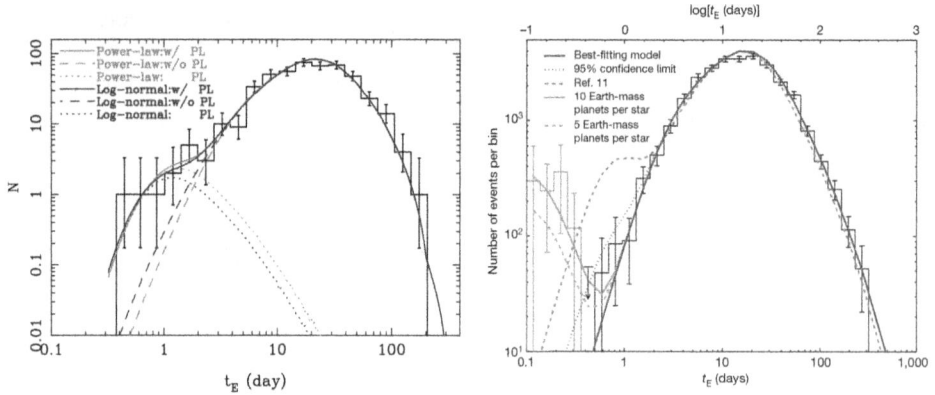

Figure 3.31. The distributions of the Einstein crossing time t_E as measured from 474 events from the MOA-II survey during 2006–2007 (black line; left panel), from Sumi et al. (2011), and from 2617 events from the OGLE-IV survey during 2010–2015 (black line; right panel), from Mróz et al. (2017). The MOA-II result, which was published prior to the start of K2C9, indicated a possible population of Jupiter-mass SPlaMOs, outnumbering main sequence stars in the Milky Way by nearly 2:1 (see the relative excess at $t_E \sim 1$ day of the observations above the predictions from a variety of Galactic models shown by the blue and red solid and dashed lines). However, the OGLE-IV result, which was published after the conclusion of K2C9 observations, found no such excess at $t_E \sim 1$ day, instead noting a potential reservoir of Earth-mass SPlaMOs, with $t_E \sim 0.1$ day. Both images reprinted by permission from Springer Nature: Nature.

3.11.3 The Overall Question

The confluence of the increased event detection rates and improved avenues of lens characterization brought about by new microlensing survey facilities presented an opportunity to explore the demographics of both bound and free-floating exoplanets. Following the failure of the second reaction wheel on board the Kepler spacecraft, and roughly concurrent with the proposal for the pilot satellite parallax campaign with Spitzer, Gould & Horne (2013) put forth the idea of conducting an automated microlensing survey with a wide-field satellite in a solar orbit. As they note, while Kepler itself is not optimally designed for such an endeavor—indeed, its primary science goal of determining $\eta_\oplus$ using the transit method sets very different design requirements—its aperture and FoV would allow for monitoring anywhere between 300–30,000 microlensing events, depending on the exact experimental design. The ability of Kepler to observe *the same* events as those detected separately by ground-based surveys, the fraction of which increases the closer Kepler points to the ecliptic, would lead to a substantial number of events for which π_E can be measured by leveraging the separation between Kepler and the Earth to observe the satellite parallax effect, ultimately providing a unique solution for the mass and distance of an ensemble of lens systems (see Figure 3.32). Moreover, this approach appeared to have several advantages as compared with the then nascent Spitzer methodology, namely that there would be no bias introduced into the observed sample due to human selection effects governing event-by-event target inclusion decisions, and that short-timescale events, lasting ~5 or fewer days, would be accessible, as there would be no lag between target selection and target observation.

Furthermore, fresh results from ground- and space-based microlensing experiments underscored the importance of conducting an automated and simultaneous microlensing survey from Earth and from space. Udalski et al. (2015b) demonstrated that a satellite parallax measurement of π_E could have a precision that is nearly an order-of-magnitude better than an orbital parallax constraint, and would also provide a cross-check on the ground-based determination. Additionally, as noted above, the targeted follow-up nature of the Spitzer microlensing campaign prevents events with timescales shorter than a few days from being observed (cf Figure 1 from Udalski et al. 2015b). Thus, an automated survey from space presented the best chance to attempt to investigate the claim of the existence of ~2 Jupiter-mass SPlaMOs per star in the Galaxy (Sumi et al. 2011). This is especially critical, given the tension of the MOA-II result with other observational constraints and expectations from simulations. Explorations of the substellar mass functions in young clusters, in particular in σ Orionis (Peña Ramírez et al. 2012) and NGC 1333 (Scholz et al. 2012), found that frequency of SPlaMOs is over an order-of-magnitude below what was claimed by the MOA-II result of Sumi et al. (2011). Meanwhile, Veras & Raymond (2012) showed that planet–planet scattering cannot produce such a large reservoir of Jupiter-mass SPlaMOs, and simulations by Pfyffer et al. (2015) of the formation and evolution of planetary systems could not eject enough planetary mass to explain the MOA-II result.

3.11.4 Experiment Design

Given these open scientific questions related to microlensing, the K2 Project Scientist ultimately decided that for Campaign 9, 85% of the downlinkable area of the K2 FoV would be dedicated solely to conducting an automated microlensing survey toward the Galactic center, spanning 86 days from 2016 April 7–July 2. The remaining ~15% of the downlinkable area would be devoted to the K2's Director's Discretionary Target program. In a departure from the standard Guest Observer procedure, through which teams of investigators submit proposals to observe a "postage stamp" of pixels centered on each object in a list of targets of interest, the K2 project solicited a call for individuals to submit proposals to join the Microlensing Science Team (MST), which would ultimately govern the scientific decision-making for K2's Campaign 9 (K2C9). Seven total individuals were selected to join the MST, with proposals primarily focused on the development of novel photometric analysis methodology and the accrual and coordination of simultaneous ground-based observing resources.

Pixel Selection
Although microlensing events are rare and generally difficult-to-impossible to predict, several ground-based surveys, including OGLE and MOA, have empirically measured the microlensing optical depth and spatially mapped the event rate toward the center of the Milky Way (see, e.g., Udalski et al. 1994; Alcock et al. 1997, 2000; Sumi et al. 2003, 2013; Sumi & Penny 2016; Navarro et al. 2018; Mróz et al. 2019). These studies have revealed a central window in the Galactic bulge with relatively

(3 Jul) Earth-*K2* Projected Separation $D_\perp$: 0.82 AU

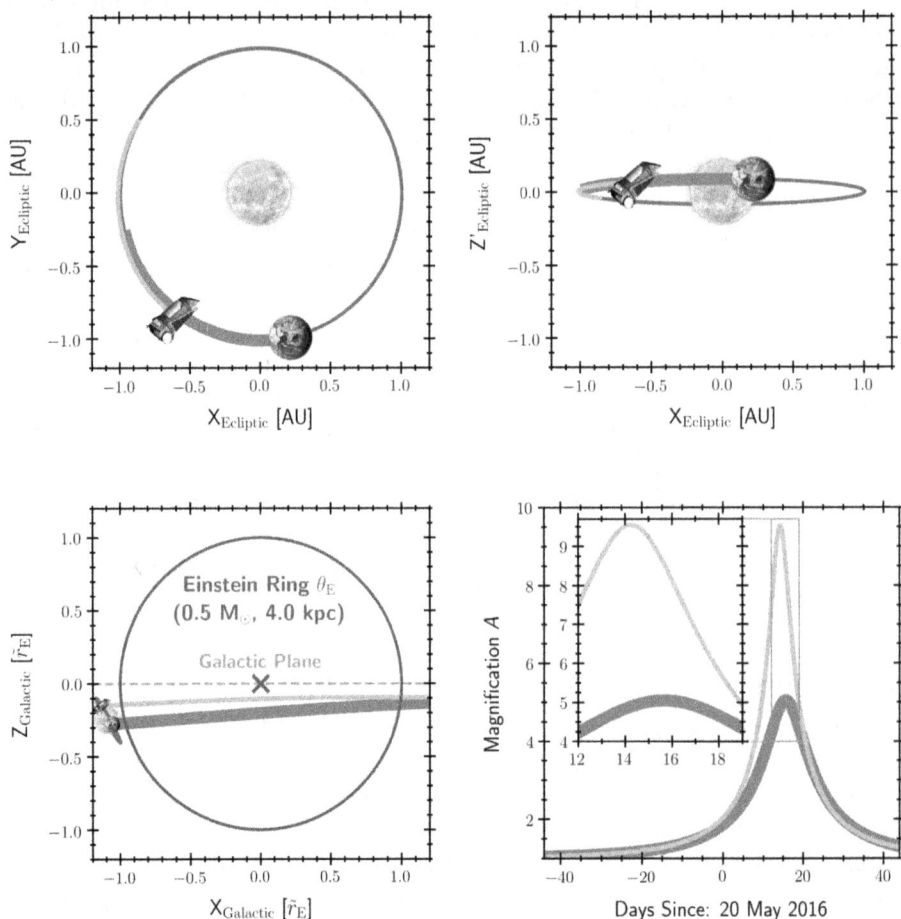

Created by Calen B. Henderson (Caltech/IPAC) using JPL's Horizons

Figure 3.32. A video visualization for the satellite parallax effect as observed from the Earth and K2 for an isolated lens with a mass of $M_\ell = 0.5\ M_\odot$ located at a distance of $D_\ell = 4.0$ kpc. The top left (right) panel shows the solar system as viewed from the top down (the center of the K2C9 survey superstamp; cf Figure 3.33), the bottom left panel is in a reference frame co-moving with the lens–source node, and the bottom right shows the resulting light-curves. Here $D_\perp$ denotes the projected separation of the Earth and K2 as viewed from the center of the K2C9 survey superstamp. (This is a modification of Figure 6 from Henderson et al. 2016. Credit: C. Henderson, Caltech/IPAC.) Video available online at http://iopscience.iop.org/book/978-0-7503-2296-6.

low extinction and a high concentration of background sources and foreground lenses that collectively produce a rate of events of $\sim 1 \times 10^{-6}$ per star per year. Poleski (2016) went on to show that a phenomenological model, formulated by a modified product of the surface density of red clump stars and the surface density of all stars brighter than the OGLE-III survey completeness limit, correlated strongly with event rate.

Limitations to on-board storage and down-link bandwidth made it impossible to retain the pixel data from the imager's full 105 deg^2 FoV at the cadence required to detect and characterize microlensing events and any planetary perturbations. As an alternative strategy, the K2 Project and the MST worked in conjunction to determine that the survey should be conducted toward a "superstamp" region of contiguous pixels. The model of Poleski (2016) was used to optimize the selection of which exact pixels would constitute the K2C9 survey superstamp, the precise geometry of which is shown in Figure 3.33. In total, the K2C9 survey superstamp comprised 3.06 million pixels, or roughly 3.7 deg^2.

The constraints on the data down-link bandwidth through the DSN led to the additional decision that the K2C9 survey would be split into two campaign halves, C9a and C9b, to help maximize the number of microlensing events for which π_E would be measured, from the ground and from space. Knowing this, the K2 Project provided two additional opportunities for target selection. The first was immediately prior to the start of C9a operations, and the second occurred during the mid-Campaign break, when the spacecraft temporarily interrupted observations in order to down-link pixel data from the first half of the Campaign. For each of these, the MST selected postage stamps for microlensing events known to be ongoing from ground-based surveys and uploaded them to the full target list of pixels to be downloaded. In total, 34 ongoing events were added for C9a and 61 were added for C9b, which totaled 70 unique events and complemented the ~100 events predicted to peak within the survey superstamp over the course of C9 observations (Henderson et al. 2016).

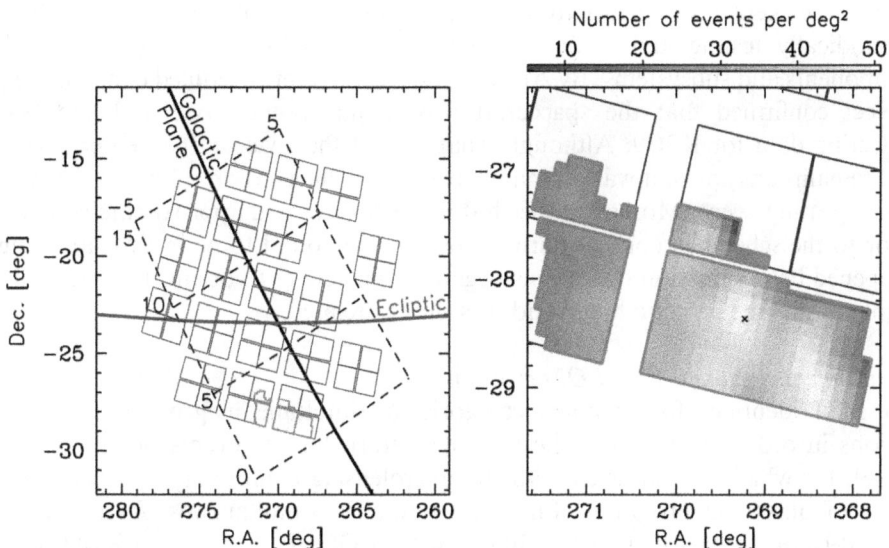

Figure 3.33. The field of view of the K2C9 survey superstamp (red) overlaid on the full Kepler detector array (blue). The superstamp covers 3.06 million pixels, or 3.7 deg^2. Reproduced from Henderson et al. (2016). © 2016. The Astronomical Society of the Pacific. All rights reserved. CC BY 3.0.

The MST maintained a running review of ongoing events detected by ground-based surveys within the K2 imager footprint, with collated information updated in real-time via NASA's ExoFOP system.[12] "Postage stamps" of pixels around selected events of interest were included in the observed pixel data.

Spacecraft Reorientation—And an Unexpected Emergency!

In order to ensure that the K2C9 space-based microlensing would occur when the Galactic bulge is simultaneously visible from the ground, the spacecraft needed to be reoriented to point along its velocity vector (+VV). During a standard, scheduled contact with the spacecraft on Thursday, 2016 April 7, mere hours after this re-pointing operation was expected to take place, the mission operations engineers found the Kepler spacecraft in emergency mode (EM). This marked the first time that Kepler had entered EM, which is its lowest operational mode and is extremely fuel-intensive. The mission quickly declared a spacecraft emergency and, in order to limit fuel use, teams at NASA Ames, Ball Aerospace, the Laboratory for Atmospheric and Space Physics at University of Colorado in Boulder, and the Deep Space Network (DSN) began working around-the-clock in order to diagnose the cause and safely return the spacecraft to scientific operations. At this point in time, Kepler was ~75 million miles from Earth, meaning the communications delay due to the light travel time was 13 minutes (roundtrip). Especially in light of this, the DSN was extremely responsive to the project's need for 70 m DSN dish time, even bringing a station out of maintenance to cover a 3 hour gap.

Incredibly, within 8 hours the teams were able to transition the spacecraft out of the fuel-heavy emergency mode and into safe mode. They then began work to restore the Kepler spacecraft to a stable science operations mode, diligently and methodically testing and recovering a wide range of engineering and scientific components and subsystems. By April 22, during another scheduled contact, the K2 project confirmed that the spacecraft was in fine point mode and had begun collecting data for K2C9. Although what caused the spacecraft to enter EM was and remains unknown, it was ultimately believed to have arisen from a transient and non-repeating event. Moreover, the shift into EM occurred approximately 36 hours prior to the scheduled contact during which it was found to be in EM, meaning it happened before mission operations began the maneuver to orient the spacecraft to point toward the galactic bulge and was thus unrelated.

Simultaneous Ground-based Observations

The MST identified four primary categories of simultaneous ground-based obser-vations in order to maximize the scientific return for the events observed during K2C9, for which they worked with the microlensing community and organized a series of observing programs. First and foremost, simultaneous observations by ground-based (optical) survey facilities, such as OGLE, MOA, and KMTNet, as well as Las Cumbres Observatory (LCO),[13] facilitate measuring π_E via the satellite

---

[12] See the Exoplanet Follow-up Observing Program, NASA Exoplanet Archive, doi: 10.26134/ExoFOP2.

[13] Formerly Las Cumbres Observatory Global Telescope (LCOGT).

parallax effect in tandem with K2 and also are critical for constructing a master database of known microlensing events within the survey superstamp, conducting near-real-time analysis, and prioritizing ground-based follow-up resources. To this end, OGLE-IV in particular modified its observing strategy to maximize coverage of the K2C9 superstamp. Second are yet higher-cadence targeted follow-up programs, conducted by a variety of teams, including LCO, to observe individual ongoing microlensing events designated as high priority, whether due to possible planetary signatures or other features of interest. A third component is multi-band photo-metric monitoring, in order to constrain the colors of the source star and ultimately measure θ_E, as described in Section 3.11.2, and achieved using MegaCam on the Canada–France–Hawaii Telescope (CFHT), among other facilities. A final, fourth feature is near-infrared (NIR) source flux measurements observations, taken in JHK_s, to provide another mass–distance relation for the lens systems, as mentioned in Section 3.11.2 and as primarily carried out using the wide-field NIR camera on the United Kingdom Infrared Telescope (UKIRT). Figure 3.34 shows a color-coded map of the facilities and instruments dedicated to each of these four observational goals, with the details of each observing program described in detail in Henderson et al. (2016).

3.11.5 Analysis Challenges and Pipeline Development

Even with the abundance and quality of ground- and space-based data, three factors inherent to the K2C9 microlensing survey necessitated the development of novel

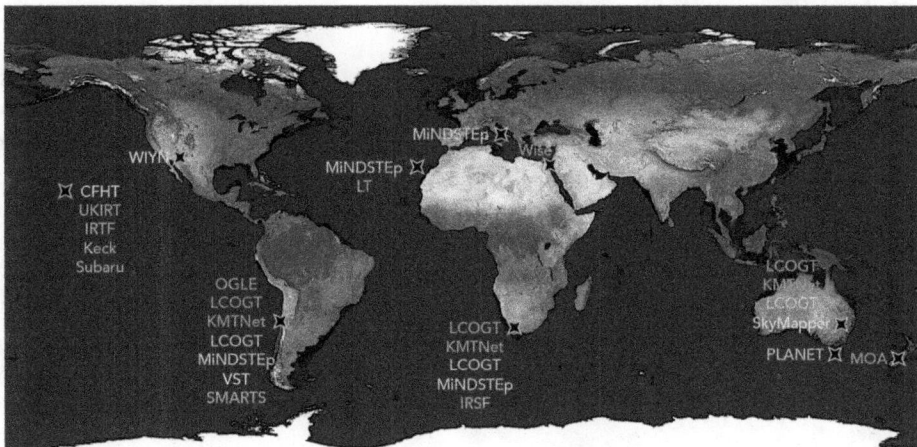

Figure 3.34. The map of all ground-based observing facilities marshaled by the Microlensing Science Team in support of the K2C9 microlensing survey, reproduced from Henderson et al. (2016). © 2016. The Astronomical Society of the Pacific. All rights reserved. CC BY 3.0. They are color-coded according to their primary science driver—automated survey (blue); high-cadence follow-up (orange); multi-band source color monitoring (yellow); and near-infrared source flux measurements (purple). See Section 3.11.4 for a complete description.

photometric pipelines for the unique challenge of extracting microlensing signals from the crowded star fields of the central Galactic bulge using the Kepler camera: (1) the stellar surface density within the K2C9 survey superstamp, which was nearly three stars with $I < 20$ per K2 pixel (cf Figure 8 of Henderson et al. 2016); (2) the drift of stars across the focal plane, which occurred at a rate of ~1 pixel each hour; and (3) the fact that, for all but the shortest-duration microlensing events, the timescale of the astrophysical variability was at least as long as a given campaign half.

To address these concerns, Zhu et al. (2017b) developed an approach whereby raw light-curves and astrometric solutions were derived using the methods of Soares-Furtado et al. (2017) and Huang et al. (2015), and then the color of the source stars in the Kepler bandpass were predicted using color–color relations. Another technique involved generating a causal pixel model (CPM) using a patch of reference pixels nearby to but at least several times the spacecraft's pixel response function (PRF) away from a given pixel of interest (Wang et al. 2017). In principle the model would include all photometric variability arising from the instrument and spacecraft, indicating that any remaining signal would be astrophysical in nature. However, the initial version of the CPM method was tested on and applied to only the least crowded regions of the K2C9 superstamp, and so Poleski et al. (2019) extended this approach to the densest regions of the superstamp via a modified CPM (MCPM) technique. A third approach used multi-band photometry taken with CFHT during C9 to facilitate measurement of the colors of the source stars (Zang et al. 2018), which aids to improve the precision of the derived values of the microlens parallax π_E (cf Section 3.11.2).

3.11.6 Results

The goal of the K2C9 microlensing survey was to enable a variety of avenues for lens characterization and demographic exploration. However, the development of photometric pipelines proved more challenging even than initially surmised, and the intrinsic Galactic frequency of Jupiter-mass solivagant planetary-mass objects was shown to be much lower than expected by Mróz et al. (2017), who examined the OGLE-IV timescale distribution and found no evidence for a substantial population of such planets. Nevertheless, there have been a handful of results for both isolated and two-body lens systems that showcase the scientific utility of this experiment, including the community-driven approach for sharing observational resources, while also emphasizing the difficulty of conducting crowded-field photometry with a large PRF to search for signals whose durations are comparable to the observing window.

Isolated Lenses
Zhu et al. (2017b) applied their approach to two events that were detected by OGLE-IV and that did not show any anomalous features indicating additional bodies in the lensing system. OGLE-2016-BLG-0980 (OB160980) was a faint event, with an I-band source magnitude of $I_s = 19.0$, while OGLE-2016-BLG-0940 (OB160940) had a much brighter source with $I_s = 17.3$. For both OGLE-2016-

BLG-0940 and OB160980 they were able to extract photometry from K2 and reliably measure non-zero values of the satellite parallax π_E. Unfortunately, since these events did not show finite-source effects, it was not possible to measure ρ, and so, even with color information, the mass of and distance to the lensing objects are not able to be uniquely determined.

Also using the methodology developed by Zhu et al. (2017b), Zang et al. (2018) analyzed the short-timescale event OGLE-2016-BLG-0795 (OB160795). An analysis using only OGLE and CFHT data yields an Einstein crossing time of $t_E < 5$ day, indicating that the lensing object has a lower mass and/or is more distant, making it a candidate SPlaMO. While this event suffers from the four-fold satellite parallax degeneracy (cf Refsdal 1966; Gould 1994), they were able to find a most likely lens mass of $M_\ell \sim 0.5\ M_\odot$ at a distance of $D_\ell \sim 7$ kpc.

The event MOA-2016-BLG-290 (MB16290) became the first-ever microlensing event for which photometric signals were extracted from three different observer locations separated by ~1 au. By combining K2C9 data with observations from Spitzer as well as ground-based facilities, Zhu et al. (2017a) were able to resolve the aforementioned four-fold satellite parallax degeneracy for this event. With a time-scale of $t_E \sim 6.4$ d, MOA-2016-BLG-290 is another candidate SPlaMO, the full light-curve of which is shown in Figure 3.35. They find that MB16290L has a mass of $M_\ell = 77^{+34}_{-23}$ M_{Jup} and a distance of $D_\ell = 6.8 \pm 0.4$ kpc, making it a brown dwarf or extremely low-mass star located in the galactic bulge.

Three of these events, OB160795, OB160980, and MB16290, were also analyzed by (Poleski et al. 2019) using their MCPM approach. For OB160980 they obtained results largely consistent with Zhu et al. (2017b), in that the statistical differences between their respective models were comparable to the π_E uncertainties as derived for other events via the annual parallax effect, but they noted that their MCPM method produced smaller scatter in the K2 data. In the case of OB160795, their extracted photometric signal from K2 differed significantly from that of Zang et al. (2018), which was analyzed using the Zhu et al. (2017b) approach. Finally, for MOA-2016-BLG-290, they obtained negatively valued blend flux F_b, indicating that the source could be located in a spatial region wherein the surface brightness of the uniform background of unresolved stars is lower than the local average, and/or that the correct model has not been found. They do note that the agreement between the predicted and fit Spitzer fluxes for the small π_E solutions from Zhu et al. (2017b) leads to the interpretation that they are correct, or that this agreement is coincidental, such that additional modeling is necessary. The different results for these same events, as analyzed using different techniques for extracting the photo-metric signal from K2 and determining the best-fit lensing model underscore the difficulties described in Section 3.11.5. There are furthermore several ongoing studies to explore the frequency of low-mass isolated lenses, including SPlaMOs (McDonald et al. 2020, in preparation) and brown dwarfs (Poleski et al. 2020).

Planetary Lenses
In addition to results for single-body lens systems, there were multiple two-body lens systems to which the full K2C9 experiment contributed data. The event MOA-2016-

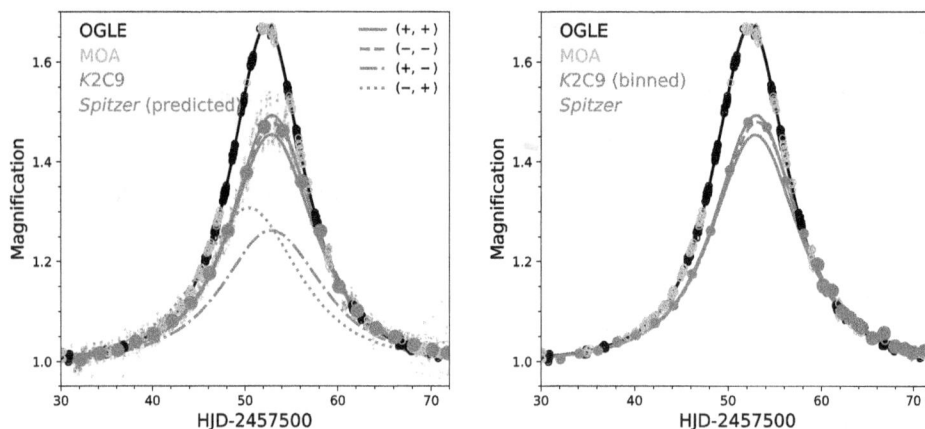

Figure 3.35. The light-curve of event MOA-2016-BLG-290 (MB16290) as viewed from the ground by OGLE-IV (black) and MOA-II (orange) and from space by K2C9 (blue) and Spitzer (red), reproduced from Zhu et al. (2017a). © 2017. The American Astronomical Society. All rights reserved. The four red lines (left panel) indicate the four competing predictions for the observed Spitzer due to the four-fold satellite parallax degeneracy. The right panel shows how this degeneracy has been reduced to a two-fold degeneracy, wherein each of the remaining two models yield a statistically indistinguishable value for the magnitude of the satellite parallax π_E.

BLG-227 (MB16227) occurred just outside the survey superstamp, for which the ground-based resources marshaled for K2C9 led to excellent coverage of the planetary deviation, ultimately revealing a super-Jupiter-mass planet located 6–7 kpc from Earth (Koshimoto et al. 2017). As noted above, K2C9 overlapped with the targeted 2016 Spitzer microlensing campaign, leading to multiple ground-breaking discoveries involving simultaneous observations at *three* different locations—K2, Spitzer, and the Earth—helping to break the satellite parallax degeneracy introduced with observations from the Earth and a single space-based observing platform.

This was leveraged for the event OGLE-2016-BLG-1190 (OB161190), which received high-cadence observations, 3–4 hr^{-1}, from OGLE, MOA, and KMTNet. The light-curve is shown in Figure 3.36, which shows that this is a long-duration (and indeed, long-timescale) event. In resolving the satellite parallax, (Ryu et al. 2018) found that the planetary companion OB161190L b has a mass of $M_p = 13.4 \pm 0.9$ M$_{\text{Jup}}$ and a semimajor axis of ~2 au, and that the lens system has a distance of $D_\ell \sim 7$kpc. This object is of particular interest given that its mass hovers right around the deuterium-burning boundary, and also because it is the first planet in the Spitzer microlensing sample that is located in the galactic bulge (Ryu et al. 2018).

Public Data Releases

In addition to the ground-breaking results for isolated and planetary microlenses from K2C9, this microlensing experiment helped usher in a watershed moment with regard to the public sharing of community data. The CFHT multi-band source color program published a data release that includes light-curves in up to all three of the *gri* filters used for 217 microlensing events (Zang et al. 2018). The full photometric

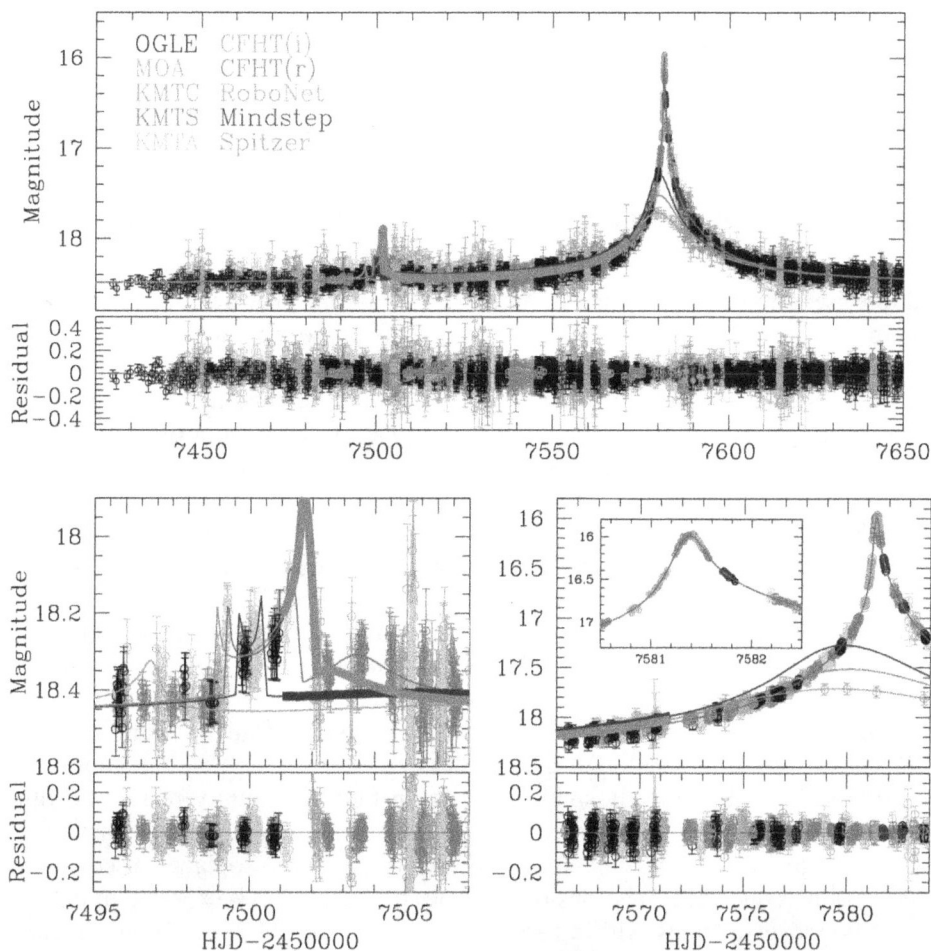

Figure 3.36. The light-curve for microlensing event OGLE-2016-BLG-1190 (OB161190), reproduced from Ryu et al. (2018). © 2018. The American Astronomical Society. All rights reserved. The thick blue and red lines show the K2C9 light-curve for two competing models. Ultimately the authors were able to use the observations from K2 to break the satellite parallax degeneracy and derive a unique mass for the planet OB161190L b of $M_p = 13.4 \pm 0.9$ M_{Jup}.

catalog from the UKIRT NIR survey is available online through both a curated search interface provided by the NASA Exoplanet Archive.[14] Finally, the KMTNet collaboration released their light-curves of known microlensing events occurring within three effective timescales t_{eff}, where $t_{\mathrm{eff}} \equiv u_0 t_{\mathrm{E}}$ and u_0 is the lens–source closest approach, in angular units normalized to θ_{E}, of the K2C9 observations (Kim et al. 2018). This immediately public data set includes 181 "clear" and 84 "possible" microlensing events as found by the KMTNet event finder, as well as 56 events

---

[14] Data made available via the UKIRT Microlensing Survey search interface, NASA Exoplanet Archive, doi: 0.26133/NEA7.

found by the OGLE and/or MOA survey(s) that were not found by KMTNet. This release of light-curves for over 300 definite and candidate microlensing events helps provide ground-based observations to pair with space-based data for determining π_E and providing one mass–distance relation for additional lens systems to be discovered and analyzed.

3.11.7 Lessons Learned and Looking Forward

From the outset of planning for K2C9, it was acknowledged that extracting accurate and precise photometry from the large-pixel data in crowded fields would be arguably the greatest challenge. While this proved to be the case, K2C9 also demonstrated that a photometric precision sufficient for many science cases (~few %) can be achieved by the application of specialized detrending algorithms. This paves the way for similar approaches to be applied to data from other wide-field/ large pixel surveys, for example ASAS-SN (Kochanek 2017), providing the CCD response is adequately characterized.

K2C9 also demonstrated the efficacy of coordination between different community groups to optimize the planning of the program and to ensure that all necessary observations are obtained as efficiently as possible. This is particularly important for transient science where additional follow-up normally cannot be obtained after the fact. Recognizing this, the NASA Exoplanet Archive extended its Exoplanet Follow-up Observing Program to provide a valuable public platform for this purpose that was customized to the particular needs of microlensing events.

K2C9 also served to highlight the need to expand the microlensing community, illustrating the comparatively small numbers in that community relative to other fields of exoplanet science. The need to cultivate expertize in this field is particularly urgent in light of NASA's upcoming WFIRST mission. Exoplanet discovery via microlensing is one of the missions twin science drivers, and it is expected to discover ~1400 bound exoplanets with masses greater than ~$0.1M_\oplus$ (Penny et al. 2019). Despite providing higher spatial resolution imaging, the analysis of these data present their own challenges, not least of which is the computationally intensive nature of modeling microlensing events, and it's many degeneracies. The need to build the microlensing community has been acknowledged within the field and is an ongoing aspect of preparations for the WFIRST mission.

3.12 Toward Eta-Earth

As discussed in the Introduction, the primary scientific goal of the NASA Kepler mission was to determine $\eta_\oplus$, the frequency of rocky planets in the habitable zone of Sun-like stars. Here we lay out the steps toward measuring $\eta_\oplus$, including the challenges, solutions, and publicly available products produced by the mission to enable future calculations.

3.12.1 Ingredients for Occurrence Rate Calculations

If we had perfect knowledge of our stellar and planet samples, then calculating $\eta_\oplus$ would be as simple as $\eta_\oplus = N_m/N_{det}$, where N_m is the number of true Earth-analogs in the

Kepler sample, and N_{det} is the number of stars around which true Earth-analogs would have been detected. But in reality we do not have perfect knowledge of either of these numbers. N_m can be both missing planets (incomplete), and contain signals falsely identified as planets (unreliable). N_{det} is confounded by hidden stellar multiplicity, variable stellar noise properties, and complicated observational window functions. To most accurately characterize N_m, the Kepler team measured both the completeness and reliability of the final Kepler planet candidate catalog with high fidelity—see Section 3.12.3 for details. The noise properties and window function of the Kepler light-curves were studied in some detail, and the project made a set of products available for use in occurrence rate calculations by the community, which can be found at the NASA Exoplanet Archive.[15] The stellar multiplicity of the Kepler sample is still an ongoing area of research—see the Chapter 4 for additional discussion.

3.12.2 Early Results and Challenges

Almost as soon as there were planet candidates coming out of Kepler, there were attempts to measure the underlying planet occurrence rates. Borucki et al. (2011a) analyzed the first quarter (~90 days) of data and launched the still-ongoing investigation of the frequencies of planets in the Kepler data. Table 3.1 shows a selection of the occurrence rate calculations for Earth-sized planets published in the literature;[16] in each case the occurrence rate calculation for the longest period bin considered in the publication is shown. The analyses using the early KOI lists suffered from many inadequacies. These early lists were generated manually in a non-uniform manner—see Section 3.9.2 for details. The completeness of the process (the rate at which true planet signals were correctly identified) and the reliability (the rate at which astrophysical false positives and instrumental false-alarm signals are correctly identified and removed) were unknown, and could only be estimated (Section 3.12.3). Many of the quirks of the Kepler instrument—and their subsequent impact on the KOI lists—were still coming to light. None of the above stopped scientists from trying, but they did introduce considerable uncertainty into the resulting calculations.

Compiling Table 3.1 also reveals a raft of other challenges when comparing results from different analyses. Almost all analyses used different sets of light-curves from different stellar samples, considered different period or semimajor axis ranges, and considered different planet radius ranges. Their treatment of completeness, false positive and false-alarm rates varied from sophisticated to none at all. Even the calculated result can be presented differently: either as a frequency of stars with the prescribed planet (F_p), or the number of those planets per star (NPPS). It may come as no surprise then that the estimated frequency of "solar-like stars" (variously including F, G, and K stars) hosting "Earth-size planets" ranges from 1.1%–46%, and the estimated number of planets per star from 0.01–2 (and that within one publication!).

---

[15] https://exoplanetarchive.ipac.caltech.edu/docs/Kepler_completeness_reliability.html.
[16] List of publications drawn from the NASA Exoplanet Archive at https://bit.ly/2mi0l65.

Table 3.1. A Selection from the Published Literature of Occurrence Rate Calculations of Earth-sized Planets using Kepler Data

References	Kepler Data	Stars	Period or a	Planet Radii	PComp	FPR	FAR	F_p or NPPS
Borucki et al. (2011b)	Q0–Q2	All (153,196)	<138 days	0.5–1.25 $R_\oplus$	No	Yes	No	F_p: 5%
Catanzarite & Shao (2011)	Q0–Q5	All (153,196)	0.95–1.37 au	0.8–2 $R_\oplus$	No	Yes	No	F_p: $1.1^{+0.6}_{-0.3}$%
Youdin (2011)	Q0–Q2	GK (58,041)	<50 days	>0.5 $R_\oplus$	No	No	No	NPPS: 0.7–1.4
Fressin et al. (2013)	Q1–Q6	All (156,453)	0.8–85 days	0.8–1.25 $R_\oplus$	Yes[a]	Yes	No	NPPS: 0.184 ± 0.037
Gaidos (2013)	Q1–Q8	Dwarfs (122,422)	0.68–1.95 au[b]	0.8–2 $R_\oplus$	No	No	No	F_p: 46^{+18}_{-15}%
Petigura et al. (2013a)	Q1–Q15	GK (42,557)	200–400 days	1–2 $R_\oplus$	Yes	Yes	No	F_p: 22 ± 8%
Dong & Zhu (2013)	Q1–Q6	FGK (122,328)	<250 days	1–2 $R_\oplus$	No	No	No	F_p: ~28%
Foreman-Mackey et al. (2014)	Q1–Q15	GK (42,557)	200–400 days	1–2 $R_\oplus$	No	No	No	F_p: $1.9^{+1.0}_{-0.8}$%
Mulders et al. (2015)	Q1–Q8	G	150–250 days	1.0–1.4 $R_\oplus$	Yes[c]	No	No	F_p: 13.5 ± 5.6%
Burke et al. (2015)	Q1–Q16	GK (91,567)	290–440 days	0.8–1.2 $R_\oplus$	Yes	Yes	No	NPPS: 0.01–2

Notes.
[a] A software pipeline completeness was estimated using the recovery rate of eclipsing binaries compared to the expected population.
[b] Scaled for each star by the effective temperature.
[c] Using the estimated software pipeline completeness from Fressin et al. (2013).

3.12.3 Ingredients for Accurate Occurrence Rates

Completeness: Which Planets Did We Miss?

One way to measure the completeness of a process such as planet detection is to use "ground-truth" data, where you start with the answer and can compare that to what you observe. For the Kepler planet finding pipeline and vetting process, this involved a series of experiments injecting hundreds of thousands of simulated planet and false positive signals into the calibrated Kepler pixels, processing them through the pipeline (and eventually Robovetter) as normal, and comparing with expectation. Figure 3.37 shows the results from one such injection and recovery experiment on the pipeline used to generate the final Kepler planet candidate catalog. The drop-off in sensitivity for longer period and smaller planets, which combine to produce transit signals with lower signal strength, is evident, and in fact the detection efficiency of the pipeline is fairly well approximated by a one-dimensional function of the signal strength, modulo some additional loss at very long periods where the number of observed transits falls to <5. Figure 3.38 shows this function, fit with a Γ cumulative distribution; see Christiansen (2017) for more details. The completeness plateaus at ~95%, even for the strongest signals, implying that there is no signal strength cut-off that would result in a fully complete catalog (an assumption made in several of the early analyses shown in Table 3.1).

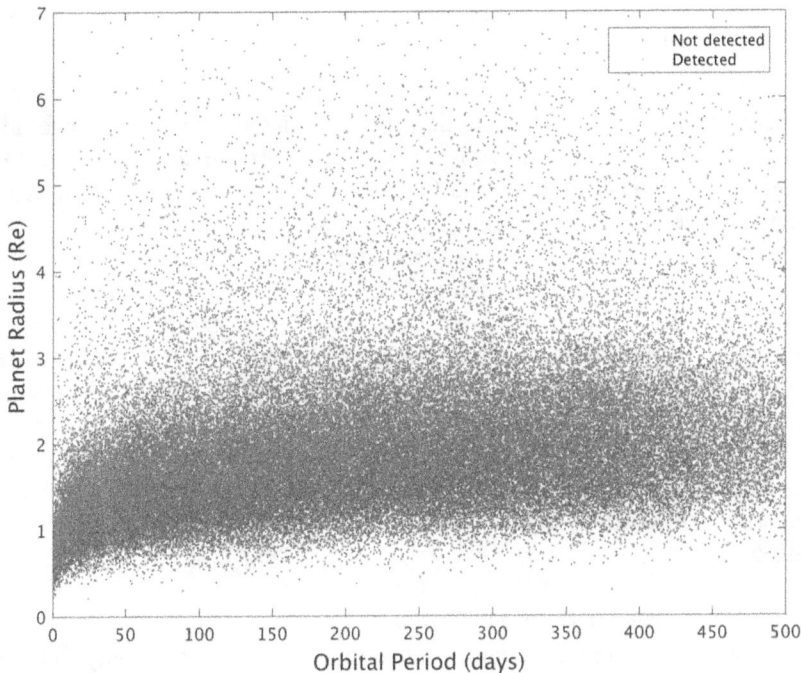

Figure 3.37. The population of simulated planets injected into the Kepler light-curves to examine the completeness of the pipeline used to generate the final Kepler planet candidate catalog. The red dots indicate planets that were successfully recovered by the pipeline; the blue dots indicate missed detections.

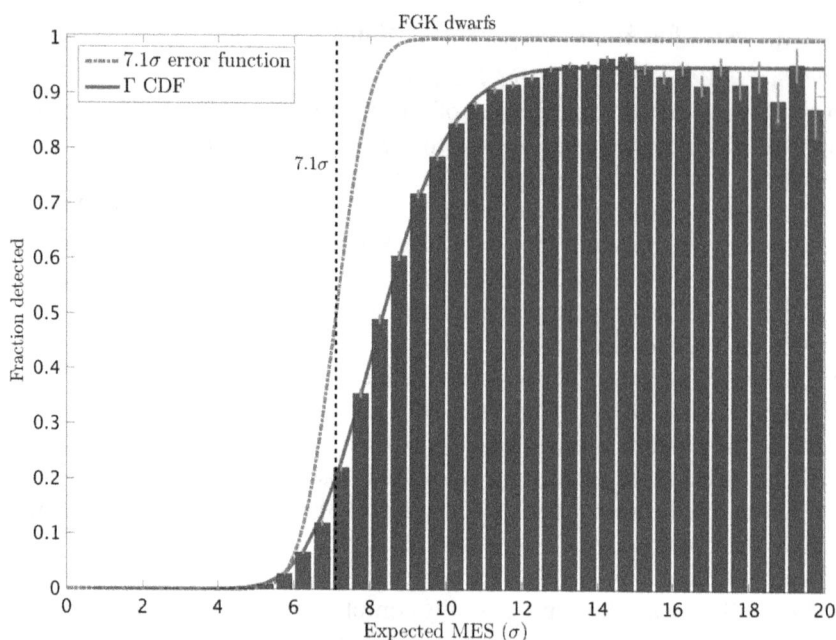

Figure 3.38. The completeness of the final Kepler pipeline as a function of the signal strength (measured here as Expected MES). The hypothetical best-case-scenario is shown as the red dashed line; the blue histogram shows the results of the injection and recovery experiment; and the blue line shows a Γ cumulative distribution function fit to the histogram.

One interesting aspect of this work was that with each evolution of the pipeline and subsequent planet candidate catalog, the completeness would change, and not necessarily always in the direction of higher completeness. Estimates of the completeness were generated for the final four Kepler planet candidate catalogs, and a concentrated community education campaign was undertaken to ensure that people using a given catalog for an occurrence rate calculation would use the associated completeness measurements.

The one-dimensional function described here is by design an average completeness over the whole stellar sample under consideration. For occurrence rate calculations, targets need to have their amenability to detection of planets assessed individually; they each have different noise properties and window functions. The Kepler mission released a set of software called KeplerPORTS[17] which takes the noise properties, window functions, and average completeness measurements to calculate the individual detection contours for each target in the occurrence rate calculation.

Reliability: Is the Signal from an Exoplanet?
When Kepler's second reaction wheel failed and it was clear that only four years of data were going to make up the hunt for Earth-size planets in the habitable zone, it

---

[17] https://github.com/nasa/KeplerPORTs.

was realized that any such transit detection was going to be very low signal to noise and made up of only three or four shallow transits. As scientists continued to examine the signals found by the transit detection algorithms they would consistently disagree on whether a signal was more likely caused by an astrophysical transit or some sort of instrumental noise. As was discussed in Section 3.12, there are several sources of false positives that do not come from any astronomical source. For accurate occurrence rates the challenge is to measure how frequently the instrument, or the astrophysical universe, can create false positives that get confused with transits. Once measured it is possible to understand the statistical reliability of the KOI catalog. The reliability against those known astrophysical events that masquerade as transits was explored with validation. However, for the lowest signal to noise events, it is the instrumental false positives that needed measured.

The final KOI catalog, because it was automated with the Robovetter (Thompson et al. 2018; Coughlin et al. 2016), could also easily evaluate simulated false positives. The problem was generating false positives at the correct rate to see how many slipped through and became planet candidates. This was done by inverting the Kepler light-curves and scrambling them by season. In this way they would produce much the same kind of false positives, but remove any coherent true astrophysical signal in the data. As a result the final KOI catalog was able to show that while the exoplanet detectors allowed a lot of false positives through, turning them into TCEs, the Robovetter worked perfectly over 99% of the time. Unfortunately, for long period planets since there are so few detected in the data set, even after 99% of the false positives were eliminated from the catalog, about half of the KOI planet candidates catalog at periods greater than 200 days are actually false positives.

While this is in some ways disappointing, it is exactly what was needed to refine the occurrence rates and get more accuracy on the value of eta-Earth. Any calculation of occurrence rates must be able to take into account both the completeness of the KOI catalog and the reliability of the signals. While it may not be possible to point to a single planet candidate and say whether it is really a planet, it does mean it is possible to measure how common small, long period planets are in the Galaxy.

3.12.4 Constraints on Eta-Earth

The Kepler mission was the first mission capable of detecting transiting planets the size of the Earth orbiting stars similar to the Sun at a distance where liquid water could pool on the surface of the planet. As we explored in the previous sections, correctly calculating $\eta_\oplus$, or how common these planets are, involves bringing together many pieces of information: transit detections, search completeness, catalog reliability and accurate stellar parameters. To do this calculation at the very edge of the Kepler mission's detection threshold is even more tricky. As shown in Figure 2.15, Kepler did not find many planets that fall in even a broad interpretation of Earth-like. Also, the reliability of the planet candidates it did find in this regime is around 50%. However, the fact that it was possible to detect Earth transits if they were there and the fact that slightly larger exoplanets in slightly closer orbits were confidently found, makes it possible to place some constraints on $\eta_\oplus$.

Many researchers have attempted to make the measurement of $\eta_\oplus$ using various techniques and various planet candidate catalogs created using the Kepler mission data. In all cases, the studies were done by making some assumptions about the completeness and reliability. Also, most of these studies extrapolated the results from larger planets orbiting in shorter periods. We have summarized some of these results in Figure 3.39 split between those that orbit stars like then Sun (G dwarfs) and those that orbit small cool stars known as M dwarfs.

The habitable zone for M dwarfs is much closer to the star where planets have orbital periods of around 20 days. These types of transiting planets were much easier to find than the Earth-sized planets orbiting the much larger G dwarf stars at around 365 days, however Kepler observed far fewer M dwarf stars, leaving large error bars on the M dwarf $\eta_\oplus$ value. Even with the small number of M dwarf stars searched, Dressing & Charbonneau (2015) determined the number of terrestrial planets in the habitable zone of M dwarf stars ranges from 16% to 42%.

For the G dwarf $\eta_\oplus$ value, the measurements vary wildly (from 2 to 50 Earth-size planets per 100 stars) and have large error bars. The large error in our understanding of this value is in part because of the different definitions of a habitable zone and what constitutes an Earth-size planet. It is also because each study uses a different statistical technique or a different Kepler exoplanet catalog, and all the systematic

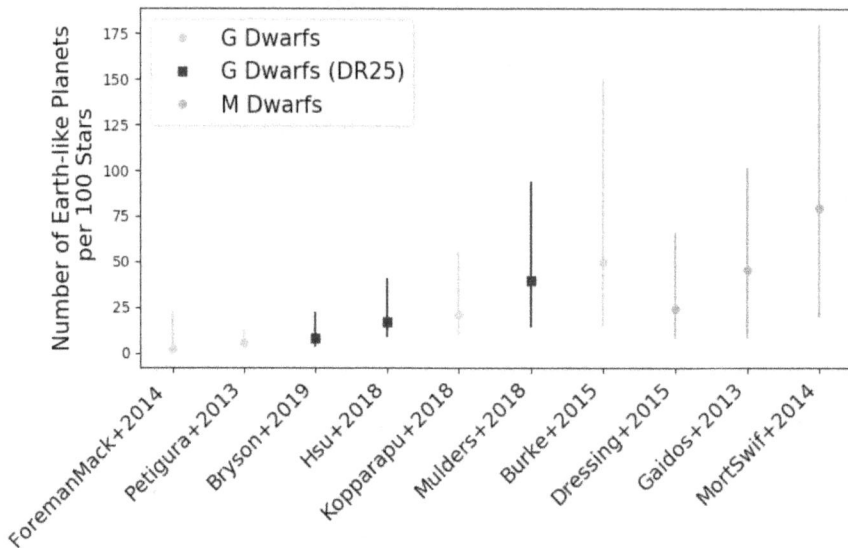

Figure 3.39. An overview of recent published values of $\eta_\oplus$ and the reported error bars for both G dwarf stars (blue) and M dwarf stars (orange). Those studies that used the Data Release 25 Kepler exoplanet catalog are labeled in dark blue and marked with a square (Foreman-Mackey et al. 2014; Petigura et al. 2013a; Bryson et al. 2020; Hsu et al. 2018; Kopparapu et al. 2018; Mulders et al. 2018; Burke et al. 2015; Dressing & Charbonneau 2015; Gaidos 2013; Morton & Swift 2014).

uncertainties must be correctly accounted for when doing the calculation. Most studies extrapolate from shorter periods (e.g., Petigura et al. 2013b) while others fit a model (e.g., Burke et al. 2015) to the observed planet population. Only three studies have used the last, full characterized Kepler catalog (known as DR25) which searched the entire 17 quarters of Kepler data (Mulders et al. 2018; Bryson et al. 2020). Both Mulders et al. (2018) and Hsu et al. (2018) decided to use the most reliable part of the catalog, which eliminates most small planet detections in or near the habitable zone. Bryson et al. (2020) is the only one that fully attempts to take the reliability into account, however, his work points out the difficulty in applying this correction. Even these three studies vary in the value of $\eta_\oplus$ from 8 to 40 planets per 100 stars, with even larger error bars. The scientific community will continue to determine the best way to constrain the value of $\eta_\oplus$ with the Kepler data by exploring which statistical technique yields the most accurate answer, by improving the measurements of the reliability of the Earth-size planets in the catalog, and by determining how to correctly account for stellar multiplicity.

Even with the large uncertainty on the value of $\eta_\oplus$, Kepler has improved our understanding of these types of planets. Prior to Kepler we did not know if every Sun-like star had an Earth, or if we were the only one. The Kepler observations have shown us that the answer is likely somewhere in between. Even if the lower value of Bryson et al. (2020) turns out to be the correct value (4 Earth-size planets per 100 G dwarf stars). This means that within 100 lt-yr of Earth we will find at least 50 Earth-size planets orbiting Sun-like stars, and we could expect to find more than 800 million in the entire Milky Way Galaxy. But if the larger numbers are correct, these numbers could easily increase to 500 within 100 lt-yr and 8 billion in the Milky Way Galaxy. Someday, with the next generation of telescopes, we may even be able to directly observe one of these nearby Earth-size planets, measuring more than just the size and surface temperature of the planet. With that information we will be far closer to understanding whether these planets are truly habitable like our planet Earth.

Exoplanets in Science Fiction

Contributed by David Lee Summers, Southern New Mexico.

There are several contenders for the very first science fiction story ever written. That said, I think a good case can be made that the first story was *Somnium*, or *The Dream*, a novelette about a voyage to the Moon written in 1608 by none other than Johannes Kepler, then published posthumously in 1634[18] In the story, Kepler imagines a witch summoning demons to take her and her astronomer son to the Moon. Once there, they are given a detailed tour and they look back and see the Earth. The book is filled with footnotes that detail Kepler's actual observations of the Moon. Although magic is

[18] A copy of Kepler's book actually appears as liner notes to Mannheim Steamroller's *Fresh Aire 5* album, which is a musical tribute to the book.

invoked to get people to the Moon, Kepler succeeded in showing the Moon as a place that could be visited and the Earth as a "planet" in the solar system.

Two and a half centuries later, Mars would capture the interest of astronomers and the imagination of science fiction writers. In 1877, during a close approach of Mars, Asaph Hall discovered Mars's two moons and Giovanni Schiaparelli mapped the planet. Schiaparelli's observations and announcement of channel-like features on the surface were popularized by Camille Flammarion[19] who both conducted his own observations of the red planet and wrote *Urania*, one of the earliest science fiction works that imagines a visit to Mars. Flammarion's work influenced Percival Lowell who would go on to observe numerous lenticular features crisscrossing the planet and conclude they must be canals built by intelligent beings. The idea of intelligent Martians fighting to survive on a desert world with two small moons hanging in the sky inspired writers such as H. G. Wells, Edgar Rice Burroughs, and Ray Bradbury. Those authors would use their words to bring Mars to life for generations of readers.

Another legacy of Lowell Observatory in the late nineteenth century was the search for planet X. The search captured the imagination of an amateur astronomer, science reporter, and science fiction writer named H. P. Lovecraft. At the time Clyde W. Tombaugh discovered Pluto in early 1930, Lovecraft was composing his novella *The Whisperer in Darkness* and the discovery seems to have influenced the story's development. The story involves the discovery of aliens called Mi-Go who came from a mysterious ninth planet discovered beyond Neptune's orbit called Yuggoth. In fact, upon Pluto's discovery, Lovecraft wrote a letter to Tombaugh suggesting Yuggoth as the planet's name. This marks an early example of a planet discovered by humans being used in a science fiction tale.

Around the time H. P. Lovecraft was imagining invaders from the outer solar system, other science fiction writers were imagining ways humans could travel to distant stars. E. E. "Doc" Smith wrote the 1928 novel *The Skylark of Space* about characters using an "atomic motor" to travel thousands of times the speed of light, dismissing Einstein's theory of relativity with a simple line of dialog. In 1931, John W. Campbell wrote about a "strain drive" in his novel *Islands of Space* which imagined ways gravitation could be harnessed to surpass the speed of light. Murray Leinster gave us an alternative to fast interstellar drives by introducing the concept of the "generation ship" in his 1935 story *Proxima Centauri*.

Starting with these stories from the earliest days of science fiction's so-called Golden Age which ran from the 1930s through the 1960s and continuing into science fiction's New Wave which ran from the 1960s through the turn of the century, science fiction writers have been creating planets for characters to travel to. Many have been diligent, keeping up with scientific research about the kinds of stars likely to have habitable planets. Often, these writers imagined distant, exotic Earths. Sometimes they imagined domed cities on less friendly worlds, water covered planets, or even cities hanging in the clouds of gas giants. Still, until the first exoplanets were discovered around pulsar PSR 1257+12 by Aleksander Wolszczan and Dale Frail in 1992, science fiction writers had to rely on their imaginations to create the worlds their characters visited.

Of course some of the imaged worlds have had possible counterparts, that is real exoplanets, actually discovered. Star Wars, for example, has given us a rich set of alien worlds to ponder, worlds on which all manner of life abounds: Kepler-16b orbits two

---

[19] Whose brother started a publishing company, which today, hundreds of years later, is one of the largest in France.

suns as does *Tatooine* and the water world Kepler-22b might represent a planet similar to that envisioned for *Kamino*.

From 1992 to the present, astronomers have discovered over 4000 confirmed exoplanets. Despite that, it's a challenge to find science fiction stories set on these planets or involving hypothetical journeys to these worlds. Notable examples of stories and novels featuring real exoplanets include Jake Kerr's 2011 novelette *The Old Equations* about a journey to Gliese 581d, T. Jackson King's 2016 novel *Battlestar* which is set in the Kepler 22 system, and Allen Steele's 2017 novel *Arkwright*, about a journey to Gliese 667 Ce. In 2013, I collaborated with then-Kepler project scientist Steve B. Howell to curate the anthology *A Kepler's Dozen* featuring 13 stories set around planets discovered by the Kepler Space Telescope. Howell and I collected a second anthology called *Kepler's Cowboys* in 2017.

Given that science fiction writers have long been inspired by real scientific discoveries, one might ask why it's so difficult to find examples of science fiction set around confirmed exoplanets. The 2014 anthology *Extreme Planets* edited by David Conyers, David Kerndt, and Jeff Harris perhaps suggests an answer. This anthology features stories set both in the outer reaches of our solar system and in other solar systems. The only exoplanet that's mentioned by its current usage is Epsilon Eridani b. Other authors do acknowledge that Kepler's announced discoveries of ice giants and super-Earths have inspired their stories, but they have set their stories in systems closer to Earth. The problem for science fiction writers, especially those who endeavor to write the most realistic science fiction, is that most exoplanets discovered so far are simply too far away to visit using technologies we believe can be developed in the not-too-distant future.

When Steve Howell and I collaborated on *Kepler's Cowboys*, we realized another reality. Explorers who visit the exoplanets which have been discovered so far are unlikely to use the catalog names and designations we use today. They are likely to give these worlds new names. In *Kepler's Cowboys*, we didn't require the authors to use contemporary names for their worlds, they just needed to set their stories on the kinds of worlds that Kepler has discovered, much like some of the authors in *Extreme Planets* did.

As we look ahead into the twenty-first century, it's clear that while science fiction writers aren't setting many of their stories on real exoplanets, they are using the results of exoplanet research to imagine believable places humans may visit in the distant future.

Works featuring planets discovered by the Kepler Space Telescope
- Howell, S. B. & Summers, D. L. (ed.) 2013, A Kepler's Dozen (Las Cruces, NM: Hadrosaur Productions).
- Howell, S. B. & Summers, D. L (ed.) 2017, Kepler's Cowboys (Las Cruces, NM: Hadrosaur Productions).
- King, T. J. 2016, Battlestar (Self-published).
- King, T. J. 2016, Battlegroup (Self-published).
- King, T. J. 2017, Battlecry (Self-published).

Other works cited
- Campbell, J. W. 1931, Islands of Space, Serialized in Amazing Stories Quarterly.
- Conyers, D., Kerndt, D. & Harris, J. (ed.) 2014, Extreme Planets (Ann Arbor, MI: Chaosium, Inc.).
- Flammarion, C. 1890, Urania. Translated by Augusta Rice Stetson (Boston, MA: Estes and Lauriat).

- Kepler, J. 1634, Somnium seu opus posthumum De astronomia lunari Frankfurt.
- Kerr, J. 2011, The Old equations, Lightspeed Magazine.
- Leinster, M 1935, Proxima Centauri, Astounding Stories.
- Lovecraft, H. P. 1931, The Whisperer in Darkness, Weird Tales.
- Smith, E. E. 1928, The Skylark of Space, Serialized in *Amazing Stories*.
- Steele, A. 2017, Arkwright (New York: Tor Books).

References

Adams, E. R., López-Morales, M., Elliot, J. L., Seager, S., & Osip, D. J. 2010, ApJ, 721, 1829
Adams, E. R., Jackson, B., Endl, M., et al. 2017, AJ, 153, 82
Alcock, C., Allsman, R. A., Alves, D., et al. 1997, ApJ, 486, 697
Alcock, C., Allsman, R. A., Alves, D. R., et al. 2000, ApJ, 541, 734
Almenara, J. M., Díaz, R. F., Bonfils, X., & Udry, S. 2016, A&A, 595, L5
Andrews, S. M., Rosenfeld, K. A., Kraus, A. L., & Wilner, D. J. 2013, ApJ, 771, 129
Ansdell, M., Gaidos, E., Jacobs, T. L., et al. 2019, MNRAS, 483, 3579
Armstrong, D. J., Santerne, A., Veras, D., et al. 2015, A&A, 582, A33
Barros, S. C. C., Almenara, J. M., Demangeon, O., et al. 2015, MNRAS, 454, 4267
Barros, S. C. C., Gosselin, H., Lillo-Box, J., et al. 2017, A&A, 608, A25
Batalha, N. M., Borucki, W. J., Bryson, S. T., et al. 2011, ApJ, 729, 27
Batalha, N. M., Rowe, J. F., Bryson, S. T., et al. 2013, ApJS, 204, 24
Batygin, K., Bodenheimer, P. H., & Laughlin, G. P. 2016, ApJ, 829, 114
Becker, J. C., & Adams, F. C. 2017, MNRAS, 468, 549
Becker, J. C., Vanderburg, A., Adams, F. C., Khain, T., & Bryan, M. 2017, AJ, 154, 230
Becker, J. C., Vanderburg, A., Adams, F. C., Rappaport, S. A., & Schwengeler, H. M. 2015, ApJL, 812, L18
Becker, J. C., Vanderburg, A., Rodriguez, J. E., et al. 2019, AJ, 157, 19
Benneke, B., Wong, I., Piaulet, C., et al. 2019, ApJL, 887, L14
Bennett, D. P., Rhie, S. H., Nikolaev, S., et al. 2010, ApJ, 713, 837
Berardo, D., Crossfield, I. J. M., Werner, M., et al. 2019, AJ, 157, 185
Borucki, W. J., Koch, D., Basri, G., et al. 2010, Sci, 327, 977
Borucki, W. J., Koch, D. G., Basri, G., et al. 2011a, ApJ, 728, 117
Borucki, W. J., Koch, D. G., Basri, G., et al. 2011b, ApJ, 736, 19
Borucki, W. J., Koch, D. G., Batalha, N., et al. 2012, ApJ, 745, 120
Borucki, W. J., Agol, E., Fressin, F., et al. 2013, Sci, 340, 587
Boyajian, T. S., LaCourse, D. M., Rappaport, S. A., et al. 2016, MNRAS, 457, 3988
Bryson, S., Coughlin, J., Batalha, N. M., et al. 2020, AJ, 159, 279
Bryson, S. T., Jenkins, J. M., Gilliland, R. L., et al. 2013, PASP, 125, 889
Bryson, S. T., Tenenbaum, P., Jenkins, J. M., et al. 2010, ApJ, 713, 97
Brzycki, B., Siemion, A., Croft, S., et al. 2019, RNAAS, 3, 147
Burke, C. J., Bryson, S. T., Mullally, F., et al. 2014, ApJS, 210, 19
Burke, C. J., Christiansen, J. L., Mullally, F., et al. 2015, ApJ, 809, 8
Burke, C. J., Mullally, F., Thompson, S. E., Coughlin, J. L., & Rowe, J. F. 2019, AJ, 157, 143
Cartier, K. M. S., Gilliland, R. L., Wright, J. T., & Ciardi, D. R. 2015, ApJ, 804, 97
Catanzarite, J., & Shao, M. 2011, ApJ, 738, 151

Christiansen, J. L. 2017, Planet Detection Metrics: Pixel-Level Transit Injection Tests of Pipeline Detection Efficiency for Data Release 25, Technical Report

Christiansen, J. L., Vanderburg, A., Burt, J., et al. 2017, AJ, 154, 122

Christiansen, J. L., Crossfield, I. J. M., Barentsen, G., et al. 2018, AJ, 155, 57

Ciardi, D. R., Beichman, C. A., Horch, E. P., & Howell, S. B. 2015, ApJ, 805, 16

Ciardi, D. R., Fabrycky, D. C., Ford, E. B., et al. 2013, ApJ, 763, 41

Ciardi, D. R., Crossfield, I. J. M., Feinstein, A. D., et al. 2018, AJ, 155, 10

Cloutier, R., Astudillo-Defru, N., Doyon, R., et al. 2019, A&A, 621, A49

Coughlin, J. L., Mullally, F., Thompson, S. E., et al. 2016, ApJS, 224, 12

Coughlin, J. L., Thompson, S. E., Bryson, S. T., et al. 2014, AJ, 147, 119

Crossfield, I. 2017, The Atmospheric Diversity of Mini-Neptunes in Multi-planet Systems, HST Proposal

Crossfield, I. J. M., Petigura, E., Schlieder, J. E., et al. 2015, ApJ, 804, 10

Crossfield, I. J. M., Ciardi, D. R., Isaacson, H., et al. 2017, AJ, 153, 255

Curtis, J. L., Vanderburg, A., Torres, G., et al. 2018, AJ, 155, 173

Dai, F., Masuda, K., & Winn, J. N. 2018, ApJL, 864, L38

Dai, F., Masuda, K., Winn, J. N., & Zeng, L. 2019, ApJ, 883, 79

Dai, F., Winn, J. N., Arriagada, P., et al. 2015, ApJL, 813, L9

Dai, F., Winn, J. N., Albrecht, S., et al. 2016, ApJ, 823, 115

Dalal, N., & Lane, B. F. 2003, ApJ, 589, 199

Dalal, S., Hébrard, G., Lecavelier des Étangs, A., et al. 2019, A&A, 631, A28

Damasso, M., Bonomo, A. S., Astudillo-Defru, N., et al. 2018, A&A, 615, A69

David, T. J., Petigura, E. A., Luger, R., et al. 2019a, ApJL, 885, L12

David, T. J., Hillenbrand, L. A., Petigura, E. A., et al. 2016, Natur, 534, 658

David, T. J., Cody, A. M., Hedges, C. L., et al. 2019b, AJ, 158, 79

Debes, J. H., & Sigurdsson, S. 2002, ApJ, 572, 556

Delplancke, F., Górski, K. M., & Richichi, A. 2001, A&A, 375, 701

Demory, B.-O., de Wit, J., Lewis, N., et al. 2013, ApJL, 776, L25

Dong, S., & Zhu, Z. 2013, ApJ, 778, 53

Dong, S., Udalski, A., Gould, A., et al. 2007, ApJ, 664, 862

Dong, S., Mérand, A., Delplancke-Ströbele, F., et al. 2019, ApJ, 871, 70

Doyle, L. R., Carter, J. A., Fabrycky, D. C., et al. 2011, Sci, 333, 1602

Dressing, C. D., & Charbonneau, D. 2015, ApJ, 807, 45

Fabrycky, D. C., Lissauer, J. J., Ragozzine, D., et al. 2014, ApJ, 790, 146

Feinstein, A. D., Schlieder, J. E., Livingston, J. H., et al. 2019, AJ, 157, 40

Fischer, D. A., Schwamb, M. E., Schawinski, K., et al. 2012, MNRAS, 419, 2900

Foreman-Mackey, D., Hogg, D. W., & Morton, T. D. 2014, ApJ, 795, 64

Fressin, F., Torres, G., Charbonneau, D., et al. 2013, ApJ, 766, 81

Fulton, B. J., & Petigura, E. A. 2018, AJ, 156, 264

Fulton, B. J., Petigura, E. A., Howard, A. W., et al. 2017, arXiv:1703.10375

Furlan, E., Ciardi, D. R., Everett, M. E., et al. 2017, AJ, 153, 71

Furlan, E., Ciardi, D. R., Cochran, W. D., et al. 2018, ApJ, 861, 149

Gaidos, E. 2013, ApJ, 770, 90

Gandolfi, D., Barragán, O., Hatzes, A. P., et al. 2017, AJ, 154, 123

Gaudi, B. S., Bennett, D. P., Udalski, A., et al. 2008, Sci, 319, 927

Ginzburg, S., Schlichting, H. E., & Sari, R. 2018, MNRAS, 476, 759

Gilliland, R. L., Brown, T. M., Guhathakurta, P., et al. 2000, ApJ, 545, 47

Gould, A. 1992, ApJ, 392, 442

Gould, A. 1994, ApJL, 421, L75

Gould, A., & Horne, K. 2013, ApJL, 779, L28

Gupta, A., & Schlichting, H. E. 2019, MNRAS, 487, 24

Hardegree-Ullman, K. K., Cushing, M. C., Muirhead, P. S., & Christiansen, J. L. 2019, arXiv:1905.05900

Hellier, C., Anderson, D. R., Collier Cameron, A., et al. 2012, MNRAS, 426, 739

Henderson, C. B., Poleski, R., Penny, M., et al. 2016, PASP, 128, 124401

Hog, E., Novikov, I. D., & Polnarev, A. G. 1995, A&A, 294, 287

Howard, A. W., Marcy, G. W., Bryson, S. T., et al. 2012, ApJS, 201, 15

Howell, S. B., Everett, M. E., Sherry, W., Horch, E., & Ciardi, D. R. 2011, AJ, 142, 19

Howell, S. B., Scott, N. J., Matson, R. A., Horch, E. P., & Stephens, A. 2019, AJ, 158, 113

Howell, S. B., VanOutryve, C., Tonry, J. L., Everett, M. E., & Schneider, R. 2005, PASP, 117, 1187

Hsu, D. C., Ford, E. B., Ragozzine, D., & Morehead, R. C. 2018, AJ, 155, 205

Huang, C., Wu, Y., & Triaud, A. H. M. J. 2016, ApJ, 825, 98

Huang, C. X., Penev, K., Hartman, J. D., et al. 2015, MNRAS, 454, 4159

Jenkins, J. M. 2017, Kepler Data Processing Handbook (KSCI-19081-002), Chapter 9

Jenkins, J. M., Chandrasekaran, H., McCauliff, S. D., et al. 2010, Proc. SPIE, 7740, 77400D

Jenkins, J. M., Twicken, J. D., Batalha, N. M., et al. 2015, AJ, 150, 56

Kane, S. R., & Hinkel, N. R. 2013, ApJ, 762, 7

Kane, S. R., Hill, M. L., Kasting, J. F., et al. 2016, ApJ, 830, 1

Kasting, J. F., Whitmire, D. P., & Reynolds, R. T. 1993, Icar, 101, 108

Kervella, P., Thévenin, F., Di Folco, E., & Ségransan, D. 2004, A&A, 426, 297

Kim, H. W., Hwang, K. H., Kim, D. J., et al. 2018, AJ, 155, 186

Kim, S.-L., Lee, C.-U., Park, B.-G., et al. 2016, JKAS, 49, 37

Kochanek, C. S., Shappee, B. J., Stanek, K. Z., et al. 2017, PASP, 129, 104502

Kopparapu, R. K., Ramirez, R. M., SchottelKotte, J., et al. 2014, ApJL, 787, L29

Kopparapu, R. K., Ramirez, R., Kasting, J. F., et al. 2013, ApJ, 765, 131

Kopparapu, R. K., Hébrard, E., Belikov, R., et al. 2018, ApJ, 856, 122

Koshimoto, N., Shvartzvald, Y., Bennett, D. P., et al. 2017, AJ, 154, 3

Kosiarek, M. R., Crossfield, I. J. M., Hardegree-Ullman, K. K., et al. 2019, AJ, 157, 97

Kostov, V. B., McCullough, P. R., Carter, J. A., et al. 2014, ApJ, 784, 14

Kostov, V. B., Orosz, J. A., Welsh, W. F., et al. 2016, ApJ, 827, 86

Kreidberg, L., Lopez, E., Cowan, N., et al. 2018, Taking the Temperature of a Lava Planet, Spitzer Proposal

LaCourse, D. M., & Jacobs, T. L. 2018, RNAAS, 2, 28

Lada, C. J., & Lada, E. A. 2003, ARA&A, 41, 57

Latham, D. W., Borucki, W. J., Koch, D. G., et al. 2010, ApJL, 713, L140

Lin, D. N. C., Bodenheimer, P., & Richardson, D. C. 1996, Natur, 380, 606

Lissauer, J. J., Fabrycky, D. C., Ford, E. B., et al. 2011, Natur, 470, 53

Lissauer, J. J., Marcy, G. W., Rowe, J. F., et al. 2012, ApJ, 750, 112

Livingston, J. H., Dai, F., Hirano, T., et al. 2018, AJ, 155, 115

Livingston, J. H., Dai, F., Hirano, T., et al. 2019, MNRAS, 484, 8

Lopez, E. D., & Fortney, J. J. 2014, ApJ, 792, 1

Lu, J. R., Sinukoff, E., Ofek, E. O., Udalski, A., & Kozlowski, S. 2016, ApJ, 830, 41

Lund, M. N., Knudstrup, E., Silva Aguirre, V., et al. 2019, AJ, 158, 248

Malavolta, L., Mayo, A. W., Louden, T., et al. 2018, AJ, 155, 107

Mann, A. W., Gaidos, E., Mace, G. N., et al. 2016a, ApJ, 818, 46

Mann, A. W., Newton, E. R., Rizzuto, A. C., et al. 2016b, AJ, 152, 61

Mann, A. W., Gaidos, E., Vanderburg, A., et al. 2017, AJ, 153, 64

Mann, A. W., Vanderburg, A., Rizzuto, A. C., et al. 2018, AJ, 155, 4

Mathur, S., Huber, D., Batalha, N. M., et al. 2017, ApJS, 229, 30

Mayo, A. W., Vanderburg, A., Latham, D. W., et al. 2018, AJ, 155, 136

McCauliff, S. D., Jenkins, J. M., Catanzarite, J., et al. 2015, ApJ, 806, 6

Miller-Ricci, E., Rowe, J. F., Sasselov, D., et al. 2008, ApJ, 682, 593

Miyamoto, M., & Yoshii, Y. 1995, AJ, 110, 1427

Montet, B. T., Morton, T. D., Foreman-Mackey, D., et al. 2015, ApJ, 809, 25

Morton, T. D., Bryson, S. T., Coughlin, J. L., et al. 2016, ApJ, 822, 86

Morton, T. D., & Swift, J. 2014, ApJ, 791, 10

Mróz, P., Udalski, A., Skowron, J., et al. 2017, Natur, 548, 183

Mróz, P., Udalski, A., Skowron, J., et al. 2019, ApJS, 244, 29

Mulders, G. D., Pascucci, I., & Apai, D. 2015, ApJ, 798, 112

Mulders, G. D., Pascucci, I., Apai, D., & Ciesla, F. J. 2018, AJ, 156, 24

Mullally, F., Coughlin, J. L., Thompson, S. E., et al. 2015, ApJS, 217, 31

Mullally, F., Thompson, S. E., Coughlin, J. L., Burke, C. J., & Rowe, J. F. 2018, AJ, 155, 210

Narita, N., Hirano, T., Fukui, A., et al. 2015, ApJ, 815, 47

Navarro, M. G., Minniti, D., & Contreras-Ramos, R. 2018, ApJL, 865, L5

Neveu-VanMalle, M., Queloz, D., Anderson, D. R., et al. 2016, A&A, 586, A93

Niraula, P., Redfield, S., Dai, F., et al. 2017, AJ, 154, 266

Nowak, G., Palle, E., Gandolfi, D., et al. 2017, AJ, 153, 131

Obermeier, C., Henning, T., Schlieder, J. E., et al. 2016, AJ, 152, 223

Orosz, J. A., Welsh, W. F., Carter, J. A., et al. 2012, Sci, 337, 1511

Orosz, J. A., Welsh, W. F., Haghighipour, N., et al. 2019, AJ, 157, 174

Owen, J. E., & Wu, Y. 2013, ApJ, 775, 105

Peña Ramírez, K., Béjar, V. J. S., Zapatero Osorio, M. R., Petr-Gotzens, M. G., & Martín, E. L. 2012, ApJ, 754, 30

Penny, M. T., Gaudi, B. S., Kerins, E., et al. 2019, ApJS, 241, 3

Pepper, J., & Burke, C. J. 2006, AJ, 132, 1177

Pepper, J., Gillen, E., Parviainen, H., et al. 2017, AJ, 153, 177

Perez-Becker, D., & Chiang, E. 2013, MNRAS, 433, 2294

Petigura, E. A., Howard, A. W., & Marcy, G. W. 2013a, PNAS, 110, 19273

Petigura, E. A., Marcy, G. W., & Howard, A. W. 2013b, ApJ, 770, 69

Petigura, E. A., Livingston, J., Batygin, K., et al. 2020, AJ, 159, 2

Pfyffer, S., Alibert, Y., Benz, W., & Swoboda, D. 2015, A&A, 579, A37

Poleski, R. 2016, MNRAS, 455, 3656

Poleski, R., Penny, M., Gaudi, B. S., et al. 2019, A&A, 627, A54

Poleski, R., Suzuki, D., Udalski, A., et al. 2020, AJ, 159, 261

Price, E. M., & Rogers, L. A. 2020, ApJ, 894, 8

Prieto-Arranz, J., Palle, E., Gandolfi, D., et al. 2018, A&A, 618, A116

Prša, A., Batalha, N., Slawson, R. W., et al. 2011, AJ, 141, 83

Quintana, E. V., Barclay, T., Raymond, S. N., et al. 2014, Sci, 344, 277

Rappaport, S., Barclay, T., DeVore, J., et al. 2014, ApJ, 784, 40

Rappaport, S., Levine, A., Chiang, E., et al. 2012, ApJ, 752, 1

Rappaport, S., Zhou, G., Vanderburg, A., et al. 2019a, MNRAS, 485, 2681

Rappaport, S., Vanderburg, A., Kristiansen, M. H., et al. 2019b, MNRAS, 488, 2455

Rasio, F. A., & Ford, E. B. 1996, Sci, 274, 954

Rattenbury, N. J., & Mao, S. 2006, MNRAS, 365, 792

Refsdal, S. 1966, MNRAS, 134, 315

Rice, K., Malavolta, L., Mayo, A., et al. 2019, MNRAS, 484, 3731

Rizzuto, A. C., Vanderburg, A., Mann, A. W., et al. 2018, AJ, 156, 195

Rodriguez, J. E., Vanderburg, A., Eastman, J. D., et al. 2018a, AJ, 155, 72

Rodriguez, J. E., Zhou, G., Vanderburg, A., et al. 2017, AJ, 153, 256

Rodriguez, J. E., Becker, J. C., Eastman, J. D., et al. 2018b, AJ, 156, 245

Rogers, L. A. 2015, ApJ, 801, 41

Rouan, D., Deeg, H. J., Demangeon, O., et al. 2011, ApJL, 741, L30

Rowe, J. F., Bryson, S. T., Marcy, G. W., et al. 2014, ApJ, 784, 45

Ryu, Y. H., Yee, J. C., Udalski, A., et al. 2018, AJ, 155, 40

Sako, T., Sekiguchi, T., Sasaki, M., et al. 2008, ExA, 22, 51

Sanchis-Ojeda, R., Rappaport, S., Winn, J. N., et al. 2014, ApJ, 787, 47

Sanchis-Ojeda, R., Winn, J. N., Dai, F., et al. 2015a, ApJL, 812, L11

Sanchis-Ojeda, R., Rappaport, S., Pallè, E., et al. 2015b, ApJ, 812, 112

Santerne, A., Malavolta, L., Kosiarek, M. R., et al. 2019, arXiv:1911.07355

Schmitt, A. R., Hartman, J. D., & Kipping, D. M. 2019, arXiv:1910.08034

Schmitt, J. R., Wang, J., Fischer, D. A., et al. 2014a, AJ, 148, 28

Schmitt, J. R., Agol, E., Deck, K. M., et al. 2014b, ApJ, 795, 167

Schmitt, J. R., Tokovinin, A., Wang, J., et al. 2016, AJ, 151, 159

Schneider, J. 2019a, RNAAS, 3, 108

Schneider, J. 2019b, RNAAS, 3, 141

Scholz, A., Jayawardhana, R., Muzic, K., et al. 2012, ApJ, 756, 24

Schwamb, M. E., Orosz, J. A., Carter, J. A., et al. 2013, ApJ, 768, 127

Shallue, C. J., & Vanderburg, A. 2018, AJ, 155, 94

Sinukoff, E., Howard, A. W., Petigura, E. A., et al. 2016, ApJ, 827, 78

Sinukoff, E., Howard, A. W., Petigura, E. A., et al. 2017, AJ, 153, 70

Smith, A. M. S., Cabrera, J., Csizmadia, S., et al. 2018, MNRAS, 474, 5523

Soares-Furtado, M., Hartman, J. D., Bakos, G. Á., et al. 2017, PASP, 129, 044501

Steffen, J. H., Ragozzine, D., Fabrycky, D. C., et al. 2012a, PNAS, 109, 7982

Steffen, J. H., Fabrycky, D. C., Ford, E. B., et al. 2012b, MNRAS, 421, 2342

Sumi, T., & Penny, M. T. 2016, ApJ, 827, 139

Sumi, T., Abe, F., Bond, I. A., et al. 2003, ApJ, 591, 204

Sumi, T., Kamiya, K., Bennett, D. P., et al. 2011, Natur, 473, 349

Sumi, T., Bennett, D. P., Bond, I. A., et al. 2013, ApJ, 778, 150

Teske, J. K., Wang, S., Wolfgang, A., et al. 2018, AJ, 155, 148

Thao, P. C., Mann, A. W., Johnson, M. C., et al. 2019, arXiv:1911.05744

Thompson, S. E., Mullally, F., Coughlin, J. L., et al. 2015, ApJ, 812, 46

Thompson, S. E., Everett, M., Mullally, F., et al. 2012, ApJ, 753, 86

Thompson, S. E., Coughlin, J. L., Hoffman, K., et al. 2018, ApJS, 235, 38

Torres, G., Fressin, F., Batalha, N. M., et al. 2011, ApJ, 727, 24

Torres, G., Kane, S. R., Rowe, J. F., et al. 2017, AJ, 154, 264

Tsiaras, A., Waldmann, I. P., Tinetti, G., Tennyson, J., & Yurchenko, S. N. 2019a, NatAs, 3, 1086

Tsiaras, A., Waldmann, I. P., Tinetti, G., Tennyson, J., & Yurchenko, S. N. 2019b, NatAs, 3, 1156

Udalski, A., Szymański, M. K., & Szymański, G. 2015a, AcA, 65, 1

Udalski, A., Szymanski, M., Stanek, K. Z., et al. 1994, AcA, 44, 165

Udalski, A., Yee, J. C., Gould, A., et al. 2015b, ApJ, 799, 237

Van Cleve, J. E., & Caldwell, D. A. 2009, Kepler Instrument Handbook (KSCI-19033-001)

Van Eylen, V., Agentoft, C., Lundkvist, M. S., et al. 2018, MNRAS, 479, 4786

van Lieshout, R., & Rappaport, S. A. 2018, Disintegrating Rocky Exoplanets, Handbook of Exoplanets (Cham: Springer), 1527

Vanderburg, A., Johnson, J. A., Rappaport, S., et al. 2015a, Natur, 526, 546

Vanderburg, A., Montet, B. T., Johnson, J. A., et al. 2015b, ApJ, 800, 59

Vanderburg, A., Becker, J. C., Kristiansen, M. H., et al. 2016a, ApJL, 827, L10

Vanderburg, A., Latham, D. W., Buchhave, L. A., et al. 2016b, ApJS, 222, 14

Vanderburg, A., Bieryla, A., Duev, D. A., et al. 2016c, ApJL, 829, L9

Vanderburg, A., Becker, J. C., Buchhave, L. A., et al. 2017, AJ, 154, 237

Vanderburg, A., Mann, A. W., Rizzuto, A., et al. 2018, AJ, 156, 46

Veras, D., & Raymond, S. N. 2012, MNRAS, 421, L117

Walker, M. A. 1995, ApJ, 453, 37

Wang, D., Hogg, D. W., Foreman-Mackey, D., & Schölkopf, B. 2017, arXiv:1710.02428

Wang, J., Fischer, D. A., Barclay, T., et al. 2013, ApJ, 776, 10

Wang, L., Kouwenhoven, M. B. N., Zheng, X., Church, R. P., & Davies, M. B. 2015, MNRAS, 449, 3543

Weiss, L. M., Rogers, L. A., Isaacson, H. T., et al. 2016, ApJ, 819, 83

Weiss, L. M., Deck, K. M., Sinukoff, E., et al. 2017, AJ, 153, 265

Weiss, L. M., Marcy, G. W., Petigura, E. A., et al. 2018, AJ, 155, 48

Weiss, L. M., & Petigura, E. A. 2020, ApJL, 893, L1

Welsh, W. F., Orosz, J. A., Seager, S., et al. 2010, ApJL, 713, L145

Welsh, W. F., Orosz, J. A., Carter, J. A., et al. 2012, Natur, 481, 475

Welsh, W. F., Orosz, J. A., Short, D. R., et al. 2015, ApJ, 809, 26

Yee, J. C., Gould, A., Beichman, C., et al. 2015a, ApJ, 810, 155

Yee, J. C., Udalski, A., Calchi Novati, S., et al. 2015b, ApJ, 802, 76

Yoo, J., DePoy, D. L., Gal-Yam, A., et al. 2004, ApJ, 603, 139

Youdin, A. N. 2011, ApJ, 742, 38

Yu, L., Crossfield, I. J. M., Schlieder, J. E., et al. 2018, AJ, 156, 22

Zang, W., Penny, M. T., Zhu, W., et al. 2018, PASP, 130, 104401

Zhou, G., Rodriguez, J. E., Vanderburg, A., et al. 2018, AJ, 156, 93

Zhu, W. 2020, AJ, 159, 188

Zhu, W., Udalski, A., Huang, C. X., et al. 2017a, ApJL, 849, L31

Zhu, W., Huang, C. X., Udalski, A., et al. 2017b, PASP, 129, 104501

Zuckerman, B., Melis, C., Klein, B., Koester, D., & Jura, M. 2010, ApJ, 722, 725

Chapter 4

Stellar Astrophysics with Kepler and K2

William J Chaplin, J J Hermes, Ann Marie Cody, Elliott P Horch, Rachel A Matson, Steven D Kawaler, Steve B Howell and David R Ciardi

4.1 Introduction

While the primary focus of the Kepler/K2 satellite was to identify transiting planets, the unique design of the mission opened up much new discovery space for time-domain astrophysics. Stellar astrophysics underwent a re-birth due to the NASA Kepler and K2 missions. Somewhere around the 1960s, astronomy started to move away from pursuing big topics in stellar astrophysics in favor of the newly discovered distant universe of far off galaxies and quasars. Stars were thought to be well understood and continued study was not seen to offer answers to the big questions of the day. However, the newly discovered quasars, far flung objects residing at the then known edge of the Universe held cosmological promise to discover the extent and fate of the Universe—big questions indeed.

The advent of exoplanet science and in particular the unprecedented photometric precision of Kepler and K2 light-curves of stars revealed not only new physics which had been missed or set aside in the preceding three to four decades but many new types of stellar sources and never before observed phenomena. These NASA missions have changed nearly every aspect of our knowledge and understanding of stars, truly providing the scientific world new paradigms in stellar astrophysics.

The breadth of the new found stellar knowledge that astronomers have and continue to gain based on results from the Kepler and K2 missions, could, in itself, fill volumes. In fact, nearly 2000 scientific papers related to new stellar astrophysics gleaned from the missions have been published since launch, an astounding average of nearly 200 papers per year. The topics of these papers vary widely and in this chapter we attempt to capture the major areas of research in stellar astrophysics enabled by the Kepler and K2 missions.

4.2 Asteroseismology

4.2.1 Overview

The success of the Kepler mission in finding and characterizing new planetary systems was astounding, as described elsewhere. The experimental design of Kepler was focused on detecting exoplanets through the reduced light from their primary star during transits. To accomplish the Level 1 mission goals, Kepler had to achieve an unprecedented duty cycle of much greater than 90%, sustained for a period measured in years, while maintaining a photometric precision in the parts-per-million range on relatively faint stars in relatively short intervals. To adequately resolve and time planetary transits, the time resolution/cadence needed to be as short as practical (a minute or so) for as many targets as possible.

As it turns out, these same requirements (high duty cycle, high photometric precision, good time resolution, long duration) are nearly identical to the needs of asteroseismology. Asteroseismology exploits the small variations in brightness (or surface velocity) that some stars exhibit, which are manifestations of self-excited oscillation modes. The frequencies (and to a lesser extent, amplitude) of these oscillations are determined by the internal mechanical and thermal structure of the star. Successful detection and measurement of these modes, along with spectroscopic determination of the star's surface temperature, T_{eff}, and surface gravity, allows precise determination of global properties (mean density, mass, radius, rotation, age) and internal structure (sound speed, differential rotation, composition gradients, convection/radiation boundaries).

Prior to Kepler, asteroseismic studies were mostly limited to ground-based observations, which, suffer from poor duty cycles, and daily, monthly, and yearly modulation of target accessibility. In addition, observing through the Earth's atmosphere introduces atmospheric scintillation noise that masks signals from the stars. Networks of ground-based telescopes tuned for photometric seismology produced tantalizing results but could only observe a limited number of targets for a few weeks at a time. Examples of such networks include the Whole Earth Telescope (WET; Nather et al. 1990) and the δ Scuti Network (Zima 1997). Such networks helped advance the study of compact stars (white dwarfs and hot subdwarfs) and main sequence stars respectively.

Kepler however was an absolute game-changer in asteroseimology, as the next few sections illustrate. For stars like our Sun, asteroseismology via photometry effectively began with thousands of stars yielding to asteroseismic analysis. Compact stars, such as pulsating white dwarfs and pulsating hot subdwarfs, while representing a smaller number of stars, show details in their oscillation spectra that were partially or totally hidden from ground-based studies because of the high duty cycle and extended timespan of the data. The efficiency and impact of Kepler is summed up in one remarkable statistic. Kepler observed approximately 40 compact pulsators for 1 year or more. That represents 40 target-years of observation during the 4 year primary mission. The WET typically observed two targets for two weeks each, twice per year—or 8 target-weeks per year. Therefore, Kepler's yield of data on compact pulsators was the equivalent of 260 years of operation of the WET network.

The targeted K2 mission had several programs that had white dwarf and sdB stars as targets, with well over 100 observed for 3 months or more.

The next few sections describe some of the results of the asteroseismic effort with Kepler. This has been the work of large collection of astronomers, often working together in large collaborative efforts. In an area of the sky as rich as the Kepler field and the K2 campaigns, many classical pulsators were also available. Substantial results from Kepler and K2 have been obtained for Cepheid variables (Derekas et al. 2012), RR Lyrae stars (Benkö et al. 2010, 2019) δ Scuti and γ Doradus stars (Uytterhoeven et al. 2011), and others. Molnár et al. (2016) briefly review some of the highlights from Kepler and K2 for classical pulsators, and readers are encouraged to consult the above references and explore the rich results on these stars in the literature.

4.2.2 Focus on Solar-like Oscillators

Solar-like oscillations, which are stochastically excited and intrinsically damped by near-surface convection (Chaplin & Miglio 2013); are seen in cool main-sequence, sub-giant, and red-giant stars. The excitation gives rise to a rich spectrum of overtones of potentially detectable radial and non-radial modes. Measured frequencies and frequency splittings provide exquisite constraints not only on the fundamental stellar properties, but also on the internal structures, physics, and dynamics of these stars; while observed amplitudes and damping rates provide diagnostics of convection and near-surface physics.

The convection limits the modes to intrinsically weak amplitudes. For example, the strongest observed radial modes of the Sun have amplitudes of only a few parts per million (ppm) in photometry, or $\simeq 20$ cm s^{-1} in Doppler velocity. This made extending detections to other solar-type stars very challenging and it was not until the 1990s that the first unambiguous detections were made on bright stars using large ground-based telescopes. The French-led CoRoT Mission provided multi-month, high-precision stellar photometric data, and while it initially added a handful of detections in solar-type stars its main breakthrough for solar-like oscillators was to show that red giants displayed rich spectra of radial *and* non-radial modes. However, it was the advent of Kepler that allowed the full potential of solar-like oscillations on other stars to be exploited. Why? First, because Kepler provided for the first time contiguous, micro-magnitude quality light-curves of durations up to four years, which gave access to seismic information that could not be extracted from shorter data sets, detailed information on rotation from mode frequency splittings being the most obvious example. Second, Kepler provided seismic data on unprecedented numbers of stars down to much fainter magnitudes than had previously been possible.

In the nominal mission, Kepler made asteroseismic detections in several hundred solar-type stars (by which we mean main-sequence and sub-giant targets) down to an apparent visual magnitude of $V \approx 11$ (Chaplin et al. 2011a); and in around 20,000 red giants, down to $V \approx 16$ (Mathur et al. 2016; Hon et al. 2019). In spite of the deterioration of photometric performance by a factor of about two in amplitude, K2 continued to yield excellent data on solar-like oscillators. By mission end it had added, from around the ecliptic, a few hundred detections in solar-type stars (Chaplin et al. 2015; Lund et al. 2016a) and tens of thousands in red giants (Stello et al. 2017).

Main sequence and sub-giant stars show detectable solar-like oscillations with periods of the order of minutes. This necessitated use of the Kepler 60 s short-cadence slots. With only up to 512 slots available at any one time, these data were clearly at a premium. The exploitation of Kepler for asteroseismology was performed within framework of the Kepler Asteroseismic Science Consortium (KASC). During the first four months of science operations, KASC was given the freedom to select all the short-cadence targets, to in effect conduct a survey to provide a large statistical seismic data sample for further analysis and to choose high-quality, high-value targets for extended short-cadence observations. This survey proved extremely successful for the solar-like oscillators, and has provided statistical and high-value target samples that will take many more years to fully mine.

As stars evolve, their detectable solar-like oscillations shift to longer periods (lower frequencies). The more numerous 29.4 minute long-cadence slots were sufficient for asteroseismic studies of stars at and beyond the base of the red-giant branch. Fortunately the Kepler long-cadence target selection, based on the Kepler input catalog, included many red giants from the outset. Once it became clear that these stars provided a treasure-trove of science, courtesy of their oscillations; and that it would be possible to use data on many thousands of stars, since the mode amplitudes are stronger than in their less evolved counterparts, they were retained throughout the nominal mission and successfully targeted for competitive time in K2.

Figure 4.1 shows the frequency–power spectrum of the Kepler light-curve of the main-sequence star HD 175289, which is the planet-host Kepler-410. The spectrum is plotted with logarithmic scales on both axes, to help identify the various components in the data. Components intrinsic to the star include: the oscillations; granulation, i.e., the visible manifestation of the near-surface convection which, as mentioned above, both excites and damps the modes; and low-frequency power from photometric variability due to magnetic activity. Non-stellar components include photon shot noise, low-frequency drifts, and narrow-band artifacts.

The top panel of Figure 4.2 shows the oscillation spectrum in more detail, the annotations tagging the angular degree of each mode, i.e., radial modes have $l = 0$, dipole modes $l = 1$, and quadrupole modes $l = 2$. Owing to geometric cancellation, only modes of low degree have sufficient amplitude to be detectable. The modes are predominantly acoustic in nature when cool stars are on the main sequence, meaning gradients of pressure provide the restoring force (hence we call them p modes). The spacing between consecutive overtones of the same angular degree—the so-called large frequency separation—scales to very good approximation with the square root of the bulk density of the star. The observed power in the modes is modulated in frequency by an envelope that is roughly Gaussian in shape; the frequency of maximum power has been shown to depend on near-surface properties, notably

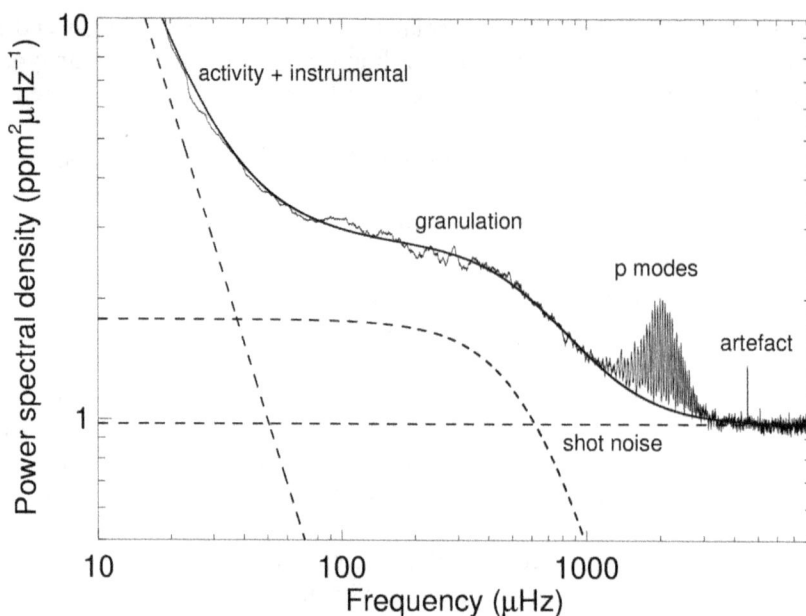

Figure 4.1. Frequency–power spectrum of the Kepler light-curve of the main-sequence star HD 175289, also known as the planet-host Kepler-410. Overlaid on the smoothed spectrum is a model fit to the background power spectral density, comprising components due to granulation, stellar activity, instrumental drifts, and photon shot noise (shown individually by the dashed lines).

effective temperature and surface gravity. Also clear from the spectrum is the finite widths of the resonant peaks, those widths depending on the rates at which the modes are damped.

The excited modes follow regular patterns in frequency when stars are on the main sequence—as shown clearly by the oscillation spectrum of HD 175289—meaning each mode can usually be trivially associated with a given angular degree. This association is required to leverage the full potential of asteroseimology. The bottom panel of Figure 4.2 shows this regular pattern in a different way. Here, the spectrum has been divided into strips that each contain one set of overtones, i.e., a width corresponding to one large frequency separation. The strips have then been stacked in ascending order of frequency, and the locations of the mode frequencies plotted with different symbols (with $l = 0$ modes as squares, $l = 1$ modes as circles, and $l = 2$ modes as triangles). This gives what is called an échelle (i.e., ladder) diagram, revealing near-vertical ridges from each set of overtones.

The solid black line in the top panel of Figure 4.2 is a multi-parameter model fit to the spectrum. This parametric model is comprised of components to represent power due to every individual mode component. The basic building-block is a Lorentzian-like profile, commensurate with the profile expected for a damped mode. Figure 4.3 is an even tighter zoom of the spectrum, showing how the non-radial modes are split into individual (azimuthal) components by the rotation.

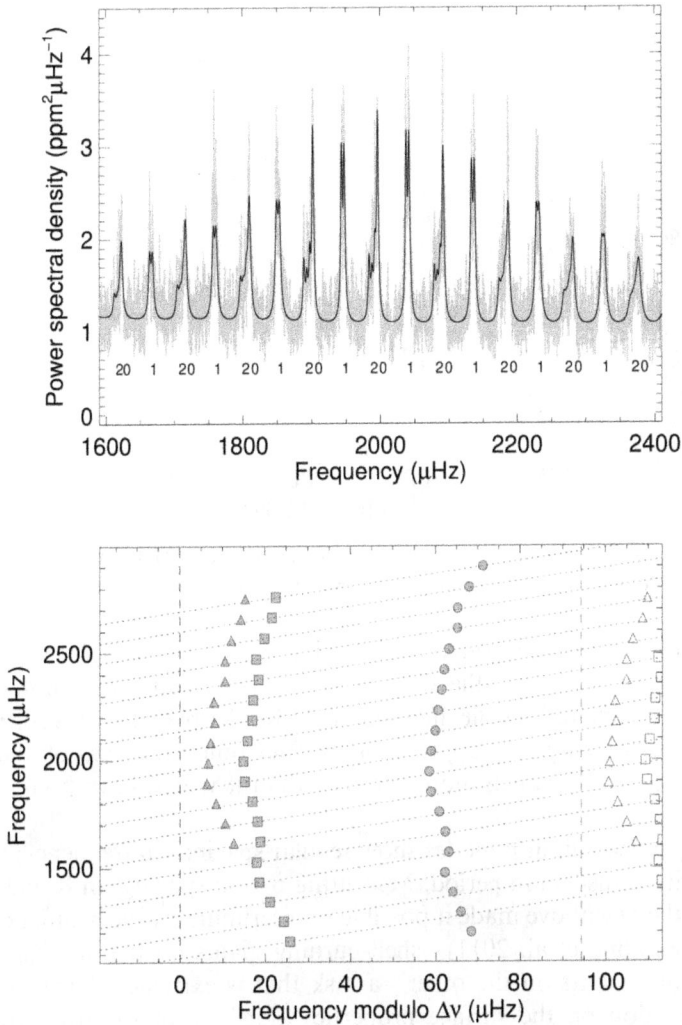

Figure 4.2. Top: Oscillation spectrum of HD 175289 (gray), showing also a parametric model fit (black). Annotations tag the angular degree of each mode, i.e., radial modes have $l = 0$, dipole modes $l = 1$, and quadrupole modes $l = 2$. Bottom: échelle diagram of the oscillation frequencies, folded at the average large frequency separation $\Delta\nu$: $l = 0$ modes plotted as squares, $l = 1$ modes as circles, and $l = 2$ modes as triangles.

As stars evolve into the sub-giant phase, following exhaustion of hydrogen burning in the core of the star, we begin to see significant changes in the nature of the observed oscillation spectrum. Changes in the cores of the stars shift the frequencies of gravity (g) modes, where buoyancy provides the restoring force, into the frequency range occupied by the detectable p modes. Non-radial p and g modes of the same degree and very similar frequency can interact—by analogy to coupled oscillators—giving modes that have mixed character: p-mode like in the envelopes of the stars, and g-mode like in the radiative interiors. Without this coupling, signatures of g modes would be too weak to detect, because pure g modes are trapped in

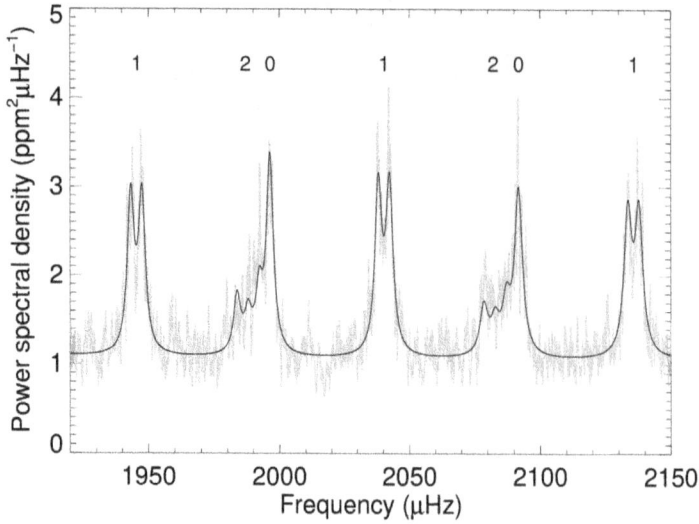

Figure 4.3. Zoom of a central region of the oscillation spectrum of HD 175289 (gray), showing also a parametric model fit (black). Annotations tag the angular degree of each mode.

cavities beneath the outer convective envelopes. Now accessible, their signatures provide exquisite probes of the deep lying layers of stars. Figure 4.4 shows the oscillation spectrum and échelle diagram of the Kepler planet host Kepler-56. This is a low-luminosity red-giant, approximately four-times the size of the Sun. The spectrum is seen to be much more complex than it would have been on the main sequence.

In the asymptotic limit p modes show regular spacings in frequency; for g modes that regularity is instead in period. Measuring period spacings in Kepler oscillation spectra like the one above made it possible to discriminate clearly different classes of red giants (Bedding et al. 2011)—shell-burning giants on the one hand and core-helium burning giants on the other—a task that is extremely hard to accomplish given information on the surface properties alone (from spectroscopy or photometric colors). That discrimination is clear in the asteroseismic data because these two classes have different structures in their deep interiors, and that difference manifests clearly in the mixed modes. The ability to robustly tag different evolutionary states has major implications for testing stellar evolutionary theory, and for studies of stellar populations and the evolution of exoplanet systems.

A wealth of Kepler data like those shown above has been used to produce catalogs of asteroseismically inferred stellar properties for solar-like oscillators (e.g., see Chaplin et al. 2014; Silva Aguirre et al. 2017). Detailed tests of the asteroseismic methods have been undertaken using independent data from the likes of interferometry (e.g., White et al. 2013), Gaia (e.g., Huber et al. 2017; Khan et al. 2019; Hall et al. 2019; Zinn et al. 2019), and stars in binaries and clusters (e.g., Gaulme et al. 2016; Miglio et al. 2016; Brogaard et al. 2018). Data on large samples and individual high-value targets have been used not only for stellar astrophysics, but also for

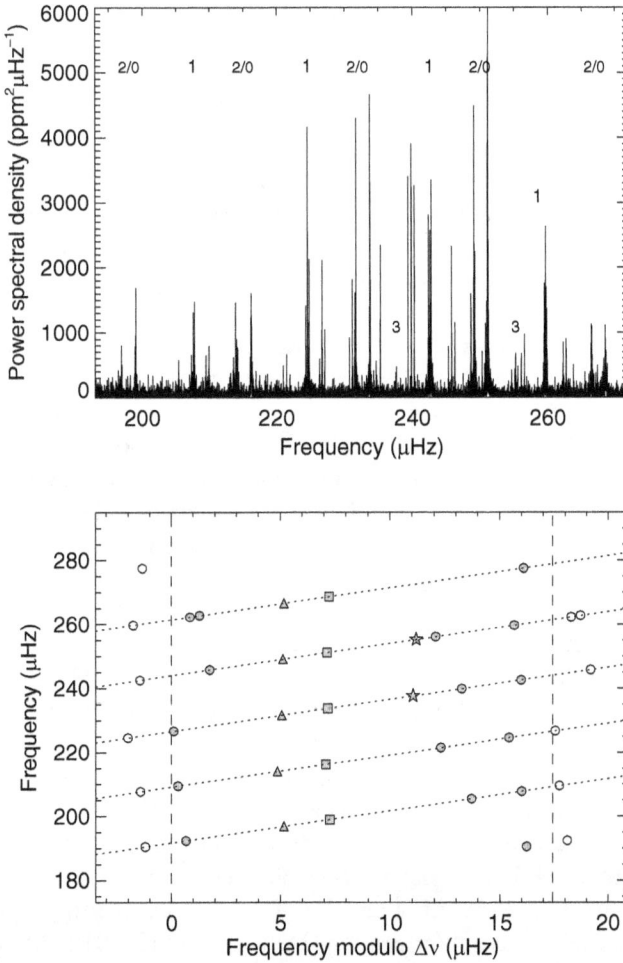

Figure 4.4. Top: Oscillation spectrum of Kepler-56. Annotations tag the angular degree of each mode. Bottom: échelle diagram of the oscillation frequencies, folded at the average large frequency separation $\Delta\nu$: $l = 0$ modes plotted as squares, $l = 1$ modes as circles, $l = 2$ modes as triangles, and $l = 3$ modes as stars.

exoplanet characterization and Galactic Archeology studies. In what follows we focus on some of the published science highlights from Kepler and K2.

Results on Stellar Structure, Physics, and Evolution
We begin with rotation. The exquisite frequency resolution afforded by the lengthy Kepler light-curves, coupled to the high signal-to-noise ratios observed in the modes, has been the key to opening detailed studies of the internal rotation of solar-like oscillators, from the main-sequence through to red giants. The frequency splittings of non-radial modes in these stars are in the range from a few μHz to a fraction of a μHz, with sub-μHz splittings being by far the most common.

On the main sequence, inference on the internal rotation of solar-like oscillators is provided by acoustic (p) modes, which provide less information on rotation in the deeper lying layers than do buoyancy dominated (g) modes accessible in more evolved stars. Nevertheless, determinations of differences in the average rotation rates of convective envelopes and radiative interiors have been made—assuming the former is constrained largely by non-seismic signatures of rotational modulation in the Kepler light-curves (due to spots and active regions)—and constrained to be no more than a factor of two in late F-type main-sequence stars (Benomar et al. 2015). This suggests these stars have rotation profiles that are, at least in a broad sense, not too dissimilar to that of our Sun, and that there must be reasonably efficient coupling between the radiative interiors and outer envelopes allowing exchange of angular momentum during evolution. Moreover, it has also been possible to place constraints on the latitudinal variation of rotation in the convective envelopes of solar-type stars. This is possible because different azimuthal components of the non-radial, rotationally split modes probe different ranges in latitude within the star. While the latitudinal resolution given by the low-degree modes accessible in distant stars is crude compared to what is possible from highly resolved helioseismic observations of the Sun—which give data on modes of much higher spatial degree—some main-sequence stars have been shown to have equatorial regions that rotate twice as fast as regions at mid-latitudes (Benomar et al. 2018). This shear is much more pronounced than in the Sun, and offers a challenge to theoretical predictions.

As stars evolve into the sub-giant phase they begin to present detectable g modes. These modes are also more lightly damped than their pure p-mode cousins. Not only do they provide much stronger constraints on the dynamics of the deep-lying layers, the rotational information is easier to extract because the mode peaks are much more clearly resolved in frequency. Results on a sample of Kepler targets with masses in the range 1.1–1.4 $M_\odot$ have shown clear evidence for spin-up of contracting cores and spin-down of expanding envelopes as stars evolve through the sub-giant phase to the base of the red-giant branch (Deheuvels et al. 2012, 2014). The ratio of core to surface rotation rates changes from ≈2.5 to more than 10. Once on the red-giant branch, Kepler has shown that the stellar cores then spin down in low-mass stars (Beck et al. 2012; Mosser et al. 2012), suggesting that transport of angular momentum via core-envelope coupling is much more efficient than in the sub-giant phase. These results have flagged the need for extra mixing mechanisms in models (Eggenberger et al. 2012) over and above the canonical mechanisms usually included.

While more massive stars (above approximately 1.5–1.6 $M_\odot$) do not show solar-like oscillations when on the main sequence, they do when they reach the red giant phase because they develop convective outer envelopes. Their solar-like oscillations then provide an excellent test of angular momentum evolution in stars that will have followed an evolution path showing more rapid rotation than at lower masses. Evidence has been found for weak radial differential rotation in a sample of intermediate-mass helium core-burning stars (so-called secondary clump stars), that have masses in range 2.2–2.9 $M_\odot$ (Deheuvels et al. 2015). The central layers of these stars rotate about three times as fast as their surfaces, suggesting efficient coupling to redistribute angular momentum, though when this occurs (e.g., in the sub-giant phase or after ignition of the Helium core) remains uncertain. These stars nevertheless seem to be rotating more slowly than expected based on models assuming angular momentum conservation and rigid rotation (Tayar & Pinsonneault 2018).

Based on what we have found with Kepler, theoretical studies suggest it is hard to reproduce the behavior of rotation in both the sub-giant and red-giant branch phase using the same mechanism of internal angular momentum transport, implying the physical nature of the underlying processes may be different (Eggenberger et al. 2019). Suffice to say, modeling and describing mixing and transport processes remains challenging.

Stars only slightly more massive than the Sun develop small convective cores as they evolve. Poor constraints on the extent and nature of overshooting of material at the edges of these cores, into the layers above, can lead to significant uncertainties in determinations of stellar ages. Asteroseismic data from Kepler have made it possible to address these issues, since certain combinations of p-mode frequencies and g-mode periods are sensitive to subtle changes in stratification brought about by these processes. We have seen that on the main sequence the modes are predominantly acoustic in character. Appropriate frequency combinations have been used to diagnose mixing processes, and to isolate the presence of small convective cores and the extent of the overshooting. The first results in this area (Silva Aguirre et al. 2013) found clear evidence for the presence of a small convective core in a star 20% more massive than the Sun. Subsequent work has uncovered tentative evidence for an increase in the extent of overshooting with increasing stellar mass in the range $\simeq 1.1$–$1.5 M_\odot$ (Deheuvels et al. 2016). Period spacings have been used in data on more evolved stars to show that canonical models of mixing underestimate overshooting at base of convective envelopes (Khan et al. 2018).

Having precise asteroseismic ages has been crucial to studies that seek to place measures of surface rotation periods from photometric variability in a proper evolutionary context (García et al. 2014); to those that seek to inform gyrochronology, i.e., the use of rotation as a "clock" to age stars; and those that aim to better understand the action of stellar dynamos (all of course related). One of the most exciting results relating to stellar activity has been evidence for a weakening of magnetic braking in older stars, which occurs via magnetized stellar winds, signaling a potentially fundamental change in the dynamo. Intriguingly, the Sun appears to lie close to or at this transition in behavior (Metcalfe et al. 2016).

Kepler has also revealed seismic stellar cycles in a growing number of solar-type stars (Salabert et al. 2016; Kiefer et al. 2017; Salabert et al. 2018; Santos et al. 2018); i.e., evidence for changes in mode frequencies and other parameters that follow systematic changes in activity. These signatures—now well established in studies of the Sun—can provide new observational constraints on stellar dynamos. Indeed, seismic data on one star—a solar analogue—have even been used to place constraints on its active latitudes, which were found to be shifted to higher latitudes compared to the Sun (Thomas et al. 2019). There is also evidence for strong attenuation of the amplitudes of solar-like oscillations in more active stars (Chaplin et al. 2011b). While very weak amplitude dipole modes have been found in evolved intermediate mass red-giant stars (Stello et al. 2016a), which did not have convective envelopes on the main-sequence. It has been proposed that this suppression is the result of dynamo action in the convective cores of these stars (Fuller et al. 2015).

Kepler data have revealed information on the structure and depth of the convective outer layers of solar-like oscillators, through so-called seismic "glitches." Abrupt structural changes—here, we mean abrupt in depth—leave distinct signatures on the oscillation frequencies, akin to those arising from placing a small mass on an oscillating string. Provided the mass does not lie at a node, it will modify the standing-wave frequencies in a systematic way, compared to those of the pristine string. The resulting differences in frequency are periodic in the overtone number, having an amplitude that decays at higher overtones. The period depends on the location of the mass with respect to the boundary; the amplitude depends on the added mass; and the rate of decay depends on the spatial extent (along the string) of the added mass. In solar-like oscillating stars, abrupt changes are brought about by the ionization of elements in the near-surface layers, notably helium; and by the marked change in stratification at the base of the convective envelope.

These observational signatures are clearly detectable in Kepler data (e.g., see Mazumdar et al. 2014; Broomhall et al. 2014). They have allowed tests of stellar evolutionary models, i.e., from measurement through the glitches of the convective envelope depths (Verma et al. 2017). And they have also provided a way to obtain estimates of envelope Helium abundances in these stars (Verma et al. 2019). These are in effect surface abundances, given the outer envelopes are well mixed, and are not accessible from spectroscopic photometric abundance determinations since effective temperatures are too low in solar-like oscillators to give photospheric absorption lines in helium.

Results on Exoplanet Systems
Strong synergies exist between asteroseismology and exoplanet studies, which follow first and foremost from the need to characterize in detail the host star in order to paint as complete and accurate a picture as possible of the discovered exoplanets, and the overall system characteristics and history. Asteroseismology of solar-like oscillators provides accurate and very precise fundamental properties that have direct application to the exoplanet analysis, e.g., most obviously stellar radii for calibration of exoplanet radii given by Kepler's transit data, and stellar masses for calibration of exoplanet masses given by ground-based spectroscopic follow-up data. We discuss other examples below.

The opportunities that asteroseismology would offer to Kepler exoplanet science were a key driver behind the formation of the Kepler Asteroseismic Science Consortium (KASC). Maximizing the yield of Kepler objects of interest (KOIs, later confirmed planet hosts) with asteroseismology required a responsive strategy to be implemented by KASC. This was because the vast majority of potential planet hosts acquired their KOI status while on observations in the long-cadence mode. As has been noted previously, this cadence was not rapid enough to detect oscillations in cool main-sequence and sub-giant stars. New KOIs were therefore switched rapidly into short-cadence mode if the probability of detecting solar-like oscillations—based on the already known properties of the star—was predicted to be 50% or higher (Chaplin et al. 2011b). This rapid-response strategy proved extremely successful in boosting the total number of asteroseismic hosts to in excess of 100 by the end of the nominal mission (Huber et al. 2013a; Lundkvist et al. 2016); allowing for exquisite characterization (Silva Aguirre et al. 2015; Davies et al. 2016). This sample was biased naturally to bright targets to give low enough shot noise to detect the oscillations, and contained many sub-giants owing to oscillation detection thresholds giving a bias to more evolved stars with larger oscillation amplitudes. While for K2 it was not possible to adopt the same responsive strategy for KOIs— only later would some fields be revisited—there were still several new asteroseismic hosts discovered, in addition to a few asteroseismic detections in already known hosts.

Asteroseismology had a key role to play in several high-profile discoveries. The oscillations detected in Kepler-22's host star helped confirm this as Kepler's first potential habitable-zone planet (Borucki et al. 2012). Kepler-36 (Carter et al. 2012) contained two planets showing gravitational interactions via transit-timing variations, which allowed for estimations of the planet masses relative to the mass of the star. Here, the very precise asteroseismically determined mass meant the mass of one of these exoplanets was at the time the most precisely known. Kepler discovered three transiting exoplanets in the Kepler-37 system, but one was at the time the smallest confirmed exoplanet. It was possible to pin-down its radius as lying between the Moon and Mercury in large part due to the very precise, accurate asteroseismic radius (Barclay et al. 2013). Kepler-444 (Campante et al. 2015) was noteworthy not only because of its ultra-compact system of small, terrestrial planets but also because of its old age, which asteroseismology found to be over 11 Gyr. This meant the system had formed when the Universe was only a fifth of its current age, suggesting that systems containing rocky terrestrial planets have been forming throughout a large fraction of the age of the Universe. Kepler has also found several shorter-period exoplanet systems with evolved hosts—low-luminosity red-giant stars— which complement the long-period, wider systems with evolved hosts discovered in long-standing ground-based Doppler velocity surveys. These systems are particularly well suited to asteroseismic study with Kepler, first and foremost because the stars show larger oscillation amplitudes than their less-evolved counterparts permitting detections to fainter magnitudes (covering a larger native sample from which to discover planets). K2 has been very successful in adding systems to this sample (e.g., see Grunblatt et al. 2016), courtesy of it having observed significant numbers of red giants in each campaign.

Kepler-56 (Huber et al. 2013b) was not only one of the first exoplanet systems with an evolved host discovered by Kepler—it has a low-luminosity red-giant host— it was also the first misaligned multi-planet system ever found. In our solar system, the orbital angular momentum vectors of the planets are all aligned, to within a few degrees, with the spin axis of the Sun. Observations made prior to Kepler had established the existence of misaligned single exoplanet systems, typically containing hot Jupiters, using the Rossiter–McLaughlin effect in Doppler velocity observations;

but no misaligned multi-planet systems had been found. Asteroseismology provided a new way to get information on spin–orbit alignment. When the rotationally split components of the non-radial modes can be resolved in the frequency–power spectrum the ratios of their observed amplitudes depend almost entirely on geometry alone, specifically the angle of inclination of the spin axis of the star. If a transiting exoplanet has been discovered, we know the orbital angular momentum vector must lie close to the plane of the sky. Hence, if for a transiting system the stellar-spin axis is found not to lie in the same plane, the system must be misaligned; if on the other hand it does, the result is consistent with expectations for an aligned system, though a misaligned system is not ruled out (i.e., while both vectors may lie in the plane of the sky, this does not mean they are necessarily parallel).

The first application of the asteroseismic technique to Kepler data was on two systems with bright solar-type host stars (Chaplin & Miglio 2013), both of which were found to show signatures consistent with alignment. The next was Kepler-56, which turned out to be an excellent system on which to apply the method since the resonant peaks of its non-radial modes were much more clearly resolved then in the aforementioned solar-type stars. The mode peaks were narrow in frequency, a typical feature of red giants, allowing the relative amplitudes of the constituent components of several non-radial modes to be well determined. The orbits of the two transiting exoplanets were found to be tipped by about 45 degrees with respect to the stellar spin axis (Huber et al. 2013b). It is believed that a distant planetary or stellar companion has torqued the system out of alignment.

Subsequent Kepler analyses of individual systems confirmed, for example, the misaligned nature of the already known HAT-P-7 (Lund et al. 2014); and also included systems with Rossiter–McLaughlin data allowing the full 3D geometry of the alignments to be solved (Benomar et al. 2014). The first ensemble study (Campante et al. 2016) of around 20 systems with solar-type hosts where astero-seismic inference on the alignments was possible found a strong preference for alignment, and no significant evidence for a difference in the underlying distributions of single- and multi-planet systems.

The highly constrained fundamental properties of the asteroseismic sample of host stars have played an important or even crucial role for other ensemble studies, and we finish this section by considering a few examples. Asteroseismic masses were used to constrain the masses, and hence bulk densities, of a sub-sample of small terrestrial Kepler planets (Marcy et al. 2014). Asteroseismic determinations of host-star surface gravities have been used as priors on spectroscopic analyses, helping to mitigate the effects of degeneracies in those analyses to hence provide more accurate spectroscopically determined parameters for subsequent use (Huber et al. 2013a). Asteroseismic estimates of the stellar densities have been used to make inference on the eccentricity of exoplanet orbits.

When exoplanet orbits are circular, it is straightforward to show that the transit duration is a function of the stellar density (this follows from Kepler's laws). In such cases, the measured transit parameters may therefore be used to estimate the density of the host star. However, when the orbit is eccentric, the velocity of the planet in its orbit will change, which affects the transit duration. A mismatch between densities inferred from transit parameters and asteroseismology is therefore a tell-tale indicator of a non-circular orbit and may be used to constrain the eccentricity (Van Eylen & Albrecht 2015). Results of using this technique suggest a difference in the underlying eccentricity distributions of single- and multi-planet systems (Van Eylen et al. 2019).

Finally, the highly-constrained radii of asteroseismic hosts provided the extra fidelity needed to fully tease out and define the nature of the so-called photo evaporation desert and valley. Close-in planets below a certain mass are vulnerable to losing a substantial fraction of their envelopes due to irradiation by the host star, leaving behind a small, rocky core. Asteroseismic radii of more than 100 hosts were used to constrain the stellar luminosities and hence levels of irradiation received by their planets, and revealed the presence of a lack of close-in planets having sizes between about 1.5 and 4 Earth radii (Lundkvist et al. 2016); this desert had previously been blurred and hidden by inferior estimates of the stellar radii. Further analyses using the asteroseismic sample have mapped out the theoretically predicted valley in the planet radius-irradiance and planet radius-orbital period planes (Van Eylen et al. 2018), refining how the stripping of planetary envelopes changes with star, planet, and orbital properties.

Results on Stellar Populations and Galactic Archeology
Studies of the detailed history and evolving structure of the Milky Way (Galactic archeology)—including by inference populations of stars within it—are inherently limited by the intrinsic quality of data available on individual stars. For example, poor precision on stellar ages significantly limits the ability to map the temporal evolution of different structures and populations. Kepler provided asteroseismic properties on large numbers of solar-like oscillators to the levels of fidelity needed to make a significant difference to these studies.

First and foremost are the asteroseismic data on red giant stars (Miglio et al. 2015). This has given masses on tens of thousands of stars to levels of precision and accuracy not previously available, out to distances beyond 10 kpc. Since to first order the mass of a red giant fixes its age, this has reduced uncertainties on ages to levels of 20% or better for individual stars. The ESA Gaia Mission has provided accurate positions, distances, and velocities; the distances may be used to constrain stellar luminosities and radii, allowing the asteroseismic data to work harder to place even tighter constraints on masses and ages. Very precise and accurate asteroseismic radii also fix strong constraints on the distances to the stars, and it is important to note that asteroseismology will outperform Gaia at distances beyond 3–4 kpc.

Asteroseismic data from the Kepler mission revealed a significant vertical age gradient perpendicular to plane of the Galaxy (Casagrande et al. 2016). They have been used to test so-called population synthesis models of the Milky Way (Sharma et al. 2016). And they have revealed more information about the origins of the Galactic disc through studies of populations of red giants that have different chemical signatures (Silva Aguirre et al. 2018). These results have then been used to test models describing the chemical evolution of the Galaxy (Spitoni et al. 2019).

K2 has provided the opportunity to leverage asteroseismic data from fields around the ecliptic that map to different galactic latitudes and longitudes (Howell et al. 2014; Stello et al. 2017), again revealing a gradient of age out of the plane of the Galaxy (Rendle et al. 2019).

Feeding into these studies have been large catalogs of asteroseismic properties, which need large sets of spectroscopic observations to provide the complementary data required to fully utilize the asteroseismic parameters. Notable here is the formal collaboration that was formed between the Apache Point Observatory Galaxy Evolution Experiment (APOGEE) consortium and KASC, which has produced APOKASC catalogs on red giants (Pinsonneault et al. 2018) and solar-type stars (Serenelli et al. 2017), the latter relevant for studies of the local solar neighborhood.

4.2.3 Focus on Compact Pulsators

White Dwarfs

The discovery of pulsations in stars at the final stages of their evolution—white dwarfs—came entirely by accident, more than 40 years before the launch of the Kepler space telescope. A meticulous cataloger of stellar brightness, Arlo Landolt at Louisiana State University, discovered flux variability in a white dwarf that was meant to be a photometric standard (Landolt 1968).

For decades, astronomers observed the brightness changes of these stellar remnants, which characteristically brighten with relatively high amplitudes (often above 1%) and periods of roughly 2–20 minutes, in order to constrain what lies beneath the relatively simple surfaces of stars that are no longer fusing in their core (e.g., Warner & Robinson 1972). However, monitoring the brightness changes from a single site proved extremely challenging, due to the cycle-count confusion caused by gaps in the data caused by sunrises and clouds (e.g., Kleinman et al. 1998). We now know that not only are white dwarfs comparable to Earth in radius, these stellar fossils also tend to rotate at roughly 1 day, making it extremely difficult to interpret their frequency patterns from a fixed point on the Earth's surface.

In the mid-1980s, astronomers led by Ed Nather at the University of Texas at Austin began laying the groundwork for a global network of coordinated astronomical observatories focused on overcoming these challenges by obtaining nearly uninterrupted time-series measurements of variable stars by handing off monitoring from one site to the next: thus was born the Whole Earth Telescope (Nather et al. 1990). In the first few years, the network produced prodigious observations of some of the brightest known pulsating white dwarfs, including PG1159-035 (Winget et al. 1991) and GD 358 (Winget et al. 1994). While very bright by white dwarf standards, PG 1159-035 has a relative magnitude of $V = 14.9$ mag, more than 100 times fainter than the planet host Kepler-410 (see Figure 4.1).

Dozens of astronomers trekked all over the world to participate in the labor-intensive campaigns of the Whole Earth Telescope, often braving extreme weather, altitude sickness, fires, earthquakes, political turmoil, and even near-fatal stabbings (Vauclair et al. 2002). The data from the global network provided unprecedented insights for many years on the interiors of white dwarf stars. Still, the looming full

Moon and systems of clouds fundamentally limited the Whole Earth Telescope, which rarely produced a continuous set of observations (either using two- and three-channel photometers and later CCDs) with more than 60% uptime over roughly two weeks.

So when the Kepler space telescope launched and could deliver good photometry at minute cadence, rapid enough to constrain most of the oscillations in compact stellar remnants, the community was ecstatic. Unfortunately, there was very little blue photometry of the original Kepler field, making target selection of blue white dwarfs a major challenge.

Early on, even the most marginal candidate compact pulsators selected from surveys with ultraviolet photometry, as well as blue objects with high proper motions, were observed with short-cadence observations by Kepler, since there were so few possible targets (Østensen et al. 2010). In the first four months of the Kepler mission, more than 110 of the precious short-cadence slots were used to search for candidate rapid variables. Several new pulsating hot subdwarfs were discovered in these early observations (Kawaler et al. 2010; Reed et al. 2010), but no pulsating white dwarfs.

After the launch of Kepler, two groups undertook a more sustained effort to cover the field with blue-broadband photometry, with a significant focus revealing more potential hot stars and especially white dwarfs in the Kepler field (Everett et al. 2012; Greiss et al. 2012). Hundreds of hours of complimentary ground-based observations were undertaken to find new pulsating white dwarfs in the Kepler mission field, including wide-field surveys (Ramsay et al. 2014). There were two early successes—a candidate selected from an expensive spectroscopic follow-up campaign (Østensen & Bloemen 2011), as well as another selected from imprecise photometric properties (Hermes et al. 2011)—but those were the only two pulsating white dwarfs known and observed in the Kepler field for the first 3 yr of the mission.

Aided by the new blue-broadband photometric surveys, it eventually became possible to select candidate pulsating white dwarfs, and by the summer of 2014 at least 10 were confirmed to pulsate from ground-based observations (Greiss et al. 2016). However, only five of these were eventually observed by the space telescope before the failure of its second reaction wheel, preventing it from staring at its familiar patch of sky in Cygnus.

But all was not lost on the pulsating white dwarfs with the end of the original Kepler mission: in fact, the failure of the second reaction wheel opened the sky to an entirely new set of targets along the ecliptic. The extended K2 mission afforded an opportunity to obtain runs far surpassing any by the Whole Earth Telescope, as 3 months of data were completely sufficient to fully reveal the pulsations of these stellar remnants. During a 9 day engineering run, before even K2 Campaign 0 began, observations of the pulsating white dwarf GD1212 demonstrated the quality of K2 observations (Hermes et al. 2014). By the end of K2, more than 85 pulsating white dwarfs were observed by the Kepler spacecraft.

The new insights we have gained from the dense, long-duration monitoring of pulsating white dwarfs observable by Kepler and K2 have justified the great efforts expended in finding them.

Perhaps the most surprising set of observations, which took the community entirely by surprise, came immediately from the first pulsating hydrogen-atmosphere (DAV) white dwarf monitored by Kepler: KIC 4552982 (Bell et al. 2015). On top of the pulsations that were first seen from ground-based discovery light-curve, the mean flux of this white dwarf was getting brighter by up to 15% for a few hours, and these brightening recurred stochastically roughly every 3 days. But it was hard to formulate a complete picture of the processes going on in the relatively faint (K_p = 17.9 mag) white dwarf.

Among the first 10 white dwarfs observed in K2, a second exhibited massive flux excursions: PG 1149+057 (Hermes et al. 2015). This star was 15 times brighter than the first, and clearly showed pulsations in outburst, demonstrating that not only was the white dwarf brightening, but the brightenings affected the pulsations. An annotated view of the outbursts in PG 1149+057 is shown in Figure 4.5.

These outbursts had never been seen from the ground before the long-baseline monitoring of the Kepler spacecraft, although one helium-atmosphere DBV, GD 358, underwent a massive, one-day brightening event in 1996 that those watching with the Whole Earth Telescope jokingly referred to as the "whoopsie" (Montgomery et al. 2010). Outbursts appear to be a feature of most of the coolest pulsating white dwarfs, especially those with dominant pulsation periods longer than 1000s (Bell et al. 2016, 2017). The simplest explanation appears to involve a rapid transfer of energy via a nonlinear resonance from excited pulsations that couple with

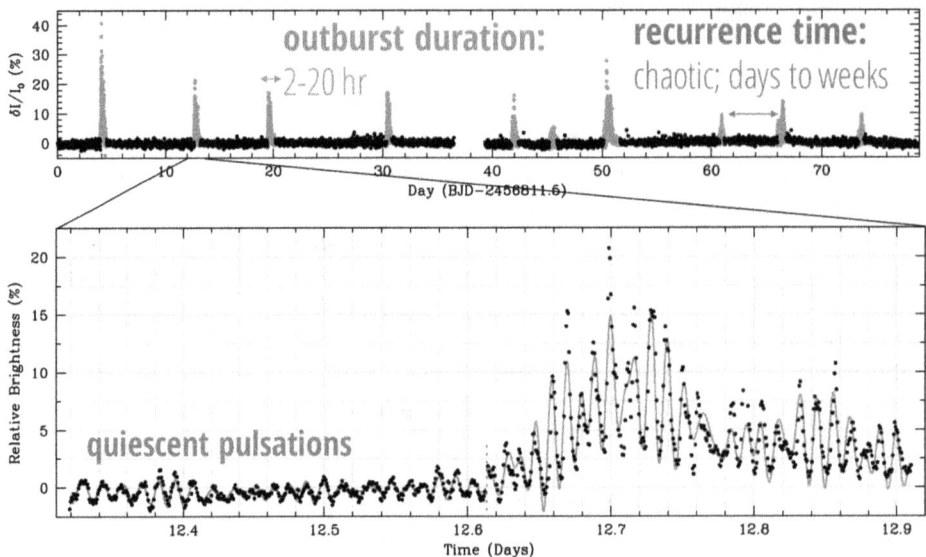

Figure 4.5. An annotated view of the K2 light-curve of the outbursting behavior observed in the white dwarf PG 1149+057, adapted from (Hermes et al. 2015). The top panel shows the full Campaign 1 light-curve, and the bottom inset shows how quiescent pulsations can grow into dramatic flux outbursts. The pulsation amplitudes rise dramatically in outburst. First discovered by Kepler, outbursts were observed in many of the coolest pulsating white dwarfs by the end of the mission and are likely due to a nonlinear resonance phenomenon. Courtesy of J. J. Hermes.

daughter modes and dump the transferred energy as heat at the surface (Hermes et al. 2015; Luan & Goldreich 2018). Interestingly, the nonlinear resonance in pulsating white dwarfs was explored in depth by Yanqin Wu in 2001 (Wu & Goldreich 2001), but all the theory lacked was a prediction of flux increases from the nonlinear resonances.

The exotic outbursting behavior was certainly not the only new discovery enabled by long-term monitoring of white dwarfs by the Kepler space telescope. Rotational multiplets were completely resolved for the majority of oscillations, for the first time allowing large-scale identification of the pulsation modes in dozens of white dwarfs. Given the unfortunate similarity between the Earth's rotation and that of a typical white dwarf, aliasing from ground-based monitoring makes it extremely difficult to observe pulsation multiplets and thus identify the modes present.

Among the first 27 DAVs observed by Kepler, roughly 40% of the more than 200 independent modes could be identified from the multiplet structure (Hermes et al. 2017b). Immediately, this mode identification was able to provide rotation rates of isolated white dwarfs. Coupled with external mass constraints from spectroscopy, Kepler allowed for the first large-scale exploration of white dwarf rotation as a function of mass, constraining the endpoints of internal angular momentum evolution in stars (Hermes et al. 2017a, 2017b). In summary, most white dwarfs evolved from 1–3 solar mass stars on the main sequence and end up rotating with a narrow range of periods between 0.5–2.2 days.

The extended space-based photometry from Kepler afforded another important insight brought about by the clean window functions in the power spectra: exploring the mode lifetimes of the longest-period modes in white dwarfs, which appear to become highly incoherent for modes with periods longer than roughly 800 s (Hermes et al. 2017b). These modes are not excited stochastically like those in the Sun, and so were not expected to show such wide linewidths in their power spectra. One possible explanation is that the outer turning point for the mode cavity for these pulsations approaches the base of the convection zone, which is always changing, causing the longest-period modes bounded by the base of the convection zone to lose phase coherence (Montgomery et al. 2019).

Kepler has also opened a new eye into the long-term phase and amplitude variability of even short-period pulsations, which in some cases can slowly wander. Although mode variability had been observed for decades (Kleinman et al. 1998), for the first time with Kepler and K2 we could nearly continuously monitor the coherence of variations over months for both hotter and cooler pulsating white dwarfs. By monitoring correlated amplitude and phase changes in pulsating white dwarfs, we see good evidence for three-mode couplings that act as nonlinear resonances to exchange energy between oscillations in many white dwarfs (Zong et al. 2016).

The richness of observational information from the hundreds of oscillations detected and well-characterized by Kepler have yet to be fully exploited for detailed asteroseismic analysis. But sophisticated tools are being improved to exploit the unprecedented data sets to extract as much interior information about white dwarfs as possible (Giammichele et al. 2017).

One of the most detailed first major explorations of a white dwarf interior using space-based photometry data centers on KIC 8626021, the first helium-atmosphere pulsating white dwarf (DBV) observed by Kepler. Asteroseismic sounding techniques have put valuable constraints on the chemical stratification of this stellar remnant as a function of radius within the star, with surprising consequences for the interior oxygen abundance set by previous epochs of nuclear burning (Giammichele et al. 2018). Figure 4.6 shows the derived mass fraction of helium, carbon, and oxygen in the interior of this white dwarf determined from the stellar pulsations, and includes updated parameterization (Charpinet et al. 2019) to address the important effects of neutrino luminosity on the structure of hot stellar remnants (Timmes et al. 2018). Importantly, the figure demonstrates the wealth of information unraveled from white dwarf oscillations, especially complimentary constraints on the complicated physics of helium burning and the highly uncertain $^{12}C\,(\alpha, \gamma)^{16}O$ reaction rate (Fields et al. 2016).

While the stars have set on additional white dwarfs observable by the Kepler space telescope, there is still ample investigation to be done of the unprecedented data sets collected on thousands of white dwarfs, nearly 80 of which pulsate. Kepler has likely provided the most extensive and comprehensive look at faint stellar remnants that will be obtained for more than a generation.

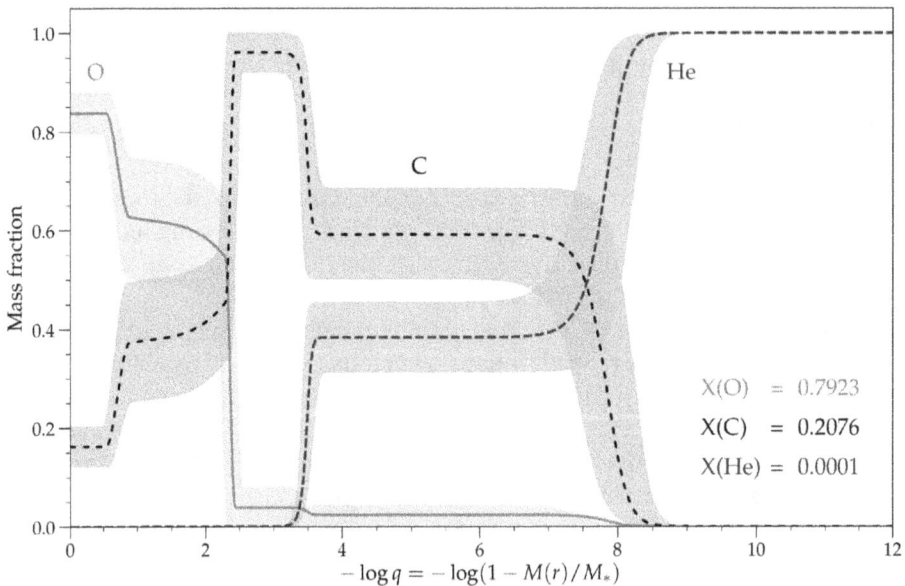

Figure 4.6. The interior chemical stratification (with 1-σ uncertainties) of the helium-atmosphere pulsating white dwarf (DBV) KIC 8626021. This figure (Charpinet et al. 2019) updates previous work (Giammichele et al. 2018) to address the important effects of neutrino luminosity on the structure of stellar remnants (Timmes et al. 2018). The run of the profile for helium, carbon and oxygen is plotted along an axis of logarithmic mass fraction; the photosphere of the star is to the right, whereas the deepest interior is at left toward 0. Reproduced from Charpinet et al. (2019). © S. Charpinet et al. 2019. CC BY 4.0.

4.2.4 Pulsating Hot B Subdwarf (sdB) Stars

A second class of compact star for which asteroseismology has had an important role are the pulsating hot subdwarf (sdB) stars. First found to be pulsators by Kilkenny et al. (1997), these stars show two broad (and in some cases overlapping) categories of pulsation—relatively short periods (500 seconds and less—the V361 Hya stars) that are attributed to p modes, and longer periods (up to a few hours—the V1093 Her stars) that are generally g modes. Several "hybrid" stars show evidence for both ranges of variability. For a good overview of these stars (and asteroseismic results) see Heber (2016). While the first decade of sdB asteroseismology saw some significant advances, both theoretically and observationally, Kepler observations completely changed the field.

This is in large part a result of the long, continuous observation, in short-cadence mode, of dozens of sdB candidates. The yield of pulsation discovery was relatively high—18 new pulsators were found over the course of the primary mission (Reed et al. 2018); out of 37 that were examined (Østensen et al. 2010; Pablo et al. 2011; Reed et al. 2012). In the K2 era, an extended guest investigator program by M. Reed and collaborators has found many more pulsators (Reed et al. 2018). Most of the new pulsating sdB stars are in the long-period V1093 Her class. Many are extremely rich, allowing determination of the asymptotic g-mode period spacing, and rotational splitting. Though this class was discovered in ground-based surveys (Green et al. 2003); their pulsation periods are similar to the nightly observing duration. As they are often multiperiodic as well, the oscillations difficult to disentangle from hours-long transparency variations at a given observing site. Kepler changed all that, providing high frequency sensitivity photometry for, in many cases, several years at 90+% duty cycle. This allowed full resolution of the frequency spectrum of many V1093 Her stars. Coupled with ground-based spectroscopic follow-up to obtain accurate T_{eff} and $\log g$, asteroseismic modeling has revealed new insights into the origin, structure, and evolution of these stars.

From early in the mission, with only a few months of data in hand, it was clear that the V1093 Her stars were rich g-mode pulsators. Longer periods in g-mode pulsators correspond to higher overtone (n, or k) pulsation, and so should tend toward an asymptotic period spacing as described in previous sections on red giant and white dwarf pulsations. Notably, the period spacings for $l = 1$ modes cluster tightly around 251 s, with a range from approximately 230 s–280 s (Reed et al. 2011). This is consistent with the period spacing of the $l = 1$ modes seen in the solar-like oscillation spectra of "clump giants" on the red giant branch. These stars have helium-burning cores, which are in some sense sdB stars shrouded by a red giant envelope.

The fact that the period spacings are relatively uniform is a bit of a surprise since these stars are expected to have relatively abrupt composition transition zones akin to what white dwarfs show, though models suggest that such regular patterns may still be seen in stratified models (Charpinet et al. 2014). Still, the lower overtone g modes as well as the p modes, along with spectroscopic constraints, have provided sufficient observational constraints to allow remarkably precise matching to asteroseismic models. One example of this analysis comes from the early observations by Kepler of KIC02697388, where the asteroseismic modeling allowed precise determination of the basic properties of this star. One of the more intriguing results was that the convective helium-burning core is much larger than any standard evolutionary models were capable of reproducing (Charpinet et al. 2011). This result has been seen for other sdB pulsators (e.g., KPD1943+4058, Van Grootel et al. 2010), suggesting that our understanding of convective helium burning in low mass stars drives as-yet poorly understood mixing mechanisms.

Beyond structural and global measures, the presence of rotationally split multiplets in the oscillation spectra of sdB stars has provided measures of surface (and perhaps internal) rotation rates for most of the Kepler and K2 sdB stars. As summarized in Reed et al. (2018) and Reed et al. (2014); the measured rotation periods range from 5 days to nearly 100 days, with a median period of about 30 days. There appears to be a weak correlation with effective temperature, with the coolers stars showing longer rotation periods. Complicating interpretation of this trend is the fact that sdBs most likely are formed during (interacting) binary evolution. Transport of angular momentum between components during prior evolution, and spin–orbit coupling, can be partially responsible for the observed rotation rate, so that what we see today may not reflect internal angular momentum evolution of a single star. Still, the measured rotation of sdBs in binaries allows us to test theories of spin–orbit coupling in close binaries. From analysis of Kepler sdB pulsators in binaries, it appears that such coupling is ineffective for orbital periods longer than a few hours. The sdB stars in these systems are rotating more slowly than synchronous (Pablo et al. 2011, 2012).

4.3 Eclipsing Binaries

While specifically designed to search for exoplanets, Kepler also enabled the discovery and monitoring of variable stars by providing high-precision nearly uninterrupted photometry of over 150,000 stars. In particular, the mission provided light-curves of thousands of eclipsing binaries with unprecedented accuracy and nearly continuous sampling for almost four years—this revolutionized the field. Kepler's exquisite photometry detected physical phenomena never before seen, such as Doppler boosting and tidal brightening/heartbeat stars, and presented challenges for theoretical models and analysis tools, which had to be adapted to comply with the unprecedented photometric precision of the data.

4.3.1 The Kepler Eclipsing Binary Catalog

Binary stars are ubiquitous throughout the Galaxy and an important source for the determination of fundamental astrophysical parameters. Eclipsing binary stars, in particular, are ideal astrophysical laboratories as the line-of-sight alignment with the orbital plane and the basic principles of classical dynamics that govern the motion of the components in a binary reduce the determination of principal parameters to a simple geometric problem (Prša et al. 2011). Photometric and spectroscopic studies of eclipsing binaries reveal fundamental stellar parameters such as masses and radii that inform our understanding of stars and constrain stellar evolutionary models. In cases where high-quality radial velocity curves exist for both stars in an eclipsing binary, the luminosities computed from the absolute radii and effective temperatures can also lead to a distance determination.

Prior to Kepler, large samples of eclipsing binaries had been generated as byproducts of large-scale automated surveys, enabling the determination of statistical properties and the discovery of rare binary systems. Kepler, however, delivered thousands of eclipsing binaries with fewer observational biases then ground-based surveys and higher detection rates (Prša et al. 2011). In fact, 1.3% of all Kepler targets were eclipsing (or ellipsoidal) binaries, which is about twice the value typical of ground-based surveys, as eclipses with smaller amplitudes and shorter fractional eclipse durations were detected (Maceroni et al. 2012). To identify and characterize the sample of eclipsing binaries detected by Kepler, an eclipsing binary catalog was developed by Prša et al. (2011) and Slawson et al. (2011); with the final catalog presented in Kirk et al. (2016), containing 2878 binaries.[1] The catalog provides the stellar parameters from the Kepler input catalog (KIC) along with measurements of the primary and secondary eclipse depth, eclipse width, eclipse separation, ephemeris, morphological classification, and principal parameters determined by geometric analysis of the phased light-curve. Raw and detrended light-curves for each system are also included in the catalog, as well as an analytic approximation of the light-curve via a polynomial chain (Prša 2008).

In addition to identifying, classifying, and conducting a preliminary analysis of the eclipsing binaries detected by Kepler, the Kepler eclipsing binary catalog enabled general population studies. For instance, it was noted that the population of eclipsing binaries was non-uniformly distributed as a function of galactic latitude. Such behavior confirms that stellar populations at lower galactic latitudes (thin disk) contain notably younger stars that are on average larger (i.e., contain more giants) than the older, sparser population of the thick disk and halo (Prša et al. 2015). As a result, the geometric probability of eclipses increases toward the galactic disk (Kirk et al. 2016).

Furthermore, the Kepler eclipsing binary catalog also serves as a valuable tool for probing the underlying distribution of binary stars. Since the primary Kepler mission ended after four years, the longest orbital periods in the catalog are ~1000 days. The low probability of eclipses at larger orbital separations and the difficulty of detecting every eclipse prevents the catalog from being complete at these longer periods. However, binaries with short periods (~1 day) should be detected due to the high geometrical probability of eclipses and the inclusion of ellipsoidal (non-eclipsing) variables. The distribution of orbital periods for all binaries in the Kepler eclipsing binary catalog are shown in Figure 4.7. The excess of systems seen at short periods occurs as close binaries can be detected due to proximity effects, even if they don't eclipse, while the gradual drop-off at longer periods is a result of

---

[1] http://keplerEBs.villanova.edu.

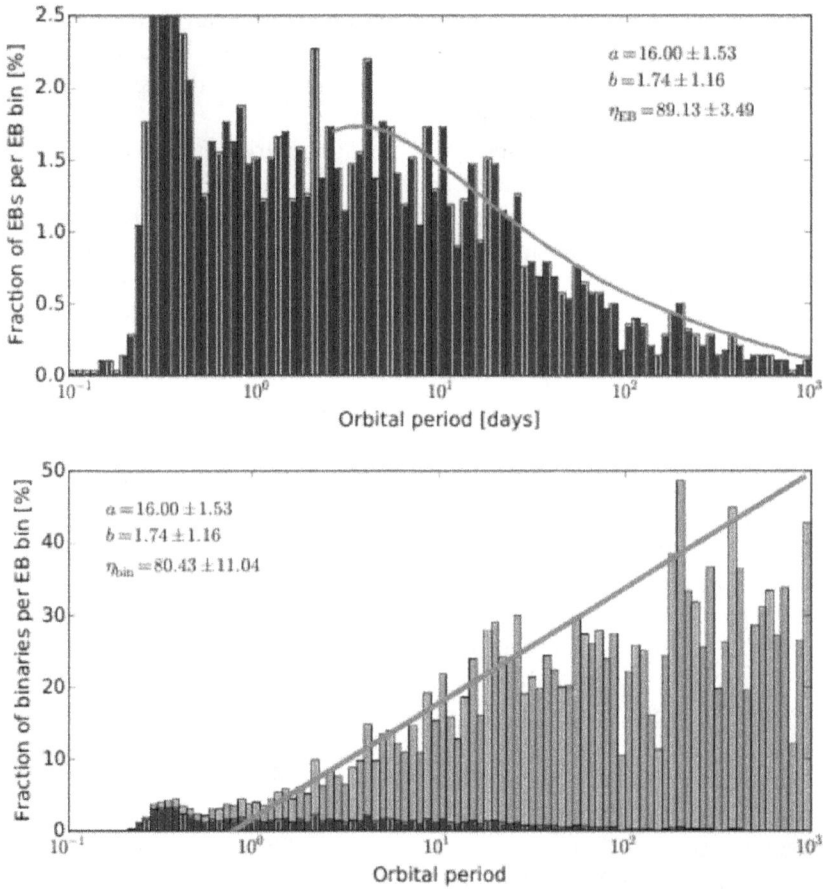

Figure 4.7. Completeness estimates for the Kepler eclipsing binary catalog (top), and projected completeness estimates for the binary star population (bottom). The observed distribution of orbital periods is depicted in blue, and the theoretical distribution of orbital periods is shown in red. The distribution in gray is the predicted occurrence rate of all binary stars based on Kepler data, compared to a best fit linear model. For details on how the catalog completeness and linear model were derived see Kirk et al. (2016). Reproduced from Kirk et al. (2016). © 2016. The American Astronomical Society. All rights reserved.

the geometrical probability of eclipses and Kepler's duty cycle. The observed distribution of orbital periods is depicted in blue, and a theoretical distribution of orbital periods, derived from a linear model of the underlying binary log-period distribution determined in Kirk et al. (2016), is shown in red. The distribution shown in gray in the bottom panel is the predicted occurrence rate of all binary stars based on Kepler data, compared to a linear model. By comparing the theoretical predictions with the observed distribution, the completeness of the Kepler eclipsing binary catalog is estimated to be 89.1% ± 3.5% for eclipsing binaries, and 80% ± 11% for all binaries (Kirk et al. 2016).

A Brief Summary of Cataclysmic Variable Stars Observed with Kepler/K2

Contributed by Paula Szkody, University of Washington, Seattle, WA.

Background

Cataclysmic variables (CVs) are generally comprised of a late main sequence or brown dwarf star that transfers mass to a white dwarf. They are very close binaries having orbital periods of a few hours, with the majority being under two hours. Their wide range of behaviors has led to various classification schemes. If they show 2–9 magnitude brightness episodes (termed outbursts) lasting for days to weeks on quasi-periodic timescales that range from weeks to decades, they are termed dwarf novae. An outburst is due to relatively low mass transfer rates which causes the disk to build up and undergo a disk instability that results in a large accretion episode onto the white dwarf. At higher mass transfer rates (and longer orbital periods), the disk remains in a hot state and does not undergo outbursts, so the CVs are termed novalikes (NLs). The NLs do show long term (weeks to months) changes in their mass transfer states, sometimes even stopping the mass transfer completely for weeks to months (termed low states). Some of these NLs contain highly magnetic white dwarfs. In these cases, the magnetic field constrains the mass transfer and accretion, causing disruption of the inner disk into accretion curtains for the objects with white dwarf fields from 1–10 MG (called intermediate polars; IPs) and removing the disk entirely for direct flow onto the magnetic poles in the higher field (10–250 MG) white dwarfs (called polars).

Because of the large range of variability both in long and short timescales, the Kepler mission was outstanding in documenting the details that have been lost with ground-based observations having to deal with weather and day/night cycles. This detailed information has provided a wealth of insight into the physical properties and changes that occur during quiescence, outburst and high/low states in these systems.

Main Results from Kepler

There are 26 known CVs located in the original Kepler field. Analysis of several dwarf novae undergoing outbursts during the three year coverage led to real understanding of how the accretion disk grows and shrinks during different types of outbursts. The short orbital period dwarf novae V344 Lyr and V1504 Cyg provided remarkable light-curves showing normal outbursts, superoutbursts (SOB), superhumps (SH) and quiescent variability. Modeling of their light-curves (Wood et al. 2011; Cannizzo et al. 2012; Kato & Osaki 2014) provided a scenario whereby the disk reached the 3:1 resonance and the tidal forces created an eccentric disk resulting in the emergence of superhumps. The light-curve of V344 Lyr was even converted to an audio version to better appreciate all the nuances that were present (Tuchton et al. 2012).

Dwarf novae were not the only CVs that provided new information from the Kepler field. An in-depth analysis of a low state that occurred in the novalike system MV Lyr (Scaringi et al. 2017) revealed the presence of mini-outburst type variability repeating on hours that was ascribed to "magnetic gating", providing the first evidence of the presence of a magnetic field in this system.

After the end of the main mission, the resurrection, K2, provided the ideal chance to obtain more observations for other dwarf novae scattered near the ecliptic plane. The resulting database contains more than 100 light-curves of CVs. Included among these is the first short period eclipsing dwarf nova that underwent a superoutburst and several normal outbursts. The eclipses in the K2 data on this 1.9 hour system, MLS0359+17, could be used to mark each orbit during the development of the SOB and its decay (Littlefield et al. 2018). The SH were observed to develop in the timespan of about two orbits, following a precursor rise and the disappearance of the prominent hot spot. During both the SOB and the normal outbursts, the eclipse times move earlier since the hot spot does not distort the eclipse as it does during quiescence. The shifting of the eclipse egress and the orbital hump to earlier phases once the SOB ends provides confirmation of the disk shrinking as predicted by the Osaki (1989) thermal-tidal disk instability model. These shifts in phases were also seen in the long period DN V729 Sgr (Ramsay et al. 2017), where the K2 data caught seven outbursts, as well as negative superhumps during quiescence. K2 was also well-suited for observing the long (hours) quasi-periods that have been previously found but not well-studied from the ground. This includes the 8 hour timescale for V406 Vir (Pala et al. 2019) that is ascribed to a new kind of thermal disk instability and the 4 hour modulation of GW Lib (Gaensicke et al. 2019) that could be related to its pulsating white dwarf.

K2 was also instrumental in providing data on polars and IPs. It managed to catch the rise from a low to high state in the polar Tau4, revealing large changes in variability as the accretion turns on (Littlefield et al. 2020). Similar large changes in the long period pre-polar V1223 Sgr were seen as indicative of mass accretion changes as part of a "magnetic syphon" (Tovmassian et al. 2018). The 69 days of monitoring of the IP FO Aqr provided the same precision timing data as decades of ground-based observations, allowing accurate spin and period measurements (Kennedy et al. 2016) revealing it was spinning down after past epochs of spin-up. The wealth of data remaining in the K2 archive will provide years of exploration and discovery.

4.3.2 Fundamental Stellar Parameters

Detached eclipsing binaries with double-lined spectra are one of the best sources of fundamental astrophysical parameters as the components are minimally distorted, can be assumed to be the same age but to have evolved independently presumably the same age, but evolved independently from one another, and the stellar parameters can be determined with high accuracy accurately determined (Torres et al. 2010). Modeling such systems with state-of-the-art modeling codes (i.e., Wilson 1979); and its derivatives ELC (Orosz & Hauschildt 2000) and PHOEBE (Prša & Zwitter 2005) that account for a range of physical circumstances allows us to determine their parameters to better than 1%. Such data can provide strict tests of stellar evolutionary models, and the distances to detached eclipsing binaries can be determined empirically to an accuracy of about 5% (Southworth 2015). The most informative comparison of stellar models with real stars is obtained when the mass, radius, temperature, and metallicity are accurately known for both stars in a binary system. If the stars differ significantly in mass and evolution, fitting both stars simultaneously for a single age provides a very stringent test of the models.

The high cadence and photometric precision of Kepler data opened new possibilities in studies of detached eclipsing binaries, as the data allowed for better precision in the derivation of light-curve-based parameters, like fractional radii or orbital inclination, and observations of various other phenomena. When combined with high-quality spectroscopic data, light-curves from Kepler enabled many systems to be analyzed and accurate masses and radii to be derived to better than 1%–3%. Such studies tended to focus on the most scientifically interesting/useful systems, including: detached eclipsing binaries in nearly circular orbits used to

calibrate M–L–R–T relationships, binaries with low-mass main sequence components to probe the larger than predicted stellar radii observed in such systems,
eclipsing binaries with intrinsic stellar pulsations which allow the comparison of
fundamental parameters derived from binarity to those derived from asteroseismic
scaling relations, and systems showing evidence of tertiary companions.

Low-mass Eclipsing Binaries

Previous studies of G–K-type eclipsing binaries have demonstrated that stars less
massive than the Sun in short period binaries exhibit major discrepancies from
standard stellar models that provide reasonable fits to similar stars in long-period
orbits, and therefore slower rotation (Torres et al. 2010). These low-mass stars are
up to 10% larger than their slowly-rotating counterparts, with their measured radii
falling above model isochrones in the mass–radius plane. The most common
explanation for this discrepancy is that tidal interactions in short-period systems
cause higher than normal rotational speeds, which in turn increases the level of
stellar activity (Bass et al. 2012).

The large sample size and photometric sensitivity of Kepler enabled the detection
of several eclipsing binaries with low-mass companions, including some in longer
period orbits. An analysis of KIC 6131659 by Bass et al. (2012) showed that the
system contains two stars with $M_1 = 0.922 \pm 0.007 M_\odot$, $R_1 = 0.880\,0 \pm 0.002\,8 R_\odot$,
and $M_2 = 0.685 \pm 0.005 M_\odot$, $R_2 = 0.639\,5 \pm 0.006\,1 R_\odot$. However, neither star is
bloated and both fit nicely on the same theoretical mass–radius relation, as seen in
Figure 4.8. In conjunction with the low stellar activity evidenced by the minimal out-
of-eclipse variability (<0.2%) in the light-curve of KIC 6131659, this system
supports the idea that the observed disparity between observational and theoretical
stellar radii is caused by stellar activity. The primary star in the Kepler-16 system,
which consists of two low-mass stars in a 41 day eclipsing binary orbited by a
circumbinary planet (Doyle et al. 2011); similarly fits on the model isochrone in the
mass–radius diagram shown in Figure 4.8, while the secondary star does not. The
out-of-eclipse modulation in the light-curve of Kepler-16 was found to be ~1%,
perhaps suggestive of an activity level threshold needed to inflate low-mass stars that
is not reached in long-period systems.

Doppler Beaming

The precision of Kepler photometry also allowed for the detection and characterization of binary systems through out-of-eclipse flux variations in the light-curve.
Ellipsoidal variations, cosine-like variations in the light-curve due to the changing
projected cross-section of a tidally distorted star, and the reflection effect, where the
light emitted by one component is scattered off of or thermally re-emitted from the
dayside of the other component, are two well-known effects caused by the proximity
of two stars in a binary system (see Figure 4.9). While these effects have long been
observed and modeled in binary systems, they could be detected with amplitudes of
10–100 parts per million by Kepler and used to search for low-mass non-transiting
stellar and planetary companions (Faigler & Mazeh 2011; Faigler et al. 2015); as
well as estimate the mass of such companions (Mazeh et al. 2012).

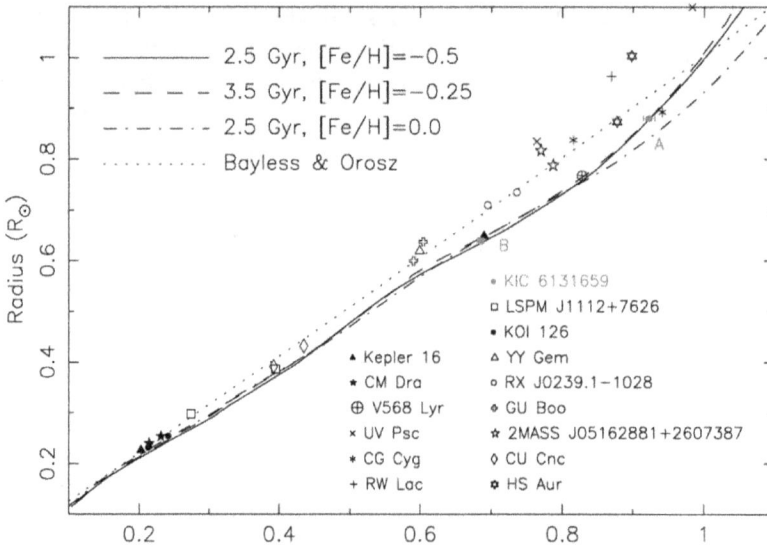

Figure 4.8. A mass–radius diagram for well-characterized low-mass stars in eclipsing binaries. Evolutionary models from Dotter et al. (2008); Bayless & Orosz (2006) and the empirical relation derived by Bayless & Orosz (2006); Dotter et al. (2008) are also shown. Most of the low-mass stars with well-measured masses and radii are above the isochrones, meaning the stars are larger than expected given their masses. The location of the components in KIC 6131659 are shown in red with the primary denoted by "A" and the secondary denoted by "B." Reproduced from Bass et al. (2012). © 2012. The American Astronomical Society. All rights reserved.

A similar effect, Doppler boosting or relativistic beaming, is the result of a primary star's spectrum being Doppler shifted as it orbits, modulating the photon emission rate and beaming the photons roughly in the direction of motion (Bloemen et al. 2011). As the star approaches and recedes in its orbit, it appears to brighten and dim with an amplitude that is proportional to its radial velocity. This effect had only been detected in a few binaries due to its small amplitude, but was predicted to be seen by Kepler (Loeb & Gaudi 2003; Zucker et al. 2007). Indeed, the first detection of Doppler boosting occurred in the light-curve of KOI-74, a 9400 K early A-star with a hot, compact companion in a 5.2 day orbit (Rowe et al. 2010). van Kerkwijk et al. (2010) measured fractional amplitudes of $A_1 = (1.082 \pm 0.013) \times 10^{-4}$ and $A_2 = (1.426 \pm 0.018) \times 10^{-4}$ in the out-of-eclipse light-curve due to Doppler boosting and ellipsoidal light variations, respectively. They used the amplitude of the beaming effect to measure a radial velocity amplitude (14.7 ± 1.0 km s^{-1}) from photometry for the first time, and inferred a secondary mass of $0.22 \pm 0.03 M_\odot$. Based on the mass of the secondary and the light-curve of KOI-74, the companion was found to be a low-mass white dwarf. While Doppler boosting has not been detected in all of them, roughly 10 other low-mass, thermally bloated, hot white dwarfs have been found orbiting F/A-type primaries using Kepler photometry (Rowe et al. 2010; van Kerkwijk et al. 2010; Carter et al. 2011; Breton et al. 2012; Rappaport et al. 2015; Matson et al. 2015; Faigler et al. 2015). These systems make up nearly half of all known short-period

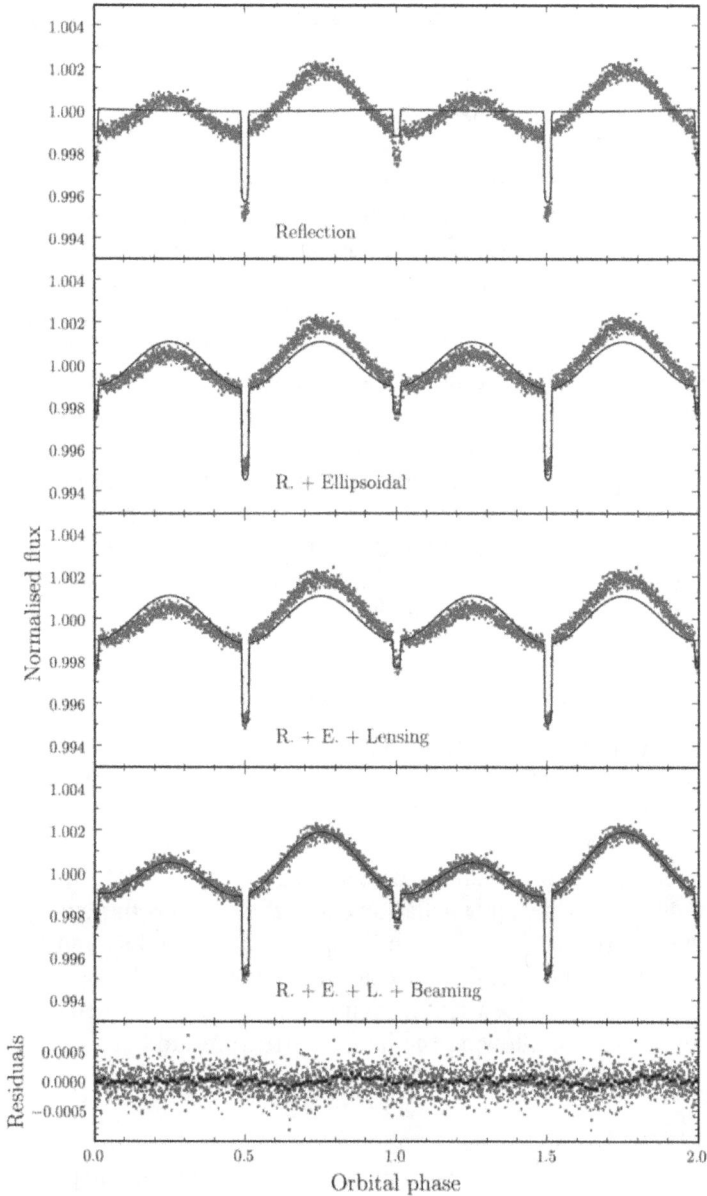

Figure 4.9. Phase-folded light-curve of KPD 1946+4340 (green, data points grouped by 30) and the best-fitting model from Bloemen et al. (2011; black). In the top panel, only the eclipses and reflection effects are modeled. In the second panel, ellipsoidal modulation is added. In the third panel, gravitational lensing is taken into account as well, which affects the depth of the eclipse at orbital phase 0.5. The bottom panels show the full model—taking into account Doppler beaming—and the residuals (grouped by 30 in green and grouped by 600 in black). Reproduced from Bloemen et al. (2011) by permission of Oxford University Press on behalf of the Royal Astronomical Society.

pre-He white dwarfs in eclipsing binary systems, and represent the end product of mostly stable mass transfer from the white dwarf progenitor on to the current primary star (Maxted et al. 2014).

Doppler beaming was also detected in KPD 1946+4340, a short-period eclipsing binary consisting of a subdwarf B-star and a white dwarf (Bloemen et al. 2011). Primary and secondary eclipses, ellipsoidal modulation, and Doppler beaming were all detected in the Kepler light-curve of this system as shown in Figure 4.9; and had to be taken into account when modeling the system. This system highlights some of the consequences of the extreme precision of Kepler photometry, as effects like Doppler beaming, reflection/scattering of light, and gravitational lensing are no longer lost in the noise, but can be measured and used to constrain stellar parameters. For instance, the observed Doppler beaming amplitude for KPD 1946+4340 was found to be in agreement with the amplitude expected from spectroscopic radial velocity measurements, resulting in masses of $0.59 \pm 0.02 M_\odot$ and $0.47 \pm 0.03 M_\odot$ for the white dwarf and subdwarf, respectively. While the theoretical estimates presented by Bloeman et al. (2011) were enough to confirm the effect of Doppler beaming and its agreement with the expected radial velocity amplitude, it clearly indicates the need to incorporate effects like Doppler beaming into rigorous eclipsing binary models in order to derive accurate stellar parameters (Prša et al. 2012).

4.3.3 Eclipsing Binaries with Pulsating Components

Most eclipsing binary light-curves in Kepler show variability in addition to their characteristic eclipses. Such variations are often due to stellar activity, but they can also be caused by intrinsic stellar pulsations. As discussed in Section 4.2 (Asteroseismology), the photometric precision, duty cycle, and long timespan of Kepler enabled the unprecedented detection and analysis of stellar pulsations. Asteroseismology uses stellar oscillations to probe below the surface of a star, testing models of stellar evolution and improving our understanding of internal rotation, convection, and chemical mixing. Fundamental stellar parameters derived from eclipsing binaries can be used to minimize parameter degeneracies and further constrain stellar models, allowing the interior structures and evolutionary status of stars to be examined in detail.

Asteroseismic Scaling Relations
In red-giant stars with solar-like oscillations, the internal structure and properties of the star can be inferred very precisely ($<5\%$) from their global oscillation modes. The most commonly used asteroseismic method is the derivation of stellar masses and radii using scaling relations based on the Sun and the assumption that stars are homologous (Rawls et al. 2016). These scaling relations for pressure modes have been used to derive stellar parameters for the 16,000 oscillating red-giants observed by Kepler. However, in order to ensure accurate and precise stellar parameters it is important to test asteroseismic methods with independent methods. As double-lined spectroscopic eclipsing binaries are the only stars for which accurate, model-independent masses and radii can be determined, eclipsing binaries with an

oscillating red-giant component are ideal test beds for asteroseismic scaling relations. Studies of individual systems, such as those by Frandsen et al. (2013) and Rawls et al. (2016); have found that the dynamical masses and radii of the binary agree with those from asteroseismic methods within uncertainties. However, Gaulme et al. (2016) showed that the stellar masses and radii for a sample of red-giant stars in eclipsing binaries were systematically overestimated when the asteroseismic scaling relations were used.

More recently, Brogaard et al. (2018) found agreement between dynamical and asteroseismic masses and radii of three eclipsing binary systems when applying a theoretical correction factor to the average frequency spacing measured for the red-giants, but significant overestimations when compared with a larger sample. Comparisons between the dynamical and asteroseismic mass and radius estimates for the ten red giant stars in double-lined spectroscopic binaries analyzed by Brogaard et al. (2018) are shown in Figure 4.10. The solid squares show their dynamical eclipsing binary measurements, while those of Gaulme et al. (2016) are shown as open squares. Open circles are masses and radii derived from simple asteroseismic scaling relations, while the other symbols represent scaling relations from Sharma et al. (2016) and Rodrigues et al. (2017) using different reference quantities. As noted, some stars show agreement between the dynamical and corrected asteroseismic measures while others suggest significant overestimates of the asteroseismic measures. Themeßl et al. (2018) similarly derived consistent stellar parameters from dynamical and asteroseismic methods when the measured oscillations were corrected for surface effects and the metallicity, temperature, and mass dependence of the scaling relations were taken into account. These results demonstrate the ongoing efforts to understand the pulsations observed by Kepler and connect them to a wide range of stellar properties in order to improve the accuracy of asteroseismic scaling relations.

Classical Pulsators

Hundreds of classical pulsators, such as γ Doradus and δ Scuti stars, have also been detected by Kepler. These A- and F-type stars can pulsate in radial and non-radial high-order gravity modes (γ Doradus), low-order gravity modes and pressure modes (δ Scuti), or both (hybrid; Uytterhoeven et al. 2011). Asteroseismology of these stars has been hindered by the difficulty of mode identification, rotational splitting, combination frequencies, and mismatches between observed modes and theoretical predictions (Gaulme & Guzik 2019). For classical pulsators in eclipsing binaries, dynamical mass and radius measurements impose powerful constraints on the pulsation properties and test the modeling of complex dynamical processes occurring in their interiors. In the case of δ Scuti stars, the measurement of a large number of frequencies in a star of known mass and radius aides in the identification of pulsation modes, as does constraining the inclination and stellar rotation (Murphy 2018). For example, Southworth et al. (2011) separated pulsation signatures in the light-curve of KIC 10661783 from variations due to binarity (eclipses, reflection effect, and ellipsoidal effect) to find 68 frequencies, with at least 55 that can be attributed to pulsation modes. Further analysis by Lehmann et al. (2013) determined

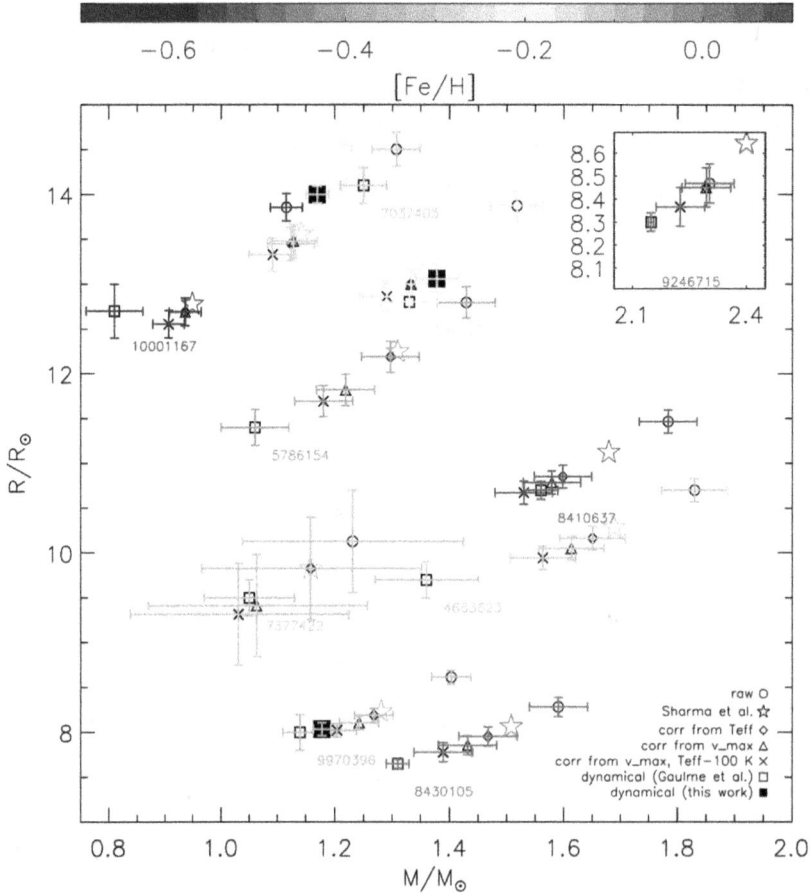

Figure 4.10. Comparisons between dynamical and asteroseismic mass and radius estimates for ten red giant stars in double-lined spectroscopic eclipsing binary systems. Big solid squares are the dynamical eclipsing binary measurements by Brogaard et al. (2018), while open squares are dynamical eclipsing binary measurements from Gaulme et al. (2016). Open circles are estimates based on the simple asteroseismic scaling relations with $f_{\Delta \nu} = f_{\nu_{max}} = 1$. Star symbols represent scaling estimates with $f_{\Delta \nu}$ from Sharma et al. (2016). Triangles represent scaling estimates with $f_{\Delta \nu}$ from Rodrigues et al. (2017) using ν_{max} as a reference, crosses are the same, but with the observed T_{eff} reduced by 100 K, while diamonds are the same corrections but using T_{eff} as the reference. Reproduced from Brogaard et al. (2011); Brogaard et al. (2018), by permission of Oxford University Press on behalf of the Royal Astronomical Society.

KIC 10661783 to be a short-period Algol-type system with the lowest mass-ratio ever observed and δ Scuti oscillations attributable to the primary.

Similarly, Hambleton et al. (2013) combined Kepler photometry and ground-based spectroscopy of KIC 4544587, a short-period eccentric eclipsing binary system showing rapid apsidal motion, to derive the absolute masses and radii of the components and show that the primary and secondary reside within the δ Scuti and γ Doradus instability regions, respectively. Frequency analysis of the pulsations revealed 31 modes, 14 in the gravity mode region, and 17 in the pressure mode region. Of the 14 gravity modes, 8 occur at integer multiples of the orbital frequency, indicating they are tidally excited oscillations or stellar pulsations that have been excited by the tidal forces of the companion star. Prior to Kepler, only a handful of observations demonstrating tidal resonance had been detected.

The analysis of Kepler light-curves also led to the discovery that many pulsating stars are γ Doradus/δ Scuti hybrids. A small number of hybrids were known previously, but Kepler lowered the detection threshold such that when Uytterhoeven et al. (2011) characterized a sample of 750 A- through F-type pulsators, 27% were δ Scuti stars, 13% were γ Doradus stars, and 23% were hybrids, while the remaining stars were other types of variables. Hybrid pulsators allow the full stellar interior to be studied, as gravity modes probe the core and pressure modes have higher amplitudes in the outer envelope.

Maceroni et al. (2014) analyzed KIC 3858884, an eclipsing binary system with deep eclipses, a complex pulsation pattern with frequencies typical of δ Scuti stars, and a highly eccentric orbit. Photometry from Kepler and high-resolution spectro-scopy yielded binary components with similar masses (1.88 and $1.86M_\odot$) and effective temperatures (6800 and 6600K), but different radii (3.45 and $3.05R_\odot$). The detection of low frequencies in the range of high order gravity modes indicated that the secondary component is a hybrid pulsator. The pulsation signature and eclipses can be seen in the section of the Kepler light-curve shown in Figure 4.11. The analysis of KIC 3858884 also highlights the challenges of the unprecedented detail of Kepler data. Due to the high eccentricity and deep eclipses of the system the eclipses and pulsations could not be disentangled with current methods, instead requiring the development of a new disentangling procedure that took into account the effect of eclipses on the observed pulsation pattern (Maceroni et al. 2014).

The probing capability of gravity and mixed modes have also enabled both surface and near-core rotation rates to be measured for an increasing number of hybrid pulsators, providing insight into their angular momentum history. For example, Schmid et al. (2015) detected rotational splitting in gravity and pressure modes in the double-hybrid binary system KIC 10080943, and used them to determine a rough estimate of the core-to-surface rotation rates of the two components. Similarly, Guo et al. (2019) characterized KIC 4142768, an eclipsing binary with two evolved A-type stars in an eccentric orbit with a period of 14 days. They determined that the surface rotation rates are only one-fifth of the pseudo-synchronous rate of the eccentric orbit, comparable to the very slow rotation rate inferred at the convective core boundary.

Heartbeat Stars
As highlighted throughout this chapter, Kepler light-curves display incredible detail that have revolutionized several areas of stellar astrophysics and discovered variability never before identified in stars. An example of the latter is the discovery of a new class of eccentric ellipsoidal binary stars or heartbeat stars. These systems show a sharp increase in brightness at the closest approach of the two binary star components due to strong gravitational distortions and heating. KOI-54 (KIC 8112039), a highly eccentric binary system composed of two A-type stars, was the

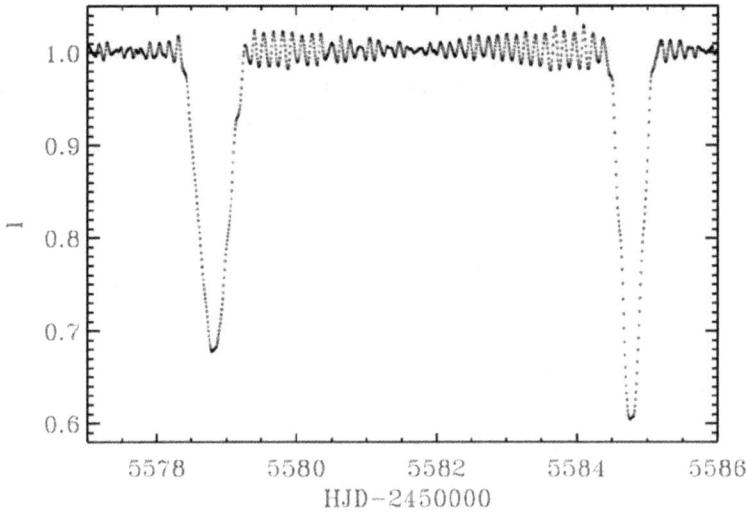

Figure 4.11. A section of the normalized and de-trended Kepler light-curve of KIC 3858884, an eccentric eclipsing binary with a hybrid δ Scuti pulsator, showing the eclipses and interlaced pulsations. Credit: Maceroni et al. (2014), reproduced with permission © ESO.

first system discovered whose light-curve is predominantly characterized by these dynamic distortions. Welsh et al. (2011) showed that the combination of the tidal ellipsoidal variation and the irradiation effect creates a periodic 0.7% brightening seen every 41.8 days. A section of the Kepler light-curve and the phase folded data can be seen in the top of Figure 4.12. The close periastron passage is also responsible for exciting stellar pulsations, with at least 30 modes identified by Welsh et al. (2011). The two largest pulsations, at the 90th and 91st harmonics of the orbital frequency, beat against each other producing the modulation envelope seen in the light-curve.

Thompson et al. (2012) identified 17 additional tidally distorted systems in the Kepler data, referring to them as heartbeat stars because the shape of the variability is reminiscent of an echocardiogram. They determined an orbital period from the light-curve and used the model of tidal distortions presented in Kumar et al. (1995) to fit the eccentricity, angle of periastron, and orbital inclination to the shape of each light-curve. Since then many more heartbeat stars have been identified, including 17 red giant heartbeat stars by Beck et al. (2014); six by Smullen & Kobulnicky (2015); and 19 by Shporer et al. (2016). While heartbeat stars are not necessarily eclipsing, their light-curves clearly indicate their binary nature and they are therefore included in the Kepler eclipsing binary catalog, which currently lists 170 heartbeat stars (Kirk et al. 2016).

Approximately 20% of heartbeat stars also demonstrate tidally induced pulsations, including KOI-54 (Welsh et al. 2011; Fuller & Lai 2012; Burkart et al. 2012; O'Leary & Burkart 2014) and KIC 4544587 (Hambleton et al. 2013). Tidally induced pulsations differ from self-excited and stochastically excited pulsations as their amplitudes and phases can be predicted from linear theory and compared with

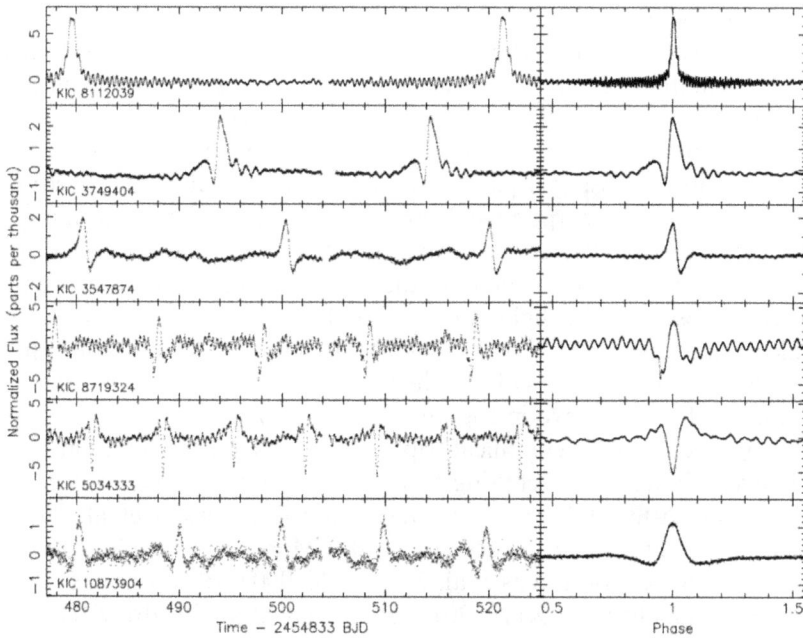

Figure 4.12. Time series of a selection of Kepler heartbeat star light-curves (left panel) and phase folded data (right panel). The phase folded data clearly depict the oscillations that are integer multiples of the orbital frequency: tidally induced pulsations. Reproduced from Kirk et al. (2016). © 2016. The American Astronomical Society. All rights reserved.

observations (Fuller 2017). Heartbeat stars can also be used to study tidal interactions between stars and estimate tidal dissipation rates. However, only observations from Kepler provide the temporal resolution to allow for a dedicated seismic analysis for these stars. Examples of Kepler light-curves of heartbeat stars are shown in Figure 4.12. Time series data is shown in the left panel, while phase folded data is shown on the right. The phase folded data shows that oscillations in the light-curve are integer multiples of the orbital frequency, and therefore due to tidally induced pulsations (Kirk et al. 2016).

4.3.4 Triples and Higher-order Multiples

Investigations of the solar neighborhood have shown that 46% of stars occur in binary or higher-order multiple star systems (Raghavan et al. 2010); with 13% having three or more stars (Tokovinin 2014). Despite being rare, such systems offer valuable insight into star formation processes and the dynamical evolution of multi-body systems. Roughly 200 triple star candidates have been discovered using Kepler data, as tertiary companions to eclipsing binaries can be detected if they transit one or both stars in the binary or if they perturb the binary enough to cause deviations in the observed times of the primary and secondary eclipses (Rappaport et al. 2016).

Multi-eclipsing Systems

Multiple star systems with eclipsing features are very rare, as they require a fortuitous orientation of the system such that the companions are seen to eclipse one another. Once again, the precision and time-span of Kepler enabled the discovery of a new category of eclipsing system through the detection of KOI-126, the first triply eclipsing triple star system. Carter et al. (2011) used a photo-dynamical model that incorporated Newton's equations of motion, along with a general relativistic correction to the orbit of the inner binary, integrated over time to determine the positions and velocities of the three bodies in order to fit the periodic, superposed eclipses in the light-curve of KOI-126. The system is a compact hierarchical triple composed of two low-mass stars (0.241 and 0.212 $M_\odot$) in a close binary orbiting a third star (1.347 $M_\odot$). The light-curve and photodynamical model of Carter et al. (2011) are shown in Figure 4.13.

Twelve other compact hierarchical triple systems have been found in Kepler/K2 data, 11 of which are triply eclipsing systems: KIC 5952403 (Derekas et al. 2011; Borkovits et al. 2013); KIC 6543674 and 7289157 (Slawson et al. 2011); KIC 2856960 (Armstrong et al. 2012; Lee et al. 2013; Marsh et al. 2014); KIC 5255552, 6964043, and 7668648 (Borkovits et al. 2015); KIC 9007918 (Borkovits et al. 2016); KIC 4150611 (Shibahashi & Kurtz 2012; Hełminiak et al. 2017b); KIC 249432662 (Borkovits et al. 2019); and HD 144548 (Alonso et al. 2015). The compact nature of these triply eclipsing systems enables the derivation of fundamental stellar and orbital parameters that constrain models of stellar evolution and allow dynamical interactions to be studied, as interactions between the stars can change the orbital elements significantly over short timescales (Borkovits et al. 2018). Hierarchical triples are believed to play an important role in close-binary formation as the outer component may remove angular momentum from the inner pair through a combination of Kozai cycles and tidal friction (Fabrycky & Tremaine 2007). Kozai cycles periodically raise the eccentricity of the close binary, while the tidal friction efficiently dissipates the orbital energy during close pericenter passages. As a result, the inner close binary undergoes an orbital period decrease and forms a tidal-locked detached binary with a short orbital period (Naoz & Fabrycky 2014). Hierarchical triples have also been hypothesized to play a role in the formation of peculiar objects, such as stellar mergers, blue stragglers, low-mass X-ray binaries, and binary pulsars (Borkovits et al. 2019).

Several higher order systems have also been detected in Kepler and K2 data, including quadrupole and quintuple star systems. Lehmann et al. (2012) used spectroscopic observations to derive fundamental parameters of KIC 4247791, an eclipsing binary with a periodic shallow dip in the light-curve, and determine that the system is likely a quadrupole system consisting of two eclipsing binaries orbiting a common center of mass. Since then, four quadrupole systems with doubly eclipsing binaries have been confirmed with Kepler/K2 data, including EPIC 220204960 (Rappaport et al. 2017) and 219217635 (Borkovits et al. 2018), which have been shown to be gravitationally bound systems. Kepler also observed KIC 4150611, a quintuple system that was observed to have eclipses at four different periods, with three stars resolvable in high-resolution imaging. Hełminiak et al. (2017a) used the

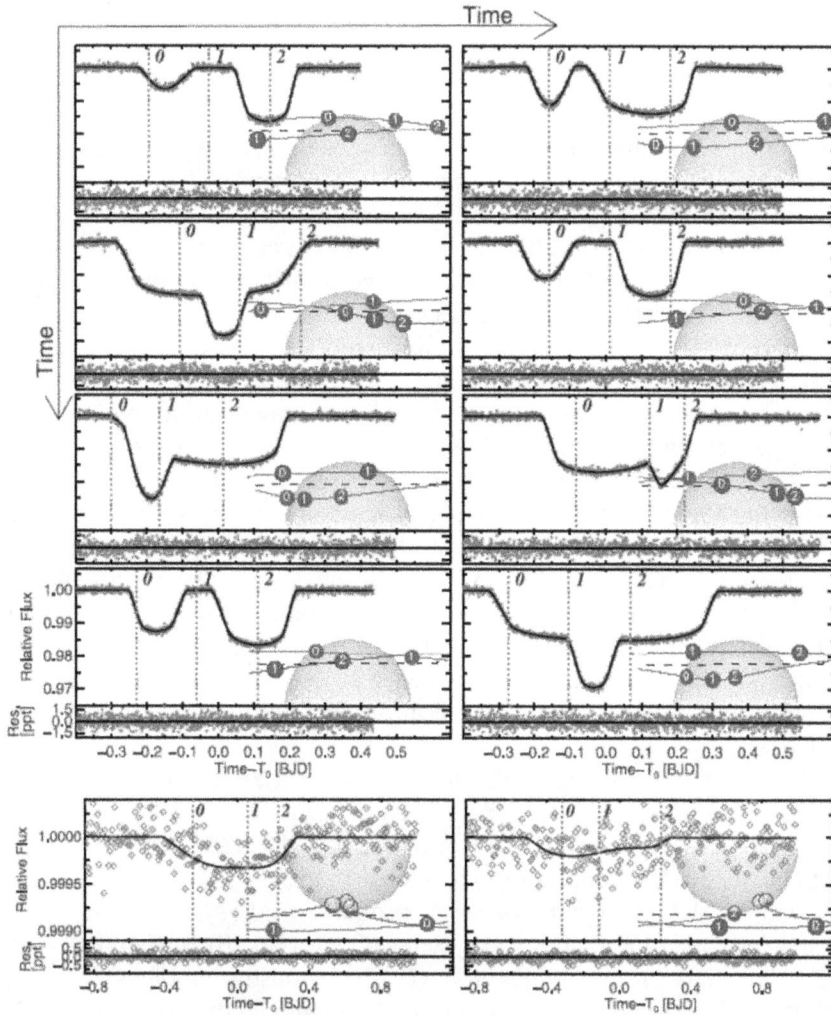

Figure 4.13. Best-fit dynamical model (solid black line) fitted to the light-curve of KOI-126 (gray points). The top eight panels are short-cadence data observed during the passage of the inner binary components in front of the more massive star from the perspective of the Kepler spacecraft. The bottom two panels are long-cadence data observed when the inner binary is behind the more massive tertiary star. The inlaid diagrams show, to scale, the orbits of the $0.241\,3M_\odot$ (green) and $0.212\,7M_\odot$ (blue) binary stars relative to the $1.347M_\odot$ tertiary (yellow). The dashed orbit is that of the center of mass of the inner binary. The numbered circles give the locations of the individual binary stars at the times indicated by the vertical dotted lines, corresponding to the numeric index (0–2). Reproduced from Carter et al. (2011).

Kepler light-curve and spectroscopic observations to determine that the system consists of a single, hybrid pulsator (A-component) orbited by an eclipsing binary of two K/M-type stars and a B-component that is itself a pair of G-type stars. An additional quintuple system, EPIC 212651213/21651234, containing two eclipsing binaries in a hierarchical quadrupole that is orbited by a single star, has been found with K2 data (Rappaport et al. 2016).

Eclipse Timing Variations

In eclipsing binaries, the time at which the eclipses occur is expected to be constant. However, apsidal motion, mass transfer, or the presence of a third body in the system can give rise to slight changes in the orbital period, which changes the time interval between consecutive eclipse events (Borkovits et al. 2016). These variations, called eclipse timing variations (ETVs), can be measured as small changes in the individual times of eclipse with respect to a linear ephemeris that assumes a Keplerian orbit. This well-known technique has been used to detect and study multi-star systems from the ground, however, the continuous observations by Kepler made it much easier to detect trends in eclipse timing variations and provided thousands of eclipsing binary light-curves to examine. This method has also been applied to detect and measure the orbits and masses of planetary systems.

When a third body perturbs the center-of-mass of a binary system the light from the system can be delayed along the light-of-sight, causing eclipses to appear earlier or later than expected. This light-travel time effect (LTTE) results in a quasi-sinusoidal variation when the observed minus calculated ($O - C$) eclipse times are plotted over time. Analyzing $O - C$ diagrams is a powerful tool for investigating triple star systems, as they can reveal evidence of long-term period changes, apsidal motion, mass transfer, starspots, and pulsations, in addition to the light-time delays associated with a third star.

The first systematic search for eclipse timing variations in Kepler data was conducted by Gies et al. (2012); who analyzed the light-curves of 41 short-period (<7 days) eclipsing binaries and identified preliminary long-term trends in 14 systems. They later extended their analysis to include all 17 quarters of Kepler data and were able to derive orbital elements for seven probable triple systems, with long term trends visible in an additional seven systems (Gies et al. 2015). Figure 4.14 depicts sample $O - C$ diagrams from Gies et al. (2015) demonstrating the effects of a third body (KIC 5513861; top left), long term trends potentially due to a third body (KIC 10736223; top right), random scatter with no visible trends (KIC 5738698; bottom left), and opposite trends in the primary and secondary eclipse times due to apsidal motion (KIC 4544587; bottom right). "Random walk" style variations due to starspots are also seen in several of the $O - C$ diagrams, and are typically more prominent in the cooler secondary component (x symbols in Figure 4.14). These variations are caused by the change in the surface intensity and center of light over the visible hemisphere.

Several subsequent studies examined larger samples of Kepler eclipsing binaries to detect possible triples and conduct preliminary statistics. Rappaport et al. (2013) searched for ETV signals in more than 2000 eclipsing binary light-curves through quarter 17, detecting 39 candidate triple systems with periods between 48–960 days. Conroy et al. (2014) similarly determined eclipse times for 1279 binaries in the Kepler eclipsing binary catalog with periods less than ~3 days. They detected 236 systems with eclipsing timing variations consistent with a third body, resulting in a

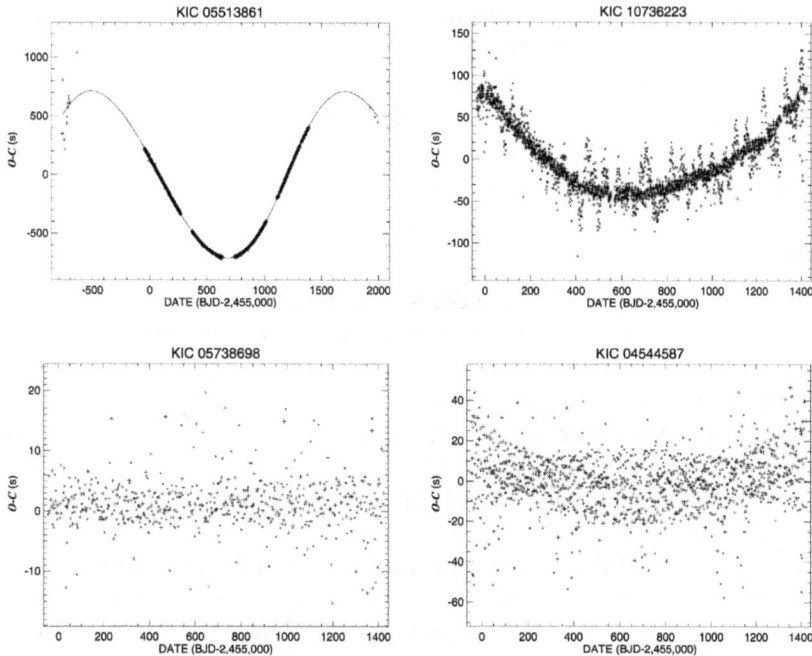

Figure 4.14. Observed minus calculated ($O - C$) eclipse times relative to a linear ephemeris of four eclipsing binaries. reproduced from Gies et al. (2015). © 2015. The American Astronomical Society. All rights reserved. The primary and secondary eclipse times are indicated by + and x symbols, respectively. Eclipse timing variations due to a short period tertiary companion (~6 yr) with a preliminary orbital solution are shown for KIC 5513861 (top left), while large parabolic variations observed in both the primary and secondary eclipses indicative of a longer period companion are seen in KIC 10738223 (top right). Variations due to starspots can also be seen in the eclipse timings of the cooler secondary component of KIC 10738223. Other eclipse timing variations appear as scatter diagrams with no trends visible in either the primary or secondary eclipse measurements, as shown in KIC 5738698 (bottom left). KIC 4544587 (bottom right) shows opposite trends for the primary and secondary eclipses as expected for apsidal motion, as well as additional scatter due to resonant pulsations.

third body rate of ~20%. Orosz (2015) also searched for ETV signals among 1322 Kepler eclipsing binaries with periods longer than ~1 day, corroborating that the occurrence rate of triples is on the order of 20%.

Borkovits et al. (2016) then performed a comprehensive study of the full sample of Kepler eclipsing binaries, finding evidence for 222 third bodies. They identified and fit eclipse timing variations for systems showing LTTE and/or dynamical effects, resulting in estimates of the masses and three-dimensional orbital characteristics for 62 systems with LTTE and dynamical delays. For the remaining 160 systems dominated by LTTE effects, they determined mass functions and orbital parameters. The sample of 222 triples have outer periods ranging from approximately 30–3000 days, with a deficit of systems that have binary periods less than ~1 day and outer periods between 50–200 days, in agreement with Conroy et al. (2014). Borkovits et al. (2016) also found a peak in the mutual inclination of the triple system and the inner binary at approximately 40°, which is consistent with predictions of models of Kozai cycles with tidal friction driven by a tertiary star. Such trends, as well as additional studies of eclipse timing variations in individual systems in the Kepler field, provide important constraints on theories of formation and evolution of hierarchical triples and close binary systems.

4.4 Star Clusters

During both the Kepler mission and the K2 mission, the Kepler spacecraft observed numerous star clusters, from young to old and open to globular. Most of these objects were targeted as part of the K2 mission, but four open clusters (NGC 6819, NGC 6791, NGC 6811, and NGC 6866) were observed in the original Kepler field. In total, the telescope observed two star-forming associations (age <10 Myr), 18 open clusters (1 Myr to 8 Gyr), and nine globular clusters ($\geqslant$ 11 Gyr). The compilation of clusters is provided in Table 4.1; along with the observing campaign and estimated age range.

For the more crowded clusters, observations were carried out in "superstamp" mode, in which all imaging data for a large (~1°), contiguous region was downloaded from the space craft. Superstamp regions include the Kepler open clusters NGC 6819 and NGC 6791, the K2 open cluster M35, the Lagoon nebula young star-forming region, as well as a number of globular clusters. For these targets, no light-curves were output by the Kepler project pipeline, and hence time series light-curves were generated by members of the astronomical community.

Regardless of imaging mode, science derived from Kepler's clusters was comprehensive and ground breaking. It encompassed many areas, from stellar to planetary science. In this section, we focus on the five key areas of stellar rotation, exoplanet discovery, eclipsing binary systems, young stellar objects, and asteroseismology.

4.4.1 Stellar Rotation

With the high precision and cadence, as well as long time baseline of Kepler photometry, rotation periods for large samples of stars are easily extracted from light-curves by identifying repeating patterns via periodogram or autocorrelation analysis (McQuillan et al. 2014). The variability is due to rotation of a non-axisymmetric distribution of spots on the stellar surface. Tens of thousands of spotted stars were observed throughout the lifetime of Kepler, enabling ensemble studies of angular momentum evolution from the pre-main sequence to intermediate ages and older. A key activity has been the comparison of theoretical models of angular momentum evolution in stars with the observed rotation sequences, as a function of both mass, age, and (on the pre-main sequence) presence or absence of a circumstellar disk.

Gyrochronology, or the determination of stellar ages based on rotation rates (Barnes 2003); was popularized over a decade ago with ground-based photometry monitoring of solar-like stars in clusters. Prior to the space-based photometry era, it

Table 4.1. Clusters and Associations Observed during the Kepler and Missions

Region	Campaign	Region Type	Age [Gyr]	Reference
NGC 6819	Kepler	Open cluster	2	Brewer et al. (2016)
NGC 6791	Kepler	Open cluster	8	García-Berro et al. (2010)
NGC 6811	Kepler	Open cluster	1	Curtis et al. (2019)
NGC 6866	Kepler	Open cluster	0.5–0.8	Kharchenko et al. (2005); Bostancı et al. (2015)
M35	K2 C0	Open cluster	0.13–0.18	Barrado y Navascués et al. (2001); Meibom et al. (2009)
NGC 2158	K2 C0	Open cluster	~2	Carraro et al. (2002)
M4	K2 C2	Globular cluster	12.5	Hansen et al. (2004)
M80	K2 C2	Globular cluster	12.5	Forbes & Bridges (2010)
Upper Scorpius	K2 C2, C15	Association	0.005–0.01	Preibisch et al. (2002); Pecaut et al. (2012)
ρ Ophiuchus	K2 C2	Open cluster	0.001–0.002	Wilking et al. (2008)
Pleiades	K2 C4	Open cluster	0.125	Bouvier et al. (2018)
Hyades	K2 C4, C13	Open cluster	0.75	Brandt & Huang (2015a)
M67	K2 C5, C16, C18	Open cluster	4	Barnes et al. (2016)
Praesepe	K2 C5, C16, C18	Open cluster	0.79	Brandt & Huang (2015b)
Ruprecht 147	K2 C7	Open cluster	3	Curtis et al. (2013)
NGC 6717	K2 C7	Globular cluster	12.5	VandenBerg et al. (2013)
NGC 6530	K2 C9	Open cluster	0.002	Prisinzano et al. (2005)
M9	K2 C11	Globular cluster	~12	Koleva et al. (2008)
M19	K2 C11	Globular cluster	~12	Forbes & Bridges (2010)
NGC 6293	K2 C11	Globular cluster	~13	Meissner & Weiss (2006)
NGC 6355	K2 C11	Globular cluster	~13	Camargo (2018)
Terzan 5	K2 C11	Globular cluster	~12	Ferraro et al. (2009)
NGC 1647	K2 C13	Open cluster	0.15	Hebb et al. (2006)
NGC 1746	K2 C13	Open cluster[a]	—	
NGC 1817	K2 C13	Open cluster	1	Donati et al. (2014)
Taurus	K2 C13	Association	0.002	Daemgen et al. (2015)
NGC 1750	K2 C13	Open cluster	0.2	Galadi-Enriquez et al. (1998)
NGC 1758	K2 C13	Open cluster	0.4	Galadi-Enriquez et al. (1998)
NGC 5897	K2 C15	Globular cluster	~12	Forbes & Bridges (2010)

Note.
[a] Basic data on the clusters and associations photometrically monitored as part of the Kepler and K2 missions, as presented in Cody et al. (2018). We note that the cluster status of NGC 1746 is debated; it may be an asterism of unrelated stars.

was commonly assumed that stellar spin-down as a function of age (t) proceeded according to t^n, where $n \sim 0.5$ as given by the Skumanich law (Skumanich 1972). However, this trend was only calibrated on solar-type stars with ages less than 1 Gyr.

An important first step in verifying and extending gyrochronology relations was to include rotation data from clusters in the 2–3 Gyr range. Meibom and collaborators took advantage of the presence of the ~2.5 Gyr old cluster NGC 6819 in the Kepler field of view (Meibom et al. 2015). They established that the gyrochronology models of Barnes (2003) hold at and beyond 2 Gyr, and hence can be used to derive ages of field stars to within ~10%.

With the onset of the K2 mission and its numerous cluster observations, further gyrochronology testing was carried out on main sequence stars in the ~4 Gyr old M67 cluster by Barnes et al. (2016). That work confirmed the same functional form for the period-mass/age plane down to G types. Attempts to further enlarge the rotation sample in this cluster have encountered challenges due to the very low amplitude of spot signals and relatively long rotation rates of order several weeks (Esselstein et al. 2018).

In the quest to explore stellar rotation trends down to lower mass, several benchmark clusters were observed during the K2 mission. These included the Hyades (~750 Myr) and Praesepe (~790 Myr). Douglas and collaborators compared rotation rate trends in the two clusters (Douglas et al. 2019). By recalibrating gyrochronology based on data points in Praesepe and the Sun, they determined a modified spin-down relation of $t^{0.62}$. Furthermore, they confirmed previous suspicions that rotation cannot be parameterized separately by age and mass for the later type K and M stars (Agüeros et al. 2018). What is instead found is that the rotational breaking seen among F and G stars stalls out at low mass (Curtis et al. 2019). This stall-out is seen even in the comparison of Praesepe and the Hyades, which are nominally only about 50 Myr apart in age (Douglas et al. 2019).

Extending studies to even younger ages, Rebull and collaborators derived rotation rates for over one thousand low-mass stars in the 5–10 Myr Upper Scorpius association, the 1–2 Myr ρ Ophiuchus cluster (Rebull et al. 2018); and the ~125 Myr Pleiades (Rebull et al. 2016a, 2016b; Stauffer et al. 2016). They identified and traced the rapidly rotating sequence of M dwarf stars, and then compared this with the low mass sequence see in the Praesepe at older ages (Rebull et al. 2017) and Upper Scorpius at younger ages. They found that this M dwarf sequence exhibits a uniform spin-up onto the main sequence from Upper Sco to the Pleiades. However, from Pleiades age to that of Praesepe, there is evidence of a mass-dependent spin-down. Full plots of the rotation sequences derived from K2 data for ρ Ophiuchus, Upper Sco, the Pleiades, and Praesepe are shown in Figure 4.15.

At young ages (e.g., <10 Myr), many stars still possess primordial circumstellar disks, and these are thought to influence stellar spin rate through accretion and magnetic locking mechanisms (Edwards et al. 1993; Koenigl 1991). Thus another subject of interest has been the comparison of rotation rates between similar age stars with and without disks. Using data from K2, Rebull et al. (2018) confirmed a

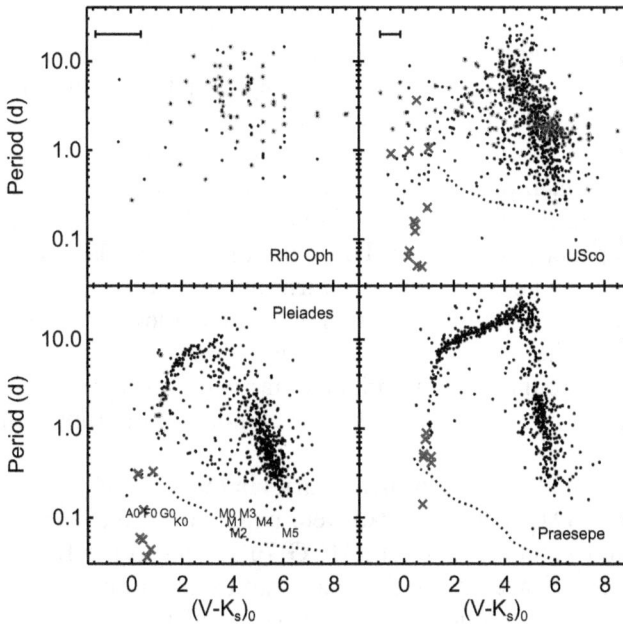

Figure 4.15. Rotation sequences for four clusters and associations observed as part of the K2 mission. Reproduced from Rebull et al. (2018). © 2018. The American Astronomical Society. All rights reserved.

correlation between infrared excess (indicating disk presence) and rotation. Specifically, stars with disks exhibit lower rotation rates than their disk-less counterparts; this correlation is particularly evident for the low-mass dwarfs with V–K colors greater than 5.0.

While indirectly related to rotation, flares in cluster members are linked to magnetic activity and hence also a topic that has been studied using K2 data. Based on examination of light-curves for stars in the Pleiades, Praesepe, and M67, Ilin and collaborators found evidence for a universal flare generation process, as indicated by similar flare frequency slopes as a function of released energy (Ilin et al. 2019).

All in all, the wealth of data from the Kepler and K2 missions has greatly expanded the number of light-curves available for active young cluster members, in particular going into the low-mass M dwarf regime. In terms of the precision, long time baseline, and uniformity across a range of masses and ages, this data set for rotation studies is unlikely to be rivaled for many years to come.

4.4.2 Eclipsing Binaries in Clusters

While rotational modulations allowed observers to trace the angular momentum evolution of a large population of stars, numerous eclipsing binaries have been uncovered in both the Kepler prime and K2 missions. These systems have been used to probe stellar structure as a function of age. They yield precise stellar radii, and often masses in cases where high resolution spectroscopic monitoring is available.

Of particular importance are eclipsing binaries in stellar clusters, as constraints on age are typically available. In combination with precisely derived mass and radius values, this information enables rigorous testing of stellar evolution models. Over a dozen eclipsing binaries have been uncovered in stellar clusters and associations targeted by the K2 mission. A list of systems is provided in Table 4.2; we only include those for which precise masses and radii are both available.

Comparison with theoretical stellar evolution tracks and isochrones has revealed numerous discrepancies between predicted and observed stellar parameters. Models typically input a mass and age, and then furnish a luminosity, effective temperature, and stellar radius. One example that highlights difficulties in matching observations to models is that of the eclipsing binary system PTFEB132.707+19.810 (Kraus et al. 2017). Kraus et al. found that the primary star is of appropriate radius but cooler and less luminous than predicted, while the secondary is appropriate luminosity but cooler and larger in radius.

Likewise, in an analysis of eclipsing binaries in the ∼125 Myr old Pleiades, David and collaborators (David et al. 2016c) were unable to fit the derived masses, radii, temperatures, and luminosities of the HCG 76 system to the Baraffe et al. (2015; BCAH15) models. However, in the mass–radius diagram, they did find broad agreement between observations and BCAH15 model predictions for a presumed cluster age of 120 Myr. The Pleiades eclipsing binary stars with masses larger than 0.5 $M_\odot$ were also consistent with 120 Myr isochrones from the PARSEC v1.2 models (Bressan et al. 2012). In contrast, several eclipsing binary systems found in K2's older Praesepe cluster field showed superior agreement with the PARSEC models (Gillen et al. 2017); particularly in their match of effective temperature to the measured masses and radii. Thus not only are there data/model discrepancies, but also model/model disagreements. Both groups (David et al. 2016c; Gillen et al. 2017) have used their K2 cluster eclipsing binary data to suggest that effective temperatures from BCAH15 are being overpredicted by several hundred Kelvin, an effect that may be related to the presence of magnetic activity and starspots.

On the pre-main sequence, an unprecedented seven eclipsing binary systems with mass and radius determinations have been discovered from K2 observations (David et al. 2016a, 2019a; Lodieu et al. 2015; Maxted & Hutcheon 2018; Kraus et al. 2015). The uniform analysis of David and collaborators indicated that traditional H–R diagram analyses of pre-main sequence stars often leads to underestimation of stellar mass by 20%–60% for stars less massive than the Sun. In general, few models are able to reproduce the observed slope of the pre-main sequence mass–radius diagram, as shown in Figure 4.16 (David et al. 2019a).

There is evidence that data/model discrepancies lessen at older ages. Torres et al. (2018) presented a G-type eclipsing binary system in the ∼2.5–3 Gyr old cluster Ruprecht 147 and found good agreement between the predictions of several stellar evolution models and the stellar properties at an age ∼2.6 Gyr. At much lower mass, an eclipsing brown dwarf companion (EPIC 219388192) was found orbiting a different Sun-like star in this same cluster, and its precisely measured radius is also consistent with the predictions of isochrones at 2.5–3 Gyr ages (Nowak et al. 2017).

Table 4.2. Parameters of Eclipsing Binaries Observed in Clusters during the K2 Mission

Star Name	Cluster/Association	M_1 ($M_\odot$)	M_2 ($M_\odot$)	R_1 ($R_\odot$)	R_2 ($R_\odot$)	Campaign number	Age (Myr)	Reference
HR 5934	Upper Sco	5.58	2.62	2.73	1.69	K2 C2	~5–10	David et al. (2019a); Maxted & Hutcheon (2018)
HD 144548	Upper Sco	0.98	0.94	1.32	1.33	K2 C2	~5–10	David et al. (2019a); Alonso et al. (2015)
USco 48	Upper Sco	0.74	0.71	1.16	1.16	K2 C2	~5–10	David et al. (2019a)
EPIC 202963882	Upper Sco	0.29	0.20	0.74	0.64	K2 C2	~5–10	David et al. (2019a)
RIK 72	Upper Sco	0.44	0.96	—	—	K2 C2	~5–10	David et al. (2019a)
UScoCTIO 5	Upper Sco	0.34	0.33	0.87	0.84	K2 C2	~5–10	David et al. (2016a, 2019a); Kraus et al. (2015)
EPIC 203710387	Upper Sco	0.12	0.11	0.43	0.42	K2 C2	~5–10	David et al. (2019a); Lodieu et al. (2015)
HCG 76	Pleiades	0.30	0.28	0.34	0.32	K2 C4	~125	David et al. (2016c)
MHO 9	Pleiades	0.41	0.17	0.46	0.32	K2 C4	~125	David et al. (2016c)
HD 23642	Pleiades	2.20	1.55	1.73	1.50	K2 C4	~125	David et al. (2016c)
AD3814	Praesepe	0.38	0.20	0.36	0.23	K2 C5	~790	Gillen et al. (2017)
AD2615	Praesepe	0.21	0.26	0.23	0.27	K2 C5	~790	Gillen et al. (2017)
AD1508	Praesepe	0.45	0.53	0.55	0.45	K2 C5	~790	Gillen et al. (2017)
WOCS 12009	M67	1.11	0.75	1.07	0.71	K2 C5	~4000	Sandquist et al. (2018)
EPIC 219394517	Ruprecht 147	1.08	1.07	1.06	1.04	K2 C7	~3000	Torres et al. (2018)
EPIC 219388192[a]	Ruprecht 147	0.99	0.03	1.01	0.10	K2 C7	~3000	Nowak et al. (2017)

Note. Kepler/K2 cluster eclipsing binary systems with mass and radius measurements. In cases where the system is not double lined, only one set of mass and radius may be listed.
[a] This system consists of a brown dwarf secondary.

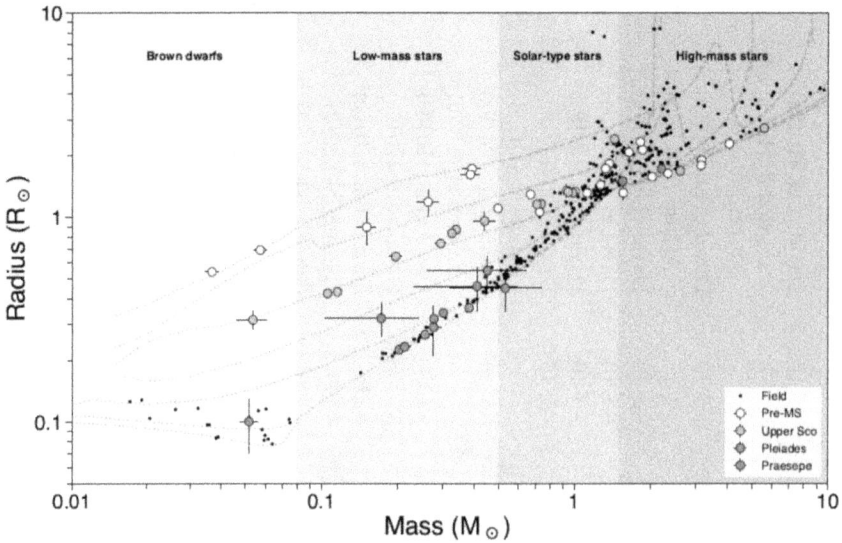

Figure 4.16. Masses and radii derived for eclipsing binary systems observed in clusters by Kepler/K2. Isochrones are taken from MIST models (dashed lines; Choi et al. 2016; Dotter 2016) and a combination of BHAC15 and COND03 (dotted lines; Baraffe et al. 2003, 2015). Reproduced from David et al. (2019a). © 2019. The American Astronomical Society. All rights reserved.

In contrast, an odd eclipsing binary system (WOCS 20009) was found in the even older K2 cluster M67 by Sandquist and collaborators (Sandquist et al. 2018). The primary star's radius was measured to be some 15% smaller than expected based on model predictions at the ~4 Gyr cluster age. In this case, however, the discrepancy does not appear to arise from an issue with the models, but rather a unique situation in which the primary is the result of a past merger of two stars. The hydrodynamics of this scenario are such that the resulting stellar merger product would appear to be only as old as the time since collision. This could explain why the radius of WOCS 20009A is more consistent with an isochronal age of 1.5 Gyr.

4.4.3 Exoplanets Discovered in Star Clusters

Smaller than the set of eclipsing binaries is the handful of transiting planets found among Kepler's clusters and associations. The Kepler mission produced the first confirmed discoveries of transiting planets in an open cluster (NGC 6811; Meibom et al. 2013), and scientists analyzing K2 data followed this up with a number of other finds. In total, 18 planets and an additional two candidates have been detected via transit; a summary of all discoveries is presented in Table 4.3.

The first two cluster exoplanets discovered with the Kepler telescope were the "mini-Neptunes" Kepler-66 b and Kepler-67 b (2.8–2.9 $R_\oplus$) in the ~1 Gyr old cluster NGC 6811 (Meibom et al. 2013). This find illustrated that relatively small planets can and do form in high-density environments subject to high-mass star radiation and dynamical perturbations. It suggested that the occurrence rate of small planets in clusters is similar to that of field stars.

Table 4.3. Confirmed and Candidate Exoplanets Discovered in Clusters Observed during the Kepler and K2 Missions

Planet Name	Cluster/ Association	R_p ($R_\oplus$)	Host SpT	Campaign number	Age (Myr)	Reference
Kepler-66 b	NGC 6819	2.8	G0	Kepler	~1000	Meibom et al. (2013)
Kepler-67 b	NGC 6819	2.9	G9	Kepler	~1000	Meibom et al. (2013)
K2-33 b	Upper Sco	5.8	M3	K2 C2	5–10	David et al. (2016b); Mann et al. (2016b)
K2-25 b	Hyades	3.4	M4.5	K2 C4	~750	David et al. (2016c); Mann et al. (2016a)
K2-309 b	Group 29	10.2	K0–K1	K2 C4	~23	David et al. (2019b)
K2-309 c	Group 29		K0–K1	K2 C4	~23	David et al. (2019b)
K2-309 d	Group 29		K0–K1	K2 C4	~23	David et al. (2019b)
K2-95 b	Praesepe	3.7	M3	K2 C5	~790	Mann et al. (2017); Pepper et al. (2017); Obermeier et al. (2016); Libralato et al. (2016)
K2-100 b	Praesepe	3.5	F8	K2 C5	~790	Mann et al. (2017); Libralato et al. (2016)
K2-101 b	Praesepe	2.0	K3	K2 C5	~790	Mann et al. (2017)
K2-102 b	Praesepe	1.3	K4	K2 C5	~790	Mann et al. (2017)
K2-103 b	Praesepe	2.2	K9	K2 C5	~790	Mann et al. (2017)
K2-104 b	Praesepe	1.9	M1	K2 C5	~790	Mann et al. (2017); Libralato et al. (2016)
EPIC 211901114 b	Praesepe	9.6	M3	K2 C5	~790	Mann et al. (2017)
K2-231 b	Ruprecht 147	2.5	G4–G5	K2 C7	~3000	Curtis et al. (2018)
K2-136 b	Hyades	1.0	K5.5	K2 C13	~750	Mann et al. (2018); Ciardi et al. (2018); Livingston et al. (2018)
K2-136 c	Hyades	2.9	K5.5	K2 C13	~750	Mann et al. (2018); Ciardi et al. (2018); Livingston et al. (2018)
K2-136 d	Hyades	1.5	K5.5	K2 C13	~750	Mann et al. (2018); Ciardi et al. (2018); Livingston et al. (2018)
HD 283869 b	Hyades	2.0	K7	K2 C13	~750	Vanderburg et al. (2018)
K2-264 b	Praesepe	2.3	M2.5	K2 C16	~790	Rizzuto et al. (2018); Livingston et al. (2019)
K2-264 c	Praesepe	2.8	M2.5	K2 C16	~790	Rizzuto et al. (2018); Livingston et al. (2019)

Note. Transiting exoplanets discovered among Kepler's observed clusters and associations.

All of the other cluster planets uncovered with Kepler were discovered via analysis of data from the K2 mission. With one exception (K2-231 b in the 3 Gyr old Ruprecht 147 cluster) they fall within the youthful age range of 5–800 Myr. Finding young planets is particularly important for constraining the processes of migration, gravitational contraction, and photo-evaporation of atmospheres. Several groups (Rizzuto et al. 2017; David et al. 2018) have sought to develop pipelines to uncover planetary transit signals among the light-curves of stellar cluster members and ultimately measure occurrence rates for comparison with older field populations.

In the assessment of planetary occurrence as a function of stellar age, a perplexing issue is that 13 planets have been discovered in the 700–800 Myr Hyades and Praesepe clusters, while *none* have been detected in the young (~125 Myr) Pleiades (Gaidos et al. 2017). Several groups (Mann et al. 2017) have argued that planet detection is more difficult among Pleiades members due to the more prominent rotation signatures there; however, the discovery of planets around the active young star K2-309 (V1298 Tau; David et al. 2019b) suggests that this is not the entire explanation.

Community-developed detection pipelines have uncovered quite a few notable exoplanetary systems. For example, Vanderburg and collaborators (Vanderburg et al. 2018) discovered a possible super-Earth planet (~2 $R_\oplus$) in the Hyades cluster. With only one transit observed, this object has an orbital period longer than 72 days and if confirmed would constitute the lowest insolation flux received by any known cluster planet.

Among K2's clusters are also three multi-planet systems. All four planets orbiting K2-309 are between the sizes of Neptune and Jupiter; no other known multi-transiting systems have four planets this large with orbits less than 300 days. In addition, the M dwarf star K2-264 hosts two mini-Neptunes orbiting with 5–20 day periods (Rizzuto et al. 2018) while K2-136 has three planets that are also in the super-Earth to mini-Neptune range (Mann et al. 2018).

At an estimated age of 5–10 Myr, K2-33 b is the youngest transiting planet found to date (David et al. 2016b; Mann et al. 2016b). This object orbits at a mere 0.04 AU from its M dwarf parent star. Such a small semimajor axis suggests *in situ* formation for Neptune-sized planets, or that some close-in planets underwent rapid migration via torques provided by a viscous protoplanetary disk.

Overall, the young to intermediate-age cluster planets discovered to date orbit mainly late-type (K and M spectral type) stars. Complementary planet-finding techniques such as radial velocity monitoring are most sensitive to planets of Jupiter mass and larger around solar-type stars. Intriguingly, no close-in giant transiting planets have been discovered in clusters; other than the warm Jupiter-sized planet found around V1298 Tau, only Neptunes and smaller have been detected. This effect may be due to the relative rarity of hot Jupiters (van Saders & Gaudi 2011).

When it comes to planetary size, several studies (Mann et al. 2017) have suggested that the radii of young planets are systematically larger than those of older planets. This may be a signature of ongoing atmospheric evaporation and/or radial contraction at sub-Gyr ages. The sample of young transiting planet radii is compared to the statistical density of field transiting planet radii in Figure 4.17. To rule out detection biases, planetary masses will eventually be required to perform a direct comparison of the densities of young versus old planets around low-mass stars. This has been carried out for one planet so far by (Barragán et al. 2019), who found that K2-100 b is a low-density cluster planet undergoing significant atmospheric loss due to radiation from its host star.

We note that not all young planet host stars reside in clusters. David et al. (2018) found a ~120 Myr planet with possible membership in the Cas-Tau association, and also reported on the discovery of a warm Jupiter-sized planet (K2-309 b) around the star V1298 Tau in the ~23 Myr Group 29 association (David et al. 2019b).

Recently, the second data release ("DR2") of the GAIA mission has provided precise proper motion and parallax data, which helps to distinguish cluster members from foreground and background stars. At least five K2-discovered planets that were at one time thought to be cluster members are now considered to be unassociated; these are K2-308 b (hot Saturn unassociated with NGC 1817; Rampalli et al. 2019), 2MASS-J06101557+2436535 b (Jupiter-mass planet or brown dwarf in the background of M35; Dholakia et al. 2019), and ESPG 008 b, ESPG 009 b, plus ESPG 010 b (1.5–2 $R_\oplus$ planet candidates near the M67 cluster field; Nardiello et al. 2016). In addition, the issue of planet K2-284 b's potential membership in the Cas-Tau association remains to be fully addressed (David et al. 2018).

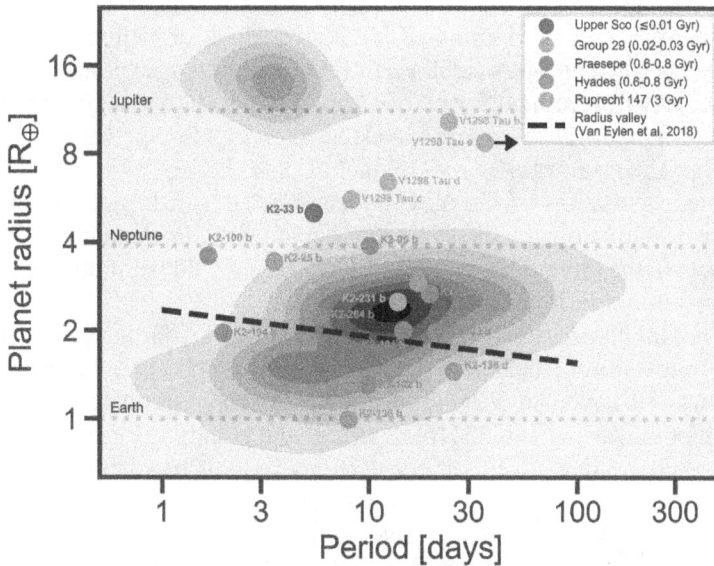

Figure 4.17. Transiting planets in young clusters and associations are shown here in the period–radius diagram, overlaid on a two dimensional histogram of the field transiting planet distribution. Young planets have a tendency toward inflated radii. Figure courtesy T. David.

4.4.4 Young Stellar Objects

Related to star and planet formation are the dusty circumstellar disks which surround the members of the youngest clusters and associations. Precision time series photometry here has led to detailed analyses of pre-main sequence accretion phenomena and obscuration by circumstellar material. Young star time domain studies have undergone a true revolution with the advent of dedicated monitoring with space telescopes. From prior ground-based studies, it was known that young stellar objects (YSOs) display variability at the 1%–100% level (e.g., Herbst et al. 1994; Cody & Hillenbrand 2010). Little more could be discerned from time series data, apart from the detection of periodicities in some objects. It was assumed that periodic behavior was caused by starspot modulation and that aperiodic behavior originated in unsteady accretion, though these were largely unconfirmed hypotheses.

With the onset of the space telescope era with *MOST* (Walker et al. 2003; Matthews et al. 2004; CoRoT Baglin et al. 2006); and finally Kepler, distinct patterns began to emerge in light-curves (e.g., Alencar & Teixeira 2010; Cody & Hillenbrand 2010; Cody et al. 2013). While no young star clusters were observed during the Kepler mission, several K2 campaign fields targeted YSOs. These included the ~1 Myr ρ Ophiuchus cluster and 5–10 Myr Upper Scorpius association in Campaign 2, the 1–2 Myr NGC 6530 cluster in the Lagoon Nebula during Campaign 9, as well as the ~3 Myr Taurus star-forming region in Campaign 13.

The bulk of K2 work published to date has focused on young stars in Campaign 2, which we highlight here. A summary of the diversity of observed YSO variability among low-mass (primarily and K and M spectral type) stars was presented by Cody & Hillenbrand (2018); and we show examples in Figure 4.18. They used metrics of periodicity and flux symmetry to determine if a given light-curve has repeating patterns and whether it is dominated by bursts, dips, or neither. Further classifications were performed by Ansdell et al. (2016) and Hedges et al. (2018).

One of the most prominent forms of variability detected was quasi-periodic flux dips of up to a magnitude at optical wavelengths. This "dipper" behavior comprised ~40% of the YSO samples in ρ Ophiuchus and Upper Scorpius; it has been interpreted as occultation of the YSO by circumstellar disk material. The precise origin of the occulting dust is the subject of debate (e.g., Bodman et al. 2017); there may be a warp in the inner disk, or alternatively there could be smaller clouds of dust elevated above the disk midplane by instabilities.

Of note, not all dippers are quasi-periodic; some display fading events that do not appear to follow any particular pattern. For example, the young star EPIC 204278916 was observed in K2 monitoring to display irregular fading events of up to 65% depth (Scaringi et al. 2016). Scaringi and collaborators suggested that the variability could be due to either occultations by a warped inner disk edge, or transits of cometary objects in orbit around the star.

Another type of YSO variability appearing prominently among K2 YSOs was transitory (day to week) increases in flux that were too symmetric and long duration to be stellar flares. This "burster" behavior (Stauffer et al. 2014; Cody et al. 2017) occurs in ~15% of low-mass stars and has been tied to unsteady gas accretion along

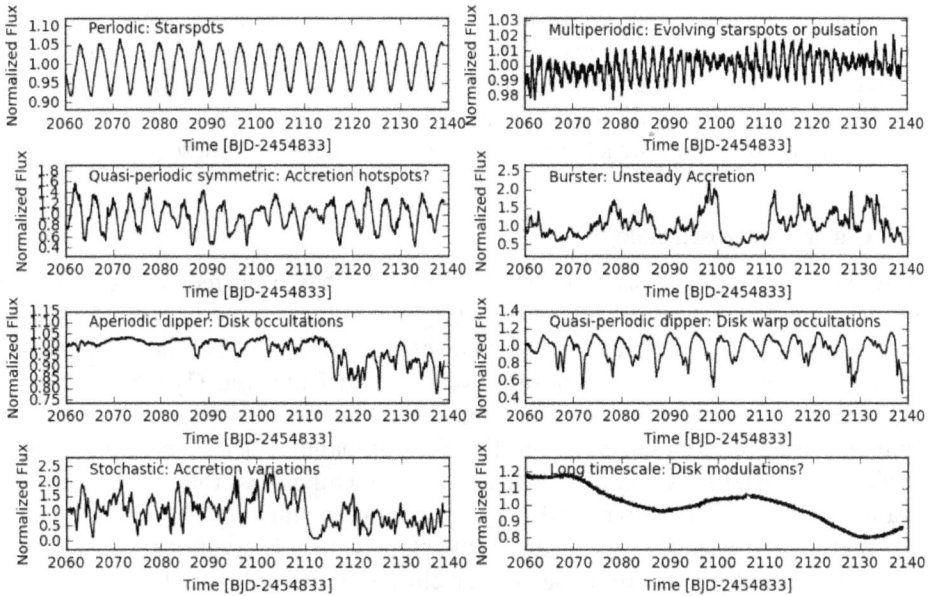

Figure 4.18. Studies on low-mass YSO light-curves obtained by K2 revealed eight different classes of variability and their hypothesized origins, as shown here. Figure courtesy of A. Cody.

magnetic field lines from the disk to stellar surface (Stauffer et al. 2014). Burstiness has been predicted by MHD simulations (e.g., Kulkarni & Romanova 2008; Romanova et al. 2013) to arise from instabilities in the inner gas disk. In some cases, light-curves with quasi-periodic bursts resembles the "propeller" regime of accretion predicted by theory; in other cases the bursts displayed no periodicity whatsoever. With K2 data, it was determined that bursters tend to have large Hα emission line widths indicative of active accretion, as well as strong infrared excesses from the inner disk (Stauffer et al. 2014).

Between the burster and dipper phenomena lie intermediate behaviors such as purely stochastic or quasi-periodic flux variations. While the origin of this variability is more uncertain, it is thought to be connected with rotational modulation of accretion hotspots on the stellar surface. Finally, stars without circumstellar material tend to display relatively simple sinusoidal type light-curves indicative of cool starspots, which have been used to reconstruct the rotational behavior of young stars across a range a masses as discussed in Section 4.4.1.

Within the framework of YSO variability categorization, several other types of flux behaviors stand out. The lowest mass young stellar cluster members—M dwarfs—typically spin rapidly, with periods around one day. A subset of them display highly structured periodic patterns, including complex wave-like undulations ("scallop shells") while others exhibit narrow ($\sim$15% of the phased pattern) triangular flux dips (Stauffer et al. 2017). The light-curve features are sufficiently unique that most cannot be explained well with a multi-starspot model. Instead, Stauffer and collaborators suggested that the behavior may be caused by eclipses of either warm coronal gas or particulate clouds.

Overall, K2 data on the youngest stars in clusters has shown that YSOs inhabit complex environments with unsteady gas accretion onto the stellar surface, as well as lumpy inner disk dust structures that change on dynamic timescales.

4.4.5 Cluster Asteroseismology

While the most vigorous work in asteroseismology has occurred for field stars, a variety of pulsators in clusters have been targeted and analyzed in both the Kepler mission and K2. These include red giants, solar type oscillators, and δ Scuti stars. Clusters receiving the most attention have been M67 and the Hyades, which are amenable to the study of solar-like oscillations. But not to be left out is the M4 globular cluster, a globular cluster in the K2 Campaign 2 field. The advantage to carrying out asteroseismology among cluster members is that their age and metallicity are already known, and vary little between stars.

Cluster asteroseismology got an early start with observations of solar-like oscillations in red giants from the Kepler cluster NGC 6819 (Stello et al. 2010). Using the first 34 days of data from the mission, Stello and collaborators were able to detect the large separation, $\Delta\nu$ in a number of red giants, confirming that oscillation amplitude scales roughly as luminosity to the 0.7 power, and showing the potential for future asteroseismology studies incorporating even longer time base-lines. Indeed, Hekker and collaborators (Hekker et al. 2011) were able to extend this study to include the entire first year of the Kepler mission, as well as two additional clusters NGC 6791 and NGC 6811, for a total of 92 oscillating red giants. They examined the $\Delta\nu$–ν_{max} relation, finding a significant dependence on mass but not on metallicity. Corsaro and collaborators (Corsaro et al. 2012) further expanded the red giant sample to 115 in these same open clusters, capitalizing on over a year and a half of Kepler data. They were able to characterize the dependence of the small separation parameters on stellar mass.

Additional success in detecting solar-like oscillations occurred for red giants in older clusters, including 3–4 Gyr old M67 (open cluster), and even the >11 Gyr old M4 globular cluster. The pulsation variability in question is very low amplitude and hence required the precision of Kepler for the first unambiguous oscillation detections here. Stello and collaborators (Stello et al. 2016b) performed asteroseismic analysis on a sample of over 30 giants and clump stars in M67. This resulted in mass derivations, a cluster age estimate, log g values, and a cluster distance estimate derived from stellar radii. All measurements were consistent with previously determined values from, e.g., H–R diagram analysis and eclipsing binary studies. Further oscillating red giants were identified in M4 (Miglio et al. 2016). In this metal-poor environment, asteroseismic masses were found to be consistent with estimates made via other methods such as isochrone fitting.

At earlier evolutionary stages, Sun-like stars in the Hyades were also found to be oscillating—the first such discovery in an open cluster (Lund et al. 2016b). During K2 Campaign 4, observations of these two fast rotating ($P < 2$ days) F stars in the

Hyades revealed excess power at 1568 and 1695 μHz in the frequency spectrum. Lund and collaborators detected the large frequency separation, $\Delta\nu$, thereby deriving asteroseismic masses, ages, and radii for these targets.

All in all, the asteroseismic masses and radii derived for oscillating cluster members are well in line with model-dependent isochrone predictions; this consistency underscores the promise of deriving asteroseismic parameters for thousands of field stars for which age and metallicity are not so well constrained.

Observing Planetary Nebula Central Stars with Kepler/K2

Contributed by George H. Jacoby, National Optical Astronomy Observatory, Tucson, AZ.

Introduction to the Science

The textbook wisdom is that normal stars like the Sun, having masses of ~1–8 $M_\odot$, will form a planetary nebula (PN) after evolving through the asymptotic giant branch (AGB) phase at the end of their nuclear fusion powered lives. During the past 15 yr, though, an alternative view has gained traction: that PNs are much more closely linked to interacting binary stars than normal quiescent stars. Several robust observations have forced that reconsideration, principally:

- Over 80% of all PNs have shapes that deviate significantly from the expected form of round spherical shells (Parker et al. 2006) that one expects from a spherical star.
- The PN phenomenon is rare: we would expect to find over 50,000 PNs in the Galaxy if all stars in the allowable mass range actually do form PNs. But, we can only account for about 11,000—which is consistent with the close binary star statistics (Moe et al. 2006).
- We find a few PNs in globular clusters, whose stars are only ~0.8 $M_\odot$, well below the mass cutoff. These can be explained as stars that have experienced significant mass transfer from a companion star during their lifetimes (Jacoby et al. 2017).
- The brightest PNs in galaxies all have very similar luminosities, but their progenitors have a wide range of masses. The absence of a correlation between progenitor mass and PN luminosity argues for another process acting to help normalize the final masses of the PN central stars (Ciardullo et al. 2005).

We wish to understand the formation process of PNs because: (1) PNs are major contributors to the chemical enrichment process in galaxies, especially for nitrogen, and so, we need to know their true rate of formation; (2) if PNs form only in binary star systems, then the Sun's future evolution is uncertain; and (3) close binaries involving white dwarfs (the descendents of PN central stars) can eventually merge to form Type Ia supernovae, whose formation process is also unclear.

Advantages Offered By Kepler/K2

The Kepler and K2 missions offered excellent opportunities to test the hypothesis that PNs form preferentially in close/interacting binary systems because we can monitor the central stars for brightness variations that serve as an indication of the presence of another star. Brightness varies in a binary star system as the secondary star orbits around the primary because of (1) radiative heating of one side of the secondary from the very hot central star (common), (2) geometrical distortion of the secondary into a teardrop shape due to the strong gravity of the central star (common), (3) eclipses of one star by the other (sometimes), (4) reflected light from the bright central star off the secondary (rare), or (5) relativistic beaming where the secondary star moves rapidly toward (or away from) us and appears brighter (or fainter) than if it is stationary (rare).

As a space-based observatory, Kepler can observe its targets all the time and at a highly predictable cadence over an extended time frame. That type of data collection is difficult from the ground (though possible: see Miszalski et al. 2009). Kepler's 30 minute cadence is especially well-matched to the variability periods of central stars, which range from a few hours to a few days.

Kepler's operating model thus enables variability studies having very high accuracy (1 part in 10,000 or better) compared to ground-based data (~1 part in 100). That level of accuracy allows us to detect secondary stars that are either very small, or much further away from the primary star, than can be detected with ground-based facilities.

Also, Kepler observes all the stars in its relatively large field of view at once, a square about 11 degrees on a side. Even though PNs are uncommon objects in the sky, Kepler's field size captures several (or many) PNs at once, though typically only ~5, in a single pointing. For fields close to the Galactic Center, the return rate can be exceptional, such as K2 Campaign 11 where over 150 PNs were in the field. That's a consequence of the high stellar density in the Galaxy's plane toward the Galactic Center. While that large number of targets is a fantastic windfall, several very serious issues arise in such dense fields that complicate the data analysis, as described in the next section.

It is important to have a large number of PNs (several dozen, at least) in order to build a healthy statistical sample for assessing whether binary central stars are the rule or the exception. Consequently, fields like K2 Campaign 11 are very attractive, despite the downsides.

A more subtle advantage of the Kepler/K2 observations is that the fields observed are selected for reasons that are totally independent of PN observations, and thus, the selection of PN targets is completely unbiased. With individually targeted observations, a selection bias can arise in which the observed PNs are more (or less) likely to have binary central stars (e.g., those with bipolar shapes compared to spherical shapes), thereby skewing the statistical conclusions on the fraction of PNs that are formed from binary stars.

Challenges Introduced With Kepler/K2
There are two very serious challenges in observing PNs central stars with Kepler/K2.
Dilution
The original Kepler mission was not designed to target faint stars in crowded fields, but rather, to study relatively bright stars among a few fainter stars. So, the large measuring aperture of ~10 arcsec was fine for deriving light-curves without contamination from neighboring field stars and other objects. PNs, however, *always* have a nebula that contaminates the measuring aperture. And, the larger the aperture, the more the contamination from the nebula. The nebula does not vary, and so, it adds a constant light source to the aperture, thereby masking variations in the brightness of the central star(s). Some tuning of the sky and target measuring masks can help slightly during post processing, but every object must be treated uniquely and improvements are modest.

Also, in crowded fields like Campaign 11, nearby field stars also significantly contaminate the measuring aperture. Figure 4.19 illustrates a serious example where the K2 light-curve from target EPIC 251248550 (the dark blue curve) is compared to a light-curve from the Kitt Peak 4 m Mayall telescope (the red curve; with a 1 arcsec

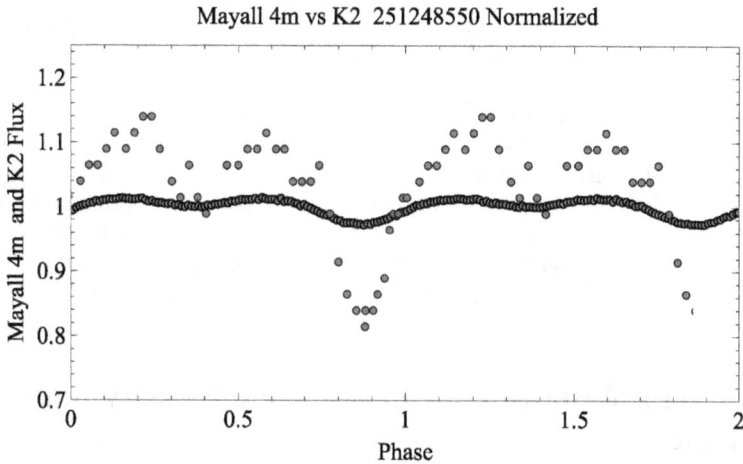

Figure 4.19. A comparison between the light-curve amplitudes derived from K2 and the Mayall 4 m for Campaign 11 target K2 251248550. The Mayall data (red dots) reveal a much larger variability amplitude than that implied by the K2 data. Figure courtesy of G. Jacoby.

measuring aperture). The amplitude of the variations is dramatically washed out in the K2 data, being a factor of ~10 smaller.

The consequences of this dilution are two-fold: (1) many targets will fall below the threshold at which variability can be detected, thereby undercounting the number of PN central stars as variable; and (2) the derived properties of the secondary star that rely on the variability amplitude will be incorrect (e.g., the size of the secondary will be underestimated).

Confusion

When multiple stars fall within the measuring aperture, a variable light-curve may originate from a field star rather than the target PN central star. Thus, each star in the aperture must be examined to validate the source of the variability.

We encountered one glaring example where the target EPIC 251248640 exhibited a light-curve that was unmistakably that of an RR Lyr variable (OGLE 28966) separated by 8.2" as noted in SIMBAD.

Mitigations

In general, all Kepler/K2 PN observations suffer from dilution and confusion problems. We can, however, (1) use "point-spread function" photometry (Cardoso et al. 2018) where each star in the measuring aperture is analyzed simultaneously to assess its contribution. Or, (2) we can use Kepler/K2 as a screening process to identify variables, and then follow-up those objects with targeted ground-based imaging.

Results to Date

De Marco et al. (2015) reported on the first PN central star results from Kepler. Of the six PN in the original field, those authors found four PN to have variable central stars. For one of the remaining two objects, dilution from the nebula likely masked the detectability of a variability signal and was discounted. Thus, 80% (4 out of 5) of the PNs in that small sample may have originated in a close binary star system, in excellent agreement with the expectations from arguments based on the shapes and number of PNs in the Galaxy. But the sample was too small to be convincing.

Subsequent K2 fields 0, 2, and 7 targeted an additional ~20 PNs, but those objects were much more compact nebulae with fainter central stars than those in the original

Kepler field. Thus, the usefulness of the observations in those three fields was limited due to dilution and brightness factors. Only 20% of those PNs exhibited a signature for binarity.

Campaign 11 provided 140 more PN targets, of which 25–30 (i.e., ~20%) are variable. This fraction is higher than ground-based estimates (~15%) but not significantly, and clearly, these are underestimates. Nevertheless, these numbers are already higher than the few percent level expected for the general population of stars having close binaries that do not merge, and are in the mass range to produce PNs. Thus, we already have determined that PN central stars strongly favor progenitors that are close binary stars. But, we don't yet know if binarity is required to form a PN.

4.5 Detection and Statistics of Kepler and K2 Binary Stars

4.5.1 Introduction

Early on in the Kepler mission, it was realized that some stars that showed transits or probable transits were themselves binary star systems. This is immediately interesting because the presence of a stellar companion will of course have consequences for the formation and evolution of planetary systems. For example, in terms of formation, a second protostar orbiting a star with a disk could truncate the disk, thereby affecting the range of periods, separations, and other orbital characteristics of the exoplanets that form from the disk. A second star would then subsequently affect, or potentially affect, the evolution of orbital parameters, cause orbital alignments or misalignments, or influence migration of planets within the disk. If good statistics of these effects can be observed, then the presence of stellar companions to exoplanet host stars could be an excellent laboratory in which to test theories of star and planet formation and evolution.

4.5.2 High-resolution Imaging

The size of the Kepler/K2 pixels on the sky was large, 4 ×4 arcsec, and therefore the problem of more than one star contributing to a particular pixel's photometric signal is quite common. If this were to be left unaccounted for, two main issues could arise. First, if a background star in the same pixel was itself an eclipsing binary star system, its faintness relative to the star under study might allow its eclipses to mimic an exoplanet transit, creating a "false positive," that is, a star thought to harbor an exoplanet that in reality does not. Second, even if the background star were to be single and constant in brightness, the increase in the total amount of light received in the pixel would dilute the transit signal of the star under study. The consequence of that would be to underestimate the radius of any exoplanet detected, thus creating the possibility that a planet that was really the size of Neptune might be thought to be more Earthlike.

To guard against these possibilities, an observational effort to obtain the best possible information of Kepler and K2 stars that exhibited transit signals was undertaken by multiple research teams. One area of activity was in taking high-resolution ground-based images in order to search for stellar companions. The primary techniques used in this regard were speckle imaging, direct infrared imaging, and adaptive optics. All of these methods offer a way to reach diffraction-limited or nearly-diffraction-limited resolution using large, single-aperture telescopes, but they have different strengths and weaknesses. For example, adaptive optics is a technique that is typically used in the near-infrared while speckle imaging is most often pursued in the visible range. In diffraction-limited imaging, the final resolution is related to the Rayleigh criterion, given by

$$\theta_R = \frac{1.22\lambda}{D},\qquad(4.1)$$

where θ_R is the smallest resolvable angular separation between two stars, λ is the wavelength of light being observed, and D is the diameter of the telescope. Thus, for the same diameter telescope, speckle imaging offers higher spatial resolution based on the smaller wavelengths of light in the visible compared to the infrared. On the other hand, if a companion star is physically bound to its primary star and both are on the main sequence, then the secondary will be redder (especially if low mass) than the primary star and therefore may appear brighter at infrared wavelengths relative to the primary star—allowing for easier detection on that basis. Thus, both wavelength ranges have something unique to offer in the realm of companion detection. This is why both wavelength ranges have been used, and often used in combination, to study the environs of exoplanet host stars.

Some Brief Technical Aspects of Speckle and AO

Contributed by Elliott Horch, Southern Connecticut State University, New Haven, CT.

Some readers may be familiar with both speckle imaging and adaptive optics as high resolution imaging techniques, but a brief technical comparison may still be useful. The air above the telescope aperture is a turbulent medium, creating variation in temperature and pressure on a scale of a few to 10s of centimeters. This leads to a variation in the index of refraction over the same scale, so that flat wavefronts from starlight above the atmosphere are disturbed or corrupted as they travel down through the air and enter the telescope itself. When this wavefront is brought to a focus by the telescope, a complex interference pattern is created, known as a speckle pattern. The speckle pattern changes rapidly in time, and if a long exposure image is taken (long means a second or more), the resulting image is the blurring of many different patterns seen in sequence. This leads to images that have so-called "seeing-limited" resolution (Figure 4.20).

Speckle imaging consists of taking many short exposures of the target ($\ll 1$ s), something that is enabled today by the existence of electron-multiplying CCD cameras. These cameras read out quickly, but as the charge collected in a given pixel is advanced toward the readout amplifier, it is done so in such a way as to create secondary electrons, thus building up the original charge collected via the photoelectric effect and amplifying it prior to read out. Thus, even a single photoelectron may be multiplied enough so that it produces a measurable signal above the read noise of the device. In this way, these cameras come very close to true photon counting (unique detection of photon events with no noise), and with the added benefit of the high quantum efficiency of modern astronomical-grade CCDs. The result is a superb detector for the task of

Figure 4.20. An image of the Differential Speckle Survey Instrument shown at the Gemini-N telescope, Maunakea, Hawaii. This is one of the primary instruments that was involved in Kepler host star vetting via high-resolution imaging. Elliott Horch built this instrument at Southern Connecticut State University in 2008. Figure courtesy of E. Horch.

taking many frames quickly. Once the sequence of frames is recorded, they must be reduced and a final reconstructed image is calculated. Generally, this is done with the technique of bispectral analysis (Lohmann et al. 1983). The image bispectrum is a four-dimensional mathematical object that is derived from the sequence of frames; it is the Fourier transform of the triple correlation function, which is defined in analogy to the autocorrelation for a single frame of the speckle sequence as:

$$T(\mathbf{x}_1, \mathbf{x}_2) = \int I(\mathbf{x})I(\mathbf{x} + \mathbf{x}_1)I(\mathbf{x} + \mathbf{x}_2)d\mathbf{x}, \qquad (4.2)$$

where $\mathbf{x}_1$ and $\mathbf{x}_2$ represent two vectors on the image plane and the function I is the irradiance recorded at that location in a speckle frame, and T is the triple correlation. It has been known since the 1980s that this function, when averaged over many frames, contains information related to both the modulus of the object's (diffraction-limited) Fourier transform and the derivative of its phase, and therefore a reconstructed image can be assembled in the Fourier domain. It is interesting to note in this context that if one thinks of two vectors on the image plane $\mathbf{x}_1$ and $\mathbf{x}_2$, then the implied vector is a third vector that would complete the triangle, namely $\mathbf{x}_2 - \mathbf{x}_1$, and there is a mathematical similarity here with the notion of closure phases in long-baseline interferometry. Regardless, the idea of speckle imaging involves some mathematical heavy-lifting after the observations are complete in order to obtain a final product. Often, a key limiting factor in the quality of the reconstructed image is the fidelity of the phase map in the Fourier domain, but because speckle imaging operates in the visible, the highest-resolution information available from ground-based follow-up of Kepler/K2 objects has come from this technique. Speckle is also an observational method with little observing overhead meaning that it can easily be used in a survey capacity, collecting data on many tens to hundreds of objects per night. An image of the speckle system that was most frequently used in Kepler follow-up is shown in Figure 4.20. New speckle instruments, NESSI, Zorro and 'Alopeke, are now in place and in routine use at the 3.5 m WIYN telescope and the two Gemini 8 m telescopes.

In contrast, adaptive optics systems have the ability to make wavefront corrections on the fly, that is, during data collection, and this allows for a high-resolution image to be built up to high signal-to-noise ratio prior to the read-out. In this case, the wavefront is measured many times a second and then changes are made in the surface of a deformable mirror so that, upon reflection off of that surface, the science wavefront is "flattened," in other words, the effect of the atmosphere turbulence (namely the speckle nature of the image) is largely removed. In this way, the image that strikes the focal plane of the detector (usually a high-grade astronomy imaging camera) is already a corrected image, and long exposures can be taken to build up the signal prior to readout. The quality of the image obtained is determined by the speed of the wavefront correction and the ability of the mirror to deform with high precision. Often, IR adaptive optics images do have fixed pattern noise in the final product due to imperfect corrections. However, they can still be extremely high in quality and exceed the ability of speckle images to detect faint red companions, particularly at separations well above the diffraction limit. This has led to the detection of some companions to Kepler/K2 stars that have been beyond the reach of speckle. Figure 4.21 shows both a speckle image reconstruction and an adaptive optics image of the same binary star and on the same scale, Ross 52 (=HIP 72896), illustrating the similarities and differences between

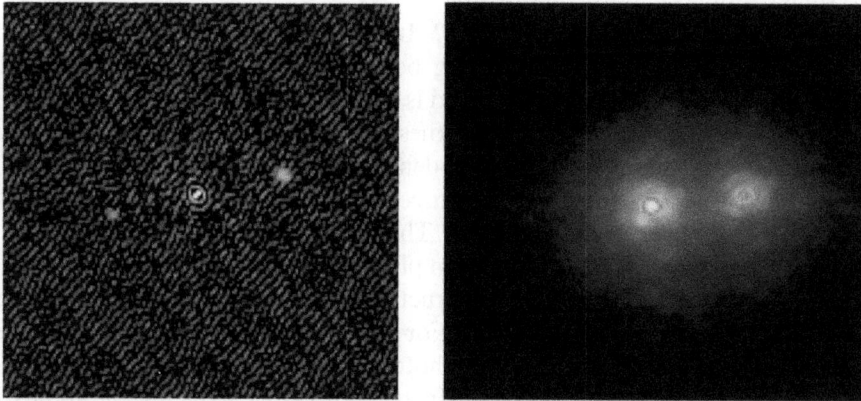

Figure 4.21. Speckle imaging and adaptive optics images of the same object. The speckle data were taken at the Discovery Channel Telescope, while the AO frame was recorded at the Palomar Observatory. Courtesy of F. Hahne, Southern Connecticut State University. Note that the speckle image is higher resolution, but that the adaptive optics image is higher signal-to-noise, as discussed in the text. The speckle image exhibits a "ghost peak" opposite the location of the secondary, which is caused by an imperfect reconstruction of the object's phase in the Fourier domain. On the other hand, a halo of irradiance is seen around each star in the AO image indicating imperfect correction by the deformable mirror.

these two techniques. The system has a composite V magnitude of 11.7 and at the time of these observations, the separation between the two stars was approximately 0.7 arc seconds.

4.5.3 Limits of Companion Detection

A fundamental question in discussing the role of companion detection for Kepler and K2 stars is the degree of completeness that ground-based high resolution follow-up observations give. A common figure of merit in this regard is to construct a detection limit curve of the faintest companion that could be seen from an observation as a function of the separation of the secondary star from the primary. For example, in a seeing-limited image where the resolution is 1 arcsec, one would expect that for separations well inside one arc second, there would be little sensitivity to the presence of a companion, even if it were comparable in brightness to the primary, due to the fact that the two stellar profiles would significantly overlap and blur together. Once the separation of the companion was on the order of one arc second, then companions of near-equal brightness to the primary star could probably be detected in the image, but if the secondary star were very faint, then the companion star could still be missed. At separations well above one arc second, even a faint companion could be detected, as the stellar profiles would be well separated. So, a typical expectation would be that such a detection limit curve starts at magnitude difference zero, and monotonically increases with separation, eventually flattening out when the two stars are separated by much more than the width of the stellar profile.

How are detection limits produced? In the case of speckle imaging, the reconstructed image is used as a starting point. The image is normalized so that the peak of the central star has value 1 and is centered in the image. Then, a series of annuli of increasing radii and equal thickness are defined and peaks and valleys in irradiance that lie inside the annulus are identified. Each local maximum within the image as treated as a possible stellar source, and the average value and standard deviation of these maxima are computed. The detection limit is then usually judged to be a peak of the same value of the mean plus 5 times the standard deviation, a so-called "5-σ" detection. However, reconstructed images made from speckle imaging data sometimes have fixed pattern noise. For example, if the star is not well-centered in the raw speckle frames, then some photons in the wings of the seeing-limited images spill off the edge of the frame. In the process of reducing the data using correlation functions and Fourier transforms, this can produce a final reconstructed image that has a non-physical faint "cross" that is centered on the primary star's diffraction-limited profile. Although this can be calibrated out to some extent, the distribution of local maxima can still be affected, making it a non-Gaussian distribution. Thus, a "5-σ" detection does indeed represent a peak that is five standard deviations above the mean, but it does not necessarily correlate to the same probability of random occurrence as in a Gaussian distribution.

With his caveat in mind, detection limit curves derived from speckle imaging and adaptive optics observations would be expected to follow a similar pattern, albeit that the separations at which companions could be easily detected would extend to smaller separations for speckle observations. Figure 4.22 illustrates this situation, for some of the major instruments used in exoplanet host star vetting, from Crossfield et al. (2016). The detected companions presented in that paper are shown with the triangle symbols, but this is not a complete accounting; for example Furlan et al. (2017) and Hirsch et al. (2017) have more complete information, but the Crossfield et al. curve reproduced here serves to give the reader a good illustration of detection capabilities. A representative comparison is given with the IR NIRI instrument K-band curve versus either of the speckle (DSSI) curves in the plot. Both NIRI and DSSI were used at the Gemini-North telescope, but the NIRI observations were at a much longer wavelength, leading to a larger diffraction limited spot size. Thus, for separations between the diffraction limit and 0.2 arcsec, speckle proves to be the more sensitive technique, but for larger separations, K-band imaging from NIRI is better.

Another issue for detection limit curves which is also illustrated in Figure 4.22 is, over what separation range is the the curve approximately valid. One might naïvely think that the limits could be extended out to the largest diameter of a circle that inscribes the final image frame that is studied. However, in high-resolution imaging, it is not so simple. In speckle imaging and in IR adaptive optics, there is the notion of

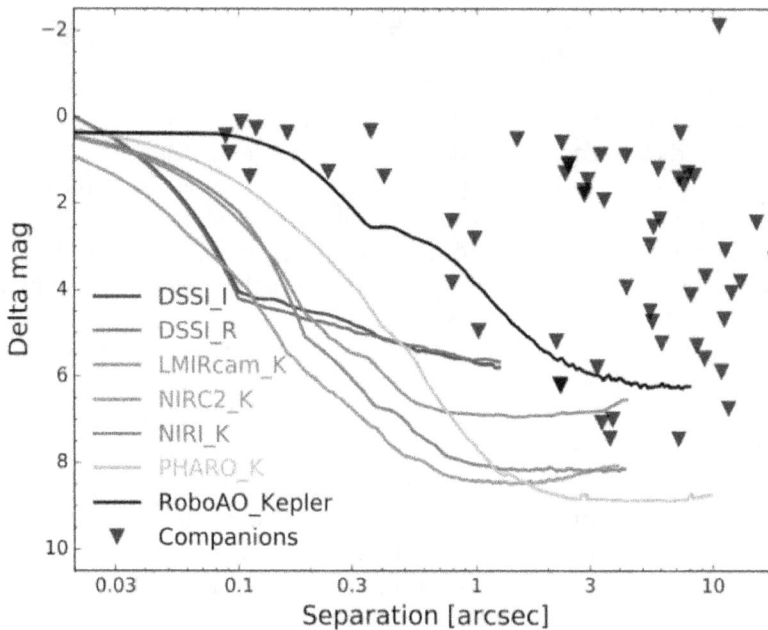

Figure 4.22. Detection limit curves for several instrument used in high-resolution imaging of exoplanet host stars. Reproduced from Crossfield et al. (2016). © 2016. The American Astronomical Society. All rights reserved.

the isoplanatic angle—the angle on the sky over which the speckle pattern that would be recorded for a point source remains constant. For stars close together on the sky, the light from both travels through the same turbulence in the atmosphere, and so the speckle patterns generated are identical in both cases, but if one imagines two stars that have a large separation, the turbulence experienced by the wavefronts of each star are not the same, and they will generate uncorrelated or partially correlated speckle patterns. In the visible, the size of the isoplanatic angle is at most a few arc seconds. However, the isoplanatic assumption begins to break down at smaller separations–as small as 1 arcsec or less depending on observing conditions. This can potentially affect the photometric statistics of a speckle image reconstruction at larger separations. This is why in Figure 4.22 the speckle curves are only drawn out to separations of 1.2 arcsec; the detection limit is harder to interpret at larger separations due to this effect. In contrast, since adaptive optics is performed in the infrared, where the size of the isoplanatic patch is larger, these curves extend to separations as large as 10 arcsec in the figure.

4.5.4 Are Close Companions Always Gravitationally Bound?

As research groups began to study high-resolution images of Kepler objects of interest, a significant fraction were shown to have stellar companions. Of course, the presence of a close companion does not mean it is necessarily physically associated with the primary star; for example, it could be a line-of-sight background star that is sufficiently bright to be relatively close in apparent magnitude to that of the Kepler star. To get a sense of the relative percentage of bound companions as compared to line-of-sight companions discovered in the Kepler data, Horch et al. (2014) took a statistical approach. They used the trilegal galaxy model (Girardi et al. 2005) to simulate the properties of stars in the Kepler field of view. With a list of simulated stars in hand, they then used statistics of binary and multiple stars found in Raghavan et al. (2010) to throw random bound companions into the sample. The trilegal model outputs celestial coordinates for each star when the simulation is run, so line-of-sight companions can also be identified in terms of their magnitude differences and separations, as if observed using a high-resolution imaging camera. Finally, the detection limits of a given camera can be imposed using the appropriate curve from, e.g., Figure 4.22, and the properties of the final list of "detected" binaries can be tallied up.

In doing this for the DSSI camera, Horch et al. (2014) found that most sub-arcsecond companion stars to the Kepler stars would be gravitationally bound. A slightly updated version of the plot in their paper that illustrates this point is shown in Figure 4.23. Here, open circles represent bound companions in the model, whereas filled circles represent line-of-sight companions. The blue and red circles are the locations of Kepler stars that had companions detected by speckle observations in the first two years of the program at the Gemini-North telescope with the DSSI camera. As can be read across the top of the figure, the model predicts that the fraction of bound companions increases as the separation decreases, and that most line-of-sight companions are found at larger magnitude differences and larger

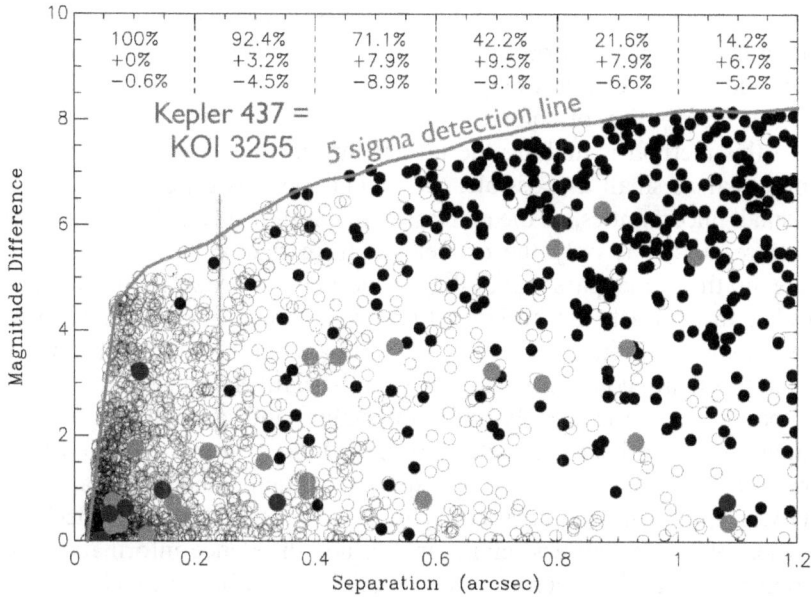

Figure 4.23. Magnitude difference as a function of separation for Kepler stars found to be binary using speckle imaging. The 5-σ detection curve for DSSI is shown in red; note that, in comparison to Figure 4.22, the x-axis is a linear scale and the y-axis of this plot is flipped. The position of Kepler437 is noted here as it will be discussed later in this chapter. Figure courtsey E. Horch.

separations. The two populations are reasonably distinct in the plot so that by simply looking at the separation and magnitude difference of a discovered companion in a high-resolution image, a reasonably good judgment can be made as to whether the companion is bound or not. Most of the companions discovered at Gemini using speckle imaging fall in the region dominated by bound companions.

Of course, just knowing whether a companion is likely to be bound may be useful in a statistical sense, but one would like to know in each specific case if the companion is truly bound or not. We outline two methods here that have been used to make more robust statements about individual companions, neither of which is completely definitive, but that can nonetheless be used to make a nearly certain statement for some Kepler binaries.

Hertzsprung–Russell (H–R) Diagram Location
The basic idea of this method posits that if two stars in a binary system are in fact gravitationally bound, then their positions on the H–R diagram will be linked by a common isochrone. In other words, the stars, to a high degree of probability may be assumed to have formed from the same cloud of gas and dust, and therefore are the same age. In most cases, this will mean that both stars are on the main sequence. On the other hand, if the secondary star is simply a line-of-sight companion, then it is not physically close to the primary star and if its absolute magnitude is derived using the distance of the primary star, a position on the H–R diagram that cannot be matched with a single isochrone will be derived. To make more certain that the H–R

diagram placement is not serendipitous (for example, a background giant that is just far enough away from the primary star to appear to be a red dwarf on the main sequence), the method can be strengthened by comparing the color of the secondary star with the distribution of colors of background stars using a galaxy model.

Key to this method is making the uncertainty in the color measurement of the companion star as small as possible; the magnitude axis generally carries comparatively small uncertainties, as the composite magnitude is usually well-known from the literature, and it is only the conversion from that value and the magnitude difference in the same pass band that is needed in order to derive individual magnitudes. In contrast, the horizontal (color) axis in the H–R diagram involves differential magnitudes in at least two different pass bands, making it less certain. If the derived secondary color is too large, then it is relatively easy to find an isochrone that will pass through the error bar of both primary and secondary in the color axis of the diagram. Nonetheless Everett et al. (2015) did a study of 18 Kepler binaries using photometry gleaned from both speckle imaging and IR adaptive optics. The analysis shows that the uncertainty in color is usually kept in check with the combination of the techniques, giving magnitude difference information from the visible range through 2.2 microns. Everett et al. were able to make unambiguous determinations of whether the secondary was gravitationally bound in six cases; five of these are bound companions and only one was a background star.

A much larger study of Kepler companions was done by Hirsch et al. (2017), where those authors studied 176 KOIs that have a stellar companion within 2 arc seconds. In addition to both speckle imaging and adaptive optics data, some targets were observed with HST and with lucky imaging. Again using H–R diagram placement, this data set shows that if the companion lies within one arc second of the primary star, there is an 80% chance of the pair being a physical system. Hirsch et al. also show that if there is an equal likelihood of an exoplanet orbiting either the primary or secondary star, then the radii of the planet will be underestimated by over 50%. This highlights the importance of robust, reliable high-resolution imaging in exoplanet characterization.

Common Proper Motion
Given the typical separations of companion stars based on the speckle results from Kepler, the orbital periods, if indeed the stellar companions are gravitationally bound, would often be in the range of hundreds or thousands of years, too long in almost all cases to determine orbital motion from the relative motion of the two stars. However, in this regime, another method for identifying bound companions is available to determine physical association: if the proper motion of both stars in the pair are the same or nearly the same, then, to a high degree of probability, one can assume that the stars are in fact at the same distance and associated. This is the basis of the method of common proper motion. Even before the advent of Gaia, many Kepler stars had measured proper motions due to *Hipparcos* and other sources, but with the release of Gaia data, we now have extremely precise proper motions for many Kepler stars that have been identified to have stellar companions. For sub-arcsecond companion stars, the Gaia measurements available to date represent the

motion of the system as a whole, not either star in the double star, but when combined with the relative positions as a function of time (which either speckle imaging or IR adaptive optics observations could provide), this permits one to solve for the individual motions and to plot those as a function of time. The first use of this technique in the literature for a Kepler star with a stellar companion is due to Wittrock et al. (2016), where those authors showed, prior to the Gaia DR2 data release, that HD 2638 was clearly a common proper motion pair. Figure 4.24 shows the motion of the two stars in this system, clearly traveling together.

A more systematic approach was taken by Hess et al. (2018). Those authors realized that speckle observations had identified a few dozen Kepler objects of interest to have stellar companions in the sub-arcsecond separation range. They took observations of these stars again years after the discovery epoch, and in many cases multiple observations over several years, in order to determine the proper motion of each companion in all of those systems. While the data set is still under study as of this writing, we show in Figure 4.25 one intriguing example from that work, Kepler 449. This 11th magnitude system, which has effective temperature just slightly cooler than the Sun, hosts two confirmed exoplanets, both with sizes between 1 and 2 Earth radii. The periods are 12.6 and 33.7 days, respectively. The astrometric data show that the pair has traveled through space together over the last eight years in terms of their absolute motion, with each star moving almost a third of an arc second in that time, in a direction just west of south. However, relative motion also appears to be present in the system along the line joining the two stars: the separation has

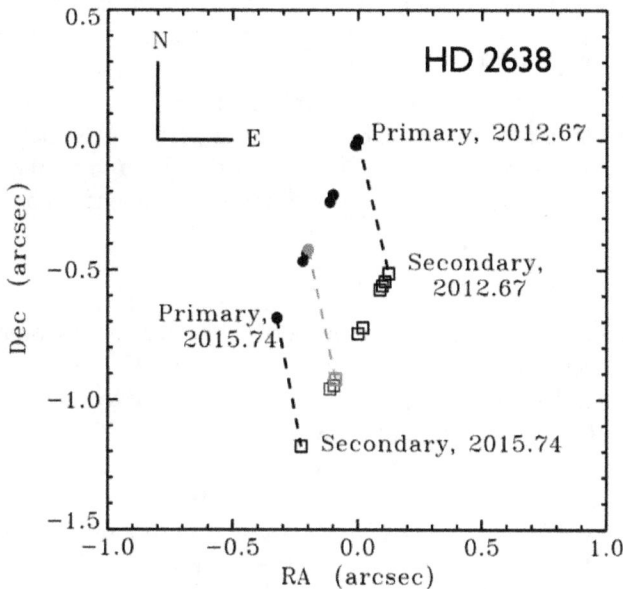

Figure 4.24. The absolute motion of the components of the binary star system HD 2638. Filled circles represent the location of the primary star and open boxes the secondary star in a sequence of observations taken between 2012 and 2015. Reproduced from Wittrock et al. (2016). © 2016. The American Astronomical Society. All rights reserved.

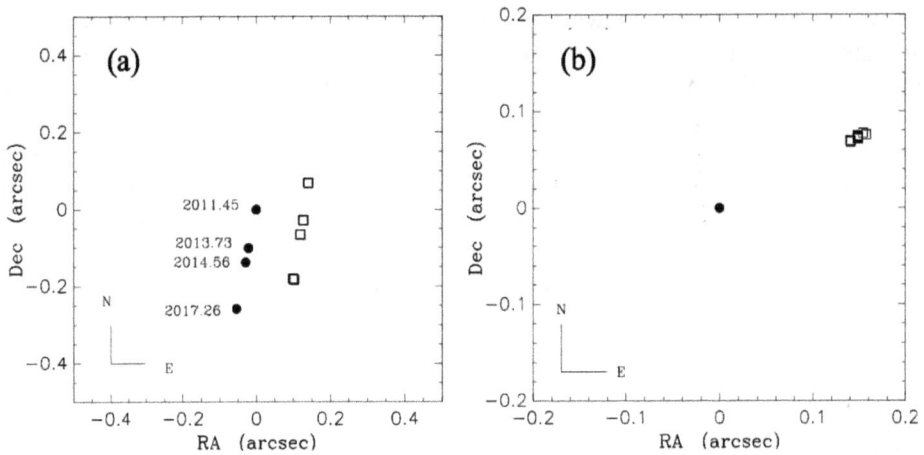

Figure 4.25. The (a) absolute and (b) relative motion of the components of binary exoplanet host Kepler 449. While overall the motion indicates the pair is traveling together, there is also relative motion along the line joining the two stars. This plot is a slightly modified version of one from Hess et al. (2018).

increased by about 20 milliarcseconds (mas) in the same time interval, suggestive of motion in the same plane as the planets are known to orbit, since they were detected via transit observations.

The Kepler ExoFOP lists the distance to the system as 260 ± 13 pc, which is relatively small for the typical KOIs. This means that the separation of the two stars in the system is about 42 au. Using the photometry of the components to estimate the mass of each star, Hess et al. concluded that the system has total mass of about 2 $M_\odot$. Combining this with the projected separation as an estimate of semimajor axis, a binary orbital period estimate of less than 200 years is obtained. Thus, the scale of the relative motion is not unreasonable for such a period—we may be viewing the stellar companion slowly orbiting in the same orbital plane as the planets. More work is needed to understand if such orbital plane alignment is common.

4.5.5 Binary Star Statistics from Kepler and K2

Efforts to understand the statistics and properties of stellar companions to exoplanet host stars have been underway for more than 10 years. Konacki et al. (2009) used high-precision radial velocities obtained via spectroscopy to rule out circumbinary exoplanets (so-called P-type exoplanets) with orbital periods below 5.3 yr around 10 known double-lines spectroscopic binary stars. They argued that the technique could be used to provide meaningful constraints on the formation of planets in a circumbinary disk if a large sample of data could be obtained. Indeed, as of this writing, some two dozen circumbinary planets are known, though these have been confirmed using a range of techniques, not simply radial velocities.

Wang et al. (2014, 2015) attempted to understand the statistics of S-type exoplanets; that is, those that orbit only one of the two stars in the system. In their 2014 paper, they suggested that there was a suppression of stellar companions in exoplanet host stars for separations less than 1500au from the primary star. However, in the 2015 work which focused on gas giant planets, the data suggested a decrease in the multiplicity rate for separations less than 20 au, and increase for the separation range of 20 to 200 au, and no statistically significant difference between the multiplicity rates for field stars and the planet-hosting sample above 200 au. The increase in stellar companions between 20 and 200 au suggests that this range of separations could be important for the formation and migration of giant planets.

As discussed in Section 4.5.4, Horch et al. (2014) also investigated the question of the multiplicity rate of planet host stars. using the trilegal galaxy model (Girardi et al. 2005) to generate a population of stars at similar distances magnitudes, and effective temperatures to the stars in the Kepler field. Given the separations of the companions that had already been detected with the speckle instrument at both the WIYN and Gemini-North telescopes, the sample of KOIs with known stellar companions below 1 arcsec in separation could be viewed as gravitationally bound. By varying the rate of bound companions in the model to match the number of companions detected in the actual speckle data versus the number of sources observed, it was determined that, in the separation range to which speckle imaging was sensitive, the data supported a multiplicity rate of exoplanet host stars of 40%–50%, comparable to the field population. Matson et al. (2018) performed a similar analysis to determine the binary star rate in the K2 fields scattered along the Ecliptic. She found a similar multiplicity rate for the K2 exoplanet host stars of 40%–50%.

Given a typical distance to the Kepler stars of about 1 kpc, the range over which speckle is sensitive is projected separations of 20–1200 au. The statistics were not robust enough to specifically investigate the range where Wang et al. described an increase in multiplicity rate over the field, but otherwise, the results between the two studies are largely consistent over the range of separation that both cover. The Wang et al. data do suggest a suppression of stellar companions in exoplanet systems that host more than one planet, something that again, the Horch et al. paper was not able to specifically address given the limitations of their statistics. However, Matson et al. (2018), using the nearer K2 binary exoplanet host stars, found no evidence for close companion suppression in planet hosting systems, observing stellar companions as close as 10–20 au.

These few examples are the beginning of what will hopefully be larger, long term efforts to obtain the statistics of exoplanets that exist in binary and multiple star systems. It remains likely that such studies will inform theories of star and planet formation along the lines touched upon in this chapter. However, one conclusion of the work to date is already something striking: it is probably a relatively common occurrence that an exoplanet orbits one star in a binary star system. An example is shown in Figure 4.26 for the exoplanet host star system Kepler437, which is a binary star of angular separation of 0.2 arcsec. While no Gaia DR2 parallax is available, the magnitude and color of the star indicates a distance of some 400 pc, leading to a stellar separation of 75 au. The exoplanet, which lies 0.3 au from its host star, appears to be in the habitable zone. This means that on Kepler437b, and on many other exoplanets, an observer would see a Sun during the day just as we do but either

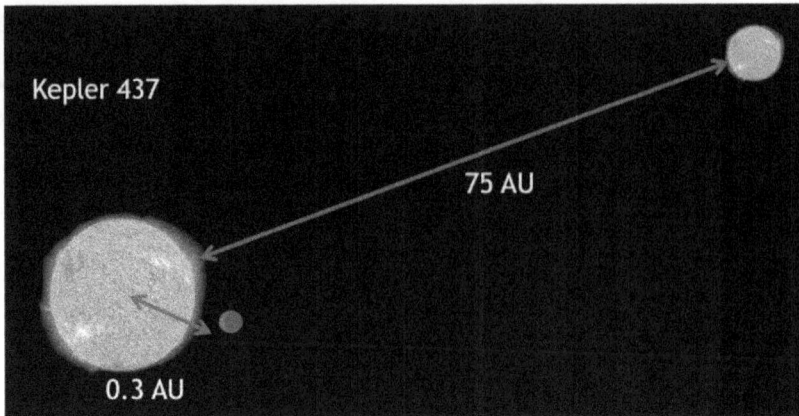

Figure 4.26. A schematic of the typical architecture for a binary system hosting an exoplanet: Kepler437. In this case, the sole known exoplanet sits at a distance of 0.3 au from the parent star, and is in or very near the habitable zone. The secondary star sits at a projected separation of 75 au, given a distance to the system of ~400 pc as listed in the Kepler ExoFOP archive. Figure courtsey E. Horch.

at night or during the day (depending on the relative positions of the alien Sun and the stellar companion on the sky), one would also see an extremely bright star, something comparable to the brightness of the full Moon in our sky, but concentrated into very small Sun-like object. What a sight that would be!

4.5.6 Postscript: Which Star Does the Exoplanet Orbit?

It has already been mentioned that in S-type exoplanet binaries, the exoplanet (or exoplanets) may orbit either of the two stars. Or indeed, each star may have one or more planets in its retinue. For a simple system of two stars and one planet, what prospects are there for being able to determine which star is the exoplanet host? This is an important question, as has already been touched on, because depending on the magnitude difference between the two stars (yielding a stellar radius difference), there could be a significant error in the derived radius of the planet if one assumes it orbits the primary when it really orbits the secondary star. The true exoplanet radius in that case would be larger than the transit signal would suggest.

In a recent paper by Howell et al. (2019), a first attempt was made to make a judgment of whether the exoplanet in a Kepler binary star system orbits the primary or secondary star. Speckle observations were made using the speckle imager 'Alopeke at the Gemini-North telescope, before, during and after a transit event of Kepler13. The data were then processed in two independent ways. Since the system has a separation of about 1.1 arcsec, the two stars were resolved throughout the sequence of observations. Thus, aperture photometry could be used to develop a time series of the magnitude difference in transit and out of transit. This analysis showed that the magnitude difference becomes smaller during the transit, indicating that the primary star was becoming fainter during the transit, and therefore the exoplanet orbits the primary star. The same data set was then used for a speckle analysis using correlation functions. The idea was to see if the same result could be obtained from high-resolution image products. The authors were able to show that a similar implied transit depth around the primary was derived from the speckle analysis, although with lower precision. To extend this result to get high-precision, high-resolution results, one would have to observe 10–20 transits—a considerable amount of telescope time. But, this paper shows that it can be done even for systems with smaller angular separations between the stars; those are situations where aperture photometry would not be available as the seeing limited images would be highly blended. Thus, speckle imaging could be a viable way forward for understanding which star in a binary system the exoplanet actually orbits.

References

Agüeros, M. A., Bowsher, E. C., Bochanski, J. J., et al. 2018, ApJ, 862, 33

Alencar, S. H. P., Teixeira, P. S., & Guimarães, M. M. 2010, A&A, 519, A88

Alonso, R., Deeg, H. J., Hoyer, S., et al. 2015, A&A, 584, L8

Ansdell, M., Gaidos, E., Rappaport, S. A., et al. 2016, ApJ, 816, 69

Armstrong, D., Pollacco, D., Watson, C. A., et al. 2012, A&A, 545, L4

Baglin, A., Auvergne, M., Barge, P., et al. 2006, in ESA Special Publication, Vol. 1306, The CoRoT Mission Pre-Launch Status—Stellar Seismology and Planet Finding, ed. M. Fridlund, A. Baglin, J. Lochard, & L. Conroy (ESA), 33

Baraffe, I., Chabrier, G., Barman, T. S., Allard, F., & Hauschildt, P. H. 2003, A&A, 402, 701

Baraffe, I., Homeier, D., Allard, F., & Chabrier, G. 2015, A&A, 577, A42

Barclay, T., Rowe, J. F., Lissauer, J. J., et al. 2013, Natur, 494, 452

Barnes, S. A. 2003, ApJ, 586, 464

Barnes, S. A., Weingrill, J., Fritzewski, D., Strassmeier, K. G., & Platais, I. 2016, ApJ, 823, 16

Barrado y Navascués, D., Stauffer, J. R., Bouvier, J., & Martín, E. L. 2001, ApJ, 546, 1006

Barragán, O., Aigrain, S., & Kubyshkina, D. 2019, MNRAS, 490, 698

Bass, G., Orosz, J. A., Welsh, W. F., et al. 2012, ApJ, 761, 157

Bayless, A. J., & Orosz, J. A. 2006, ApJ, 651, 1155

Beck, P. G., Montalban, J., Kallinger, T., et al. 2012, Natur, 481, 55

Beck, P. G., Hambleton, K., Vos, J., et al. 2014, A&A, 564, A36

Bedding, T. R., Mosser, B., Huber, D., et al. 2011, Natur, 471, 608

Bell, K. J., Hermes, J. J., Bischoff-Kim, A., et al. 2015, ApJ, 809, 14

Bell, K. J., Hermes, J. J., Montgomery, M. H., et al. 2017, in ASP Conf. Ser. 509, 20th European White Dwarf Workshop, ed. P. E. Tremblay, & B. Gaensicke (San Francisco, CA: ASP), 303

Bell, K. J., Hermes, J. J., Montgomery, M. H., et al. 2016, ApJ, 829, 82

Benkő, J. M., Kolenberg, K., & Szabó, R. 2010, MNRAS, 409, 1585

Benkő, J. M., Jurcsik, J., & Derekas, A. 2019, MNRAS, 485, 5897

Benomar, O., Masuda, K., Shibahashi, H., & Suto, Y. 2014, PASJ, 66, 94

Benomar, O., Takata, M., Shibahashi, H., Ceillier, T., & García, R. A. 2015, MNRAS, 452, 2654

Benomar, O., Bazot, M., Nielsen, M. B., et al. 2018, Sci, 361, 1231

Bloemen, S., Marsh, T. R., Østensen, R. H., et al. 2011, MNRAS, 410, 1787

Bodman, E. H. L., Quillen, A. C., Ansdell, M., et al. 2017, MNRAS, 470, 202

Borkovits, T., Hajdu, T., Sztakovics, J., et al. 2016, MNRAS, 455, 4136

Borkovits, T., Rappaport, S., Hajdu, T., & Sztakovics, J. 2015, MNRAS, 448, 946

Borkovits, T., Derekas, A., Kiss, L. L., et al. 2013, MNRAS, 428, 1656

Borkovits, T., Albrecht, S., Rappaport, S., et al. 2018, MNRAS, 478, 5135

Borkovits, T., Rappaport, S., Kaye, T., et al. 2019, MNRAS, 483, 1934

Borucki, W. J., Koch, D. G., Batalha, N., et al. 2012, ApJ, 745, 120

Bostancı, Z. F., Ak, T., Yontan, T., et al. 2015, MNRAS, 453, 1095

Bouvier, J., Barrado, D., Moraux, E., et al. 2018, A&A, 613, A63

Brandt, T. D., & Huang, C. X. 2015a, ApJ, 807, 58

Brandt, T. D., & Huang, C. X. 2015b, ApJ, 807, 24

Bressan, A., Marigo, P., Girardi, L., et al. 2012, MNRAS, 427, 127

Breton, R. P., Rappaport, S. A., van Kerkwijk, M. H., & Carter, J. A. 2012, ApJ, 748, 115

Brewer, L. N., Sandquist, E. L., Mathieu, R. D., et al. 2016, AJ, 151, 66

Brogaard, K., Hansen, C. J., Miglio, A., et al. 2018, MNRAS, 476, 3729

Broomhall, A. M., Miglio, A., Montalbán, J., et al. 2014, MNRAS, 440, 1828

Burkart, J., Quataert, E., Arras, P., & Weinberg, N. N. 2012, MNRAS, 421, 983

Camargo, D. 2018, ApJL, 860, L27

Campante, T. L., Barclay, T., Swift, J. J., et al. 2015, ApJ, 799, 170

Campante, T. L., Lund, M. N., Kuszlewicz, J. S., et al. 2016, ApJ, 819, 85

Cannizzo, J., Smale, A. P., Wood, M. A., Still, M. D., & Howell, S. B. 2012, ApJ, 747, 117

Cardoso, J. V. D. M., Barensten, G., Hedges, C. L., et al. 2018, BAAS, 23143907C

Carraro, G., Girardi, L., & Marigo, P. 2002, MNRAS, 332, 705

Carter, J. A., Rappaport, S., & Fabrycky, D. 2011, ApJ, 728, 139

Carter, J. A., Agol, E., Chaplin, W. J., et al. 2012, Sci, 337, 556

Casagrande, L., Silva Aguirre, V., Schlesinger, K. J., et al. 2016, MNRAS, 455, 987

Chaplin, W. J., & Miglio, A. 2013, ARA&A, 51, 353

Chaplin, W. J., Kjeldsen, H., Christensen-Dalsgaard, J., et al. 2011a, Sci, 332, 213

Chaplin, W. J., Bedding, T. R., Bonanno, A., et al. 2011b, ApJ, 732, L5

Chaplin, W. J., Basu, S., Huber, D., et al. 2014, ApJ, 210, 1

Chaplin, W. J., Lund, M. N., Handberg, R., et al. 2015, PASP, 127, 1038

Charpinet, S., Brassard, P., Giammichele, N., & Fontaine, G. 2019, A&A, 628, L2

Charpinet, S., Brassard, P., Van Grootel, V., & Fontaine, G. 2014, in ASP Conf. Ser. 481, On Interpreting g-Mode Period Spacings in sdB Stars, ed. V. van Grootel, E. Green, G. Fontaine, & S. Charpinet (San Francisco, CA: ASP), 179

Charpinet, S., Van Grootel, V., Fontaine, G., et al. 2011, A&A, 530, A3

Choi, J., Dotter, A., Conroy, C., et al. 2016, ApJ, 823, 102

Ciardullo, R., Sigurdsson, S., Feldmeier, J. J., et al. 2005, ApJ, 629, 499

Ciardi, D. R., Crossfield, I. J. M., Feinstein, A. D., et al. 2018, AJ, 155, 10

Cody, A. M., Barentsen, G., Hedges, C., et al. 2018, RNAAS, 2, 199

Cody, A. M., & Hillenbrand, L. A. 2010, ApJS, 191, 389

Cody, A. M., & Hillenbrand, L. A. 2018, AJ, 156, 71

Cody, A. M., Hillenbrand, L. A., David, T. J., et al. 2017, ApJ, 836, 41

Cody, A. M., Tayar, J., Hillenbrand, L. A., Matthews, J. M., & Kallinger, T. 2013, AJ, 145, 79

Conroy, K. E., Prša, A., Stassun, K. G., et al. 2014, AJ, 147, 45

Corsaro, E., Stello, D., Huber, D., et al. 2012, ApJ, 757, 190

Crossfield, I. J. M., Ciardi, D. R., Petigura, E. A., et al. 2016, ApJS, 226, 7

Curtis, J. L., Agüeros, M. A., Douglas, S. T., & Meibom, S. 2019, ApJ, 879, 49

Curtis, J. L., Wolfgang, A., Wright, J. T., Brewer, J. M., & Johnson, J. A. 2013, AJ, 145, 134

Curtis, J. L., Vanderburg, A., Torres, G., et al. 2018, AJ, 155, 173

Daemgen, S., Bonavita, M., Jayawardhana, R., Lafrenière, D., & Janson, M. 2015, ApJ, 799, 155

David, T. J., Hillenbrand, L. A., Cody, A. M., Carpenter, J. M., & Howard, A. W. 2016a, ApJ, 816, 21

David, T. J., Hillenbrand, L. A., Gillen, E., et al. 2019a, ApJ, 872, 161

David, T. J., Hillenbrand, L. A., Petigura, E. A., et al. 2016b, Natur, 534, 658

David, T. J., Conroy, K. E., Hillenbrand, L. A., et al. 2016c, AJ, 151, 112

David, T. J., Mamajek, E. E., Vanderburg, A., et al. 2018, AJ, 156, 302

David, T. J., Cody, A. M., Hedges, C. L., et al. 2019b, AJ, 158, 79

Davies, G. R., Silva Aguirre, V., Bedding, T. R., et al. 2016, MNRAS, 456, 2183

Deheuvels, S., Ballot, J., Beck, P. G., et al. 2015, A&A, 580, A96

Deheuvels, S., Brandão, I., Silva Aguirre, V., et al. 2016, A&A, 589, A93

Deheuvels, S., Doğan, G., Goupil, M. J., et al. 2014, A&A, 564, A27

Deheuvels, S., García, R. A., Chaplin, W. J., et al. 2012, ApJ, 756, 19

Derekas, A., Kiss, L. L., Borkovits, T., et al. 2011, Sci, 332, 216

Derekas, A., Szabó, G. Y. M., & Berdnikov, L. 2012, MNRAS, 425, 1312

De Marco, O., Long, J., Jacoby, G. H., et al. 2015, MNRAS, 448, 3587

Dholakia, S., Dholakia, S., Cody, A. M., et al. 2019, PASP, 131, 114402

Donati, P., Beccari, G., Bragaglia, A., Cignoni, M., & Tosi, M. 2014, MNRAS, 437, 1241

Dotter, A. 2016, ApJS, 222, 8

Dotter, A., Chaboyer, B., Jevremović, D., et al. 2008, ApJS, 178, 89

Douglas, S. T., Curtis, J. L., Agüeros, M. A., et al. 2019, ApJ, 879, 100

Doyle, L. R., Carter, J. A., Fabrycky, D. C., et al. 2011, Sci, 333, 1602

Edwards, S., Strom, S. E., Hartigan, P., et al. 1993, AJ, 106, 372

Eggenberger, P., Montalbán, J., & Miglio, A. 2012, A&A, 544, L4

Eggenberger, P., Deheuvels, S., Miglio, A., et al. 2019, A&A, 621, A66

Esselstein, R., Aigrain, S., Vanderburg, A., et al. 2018, ApJ, 859, 167

Everett, M. E., Barclay, T., Ciardi, D. R., et al. 2015, AJ, 149, 55

Everett, M. E., Howell, S. B., & Kinemuchi, K. 2012, PASP, 124, 316

Fabrycky, D., & Tremaine, S. 2007, ApJ, 669, 1298

Faigler, S., Kull, I., Mazeh, T., et al. 2015, ApJ, 815, 26

Faigler, S., & Mazeh, T. 2011, MNRAS, 415, 3921

Ferraro, F. R., Dalessandro, E., Mucciarelli, A., et al. 2009, Natur, 462, 483

Fields, C. E., Farmer, R., Petermann, I., Iliadis, C., & Timmes, F. X. 2016, ApJ, 823, 46

Forbes, D. A., & Bridges, T. 2010, MNRAS, 404, 1203

Frandsen, S., Lehmann, H., Hekker, S., et al. 2013, A&A, 556, A138

Fuller, J. 2017, MNRAS, 472, 1538

Fuller, J., Cantiello, M., Stello, D., Garcia, R. A., & Bildsten, L. 2015, Sci, 350, 423

Fuller, J., & Lai, D. 2012, MNRAS, 420, 3126

Furlan, E., Ciardi, D. R., Everett, M. E., et al. 2017, AJ, 153, 71

Gaensicke, B. T., Toloza, O., Hermes, J. J., & Szkody, P. 2019, Proc. Conf. Compact White Dwarf Binaries, ed. G. H. Tovmassian, & B. T. Gansicke, 51

Gaidos, E., Mann, A. W., Rizzuto, A., et al. 2017, MNRAS, 464, 850

Galadi-Enriquez, D., Jordi, C., & Trullols, E. 1998, A&A, 337, 125

García, R. A., Ceillier, T., Salabert, D., et al. 2014, A&A, 572, A34

García-Berro, E., Torres, S., Althaus, L. r. G., et al. 2010, Natur, 465, 194

Gaulme, P., & Guzik, J. A. 2019, A&A, 630, A106

Gaulme, P., McKeever, J., Jackiewicz, J., et al. 2016, ApJ, 832, 121

Giammichele, N., Charpinet, S., Fontaine, G., & Brassard, P. 2017, ApJ, 834, 136
Giammichele, N., Charpinet, S., Fontaine, G., et al. 2018, Natur, 554, 73
Gies, D. R., Matson, R. A., Guo, Z., et al. 2015, AJ, 150, 178
Gies, D. R., Williams, S. J., Matson, R. A., et al. 2012, AJ, 143, 137
Gillen, E., Hillenbrand, L. A., David, T. J., et al. 2017, ApJ, 849, 11
Girardi, L., Groenewegan, M. A. T., Hatziminaoglau, E., & da Costa, L. 2005, A&A, 436, 895
Green, E. M., Fontaine, G., Reed, M. D., et al. 2003, ApJL, 583, L31
Greiss, S., Steeghs, D., Gänsicke, B. T., et al. 2012, AJ, 144, 24
Greiss, S., Hermes, J. J., Gänsicke, B. T., et al. 2016, MNRAS, 457, 2855
Grunblatt, S. K., Huber, D., Gaidos, E. J., et al. 2016, AJ, 152, 185
Guo, Z., Fuller, J., Shporer, A., et al. 2019, ApJ, 885, 46
Hall, O. J., Davies, G. R., Elsworth, Y. P., et al. 2019, MNRAS, 486, 3569
Hambleton, K. M., Kurtz, D. W., Prša, A., et al. 2013, MNRAS, 434, 925
Hansen, B. M. S., Richer, H. B., Fahlman, G. G., et al. 2004, ApJS, 155, 551
Hebb, L., Wyse, R. F. G., Gilmore, G., & Holtzman, J. 2006, AJ, 131, 555
Heber, U. 2016, PASP, 128, 082001
Hedges, C., Hodgkin, S., & Kennedy, G. 2018, MNRAS, 476, 2968
Hekker, S., Basu, S., Stello, D., et al. 2011, A&A, 530, A100
Hełminiak, K. G., Ukita, N., Kambe, E., et al. 2017a, A&A, 602, A30
Hełminiak, K. G., Ukita, N., Kambe, E., et al. 2017b, MNRAS, 468, 1726
Herbst, W., Herbst, D. K., Grossman, E. J., & Weinstein, D. 1994, AJ, 108, 1906
Hermes, J. J., Mullally, F., Østensen, R. H., et al. 2011, ApJL, 741, L16
Hermes, J. J., Charpinet, S., Barclay, T., et al. 2014, ApJ, 789, 85
Hermes, J. J., Montgomery, M. H., Bell, K. J., et al. 2015, ApJL, 810, L5
Hermes, J. J., Kawaler, S. D., Romero, A. D., et al. 2017a, ApJL, 841, L2
Hermes, J. J., Gänsicke, B. T., Kawaler, S. D., et al. 2017b, ApJS, 232, 23
Hess, N. M., Thayer, P. R., Horch, E. P., et al. 2018, SPIE Proc., 10701, 107012E
Hirsch, L. A., Ciardi, D. R., Howard, A. W., et al. 2017, AJ, 153, 117
Hon, M., Stello, D., García, R. A., et al. 2019, MNRAS, 485, 5616
Horch, E. P., Howell, S. B., Everett, M. E., & Ciardi, D. R. 2014, AJ, 795, 60
Howell, S. B., Scott, N. J., Matson, R. A., Horch, E. P., & Stephens, A. 2019, AJ, 158, 113
Howell, S. B., Sobeck, C., Haas, M., et al. 2014, PASP, 126, 398
Huber, D., Chaplin, W. J., Christensen-Dalsgaard, J., et al. 2013a, ApJ, 767, 127
Huber, D., Carter, J. A., Barbieri, M., et al. 2013b, Sci, 342, 331
Huber, D., Zinn, J., Bojsen-Hansen, M., et al. 2017, ApJ, 844, 102
Ilin, E., Schmidt, S. J., Davenport, J. R. A., & Strassmeier, K. G. 2019, A&A, 622, A133
Jacoby, G. H., De Marco, O., Davies, J., et al. 2017, ApJ, 836, 93
Kato, T., & Osaki, Y. 2014, PASJ, 66, L5
Kawaler, S. D., Reed, M. D., Quint, A. C., et al. 2010, MNRAS, 409, 1487
Kennedy, M. R., Garnavich, P., Breedt, E., et al. 2016, MNRAS, 459, 3622
Khan, S., Hall, O. J., Miglio, A., et al. 2018, ApJ, 859, 156
Khan, S., Miglio, A., Mosser, B., et al. 2019, A&A, 628, A35
Kharchenko, N. V., Piskunov, A. E., Röser, S., Schilbach, E., & Scholz, R. D. 2005, A&A, 438, 1163
Kiefer, R., Schad, A., Davies, G., & Roth, M. 2017, A&A, 598, A77
Kilkenny, D., Koen, C., O'Donoghue, D., & Stobie, R. S. 1997, MNRAS, 285, 640
Kirk, B., Conroy, K., Prša, A., et al. 2016, AJ, 151, 68

Kleinman, S. J., Nather, R. E., Winget, D. E., et al. 1998, ApJ, 495, 424

Koenigl, A. 1991, ApJL, 370, L39

Koleva, M., Prugniel, P., Ocvirk, P., Le Borgne, D., & Soubiran, C. 2008, MNRAS, 385, 1998

Konacki, M., Muterspaugh, M. W., Kulkarni, S. R., & Helminiak, K. G. 2009, AJ, 704, 513

Kraus, A. L., Cody, A. M., Covey, K. R., et al. 2015, ApJ, 807, 3

Kraus, A. L., Douglas, S. T., Mann, A. W., et al. 2017, ApJ, 845, 72

Kulkarni, A. K., & Romanova, M. M. 2008, in AIP Conf. Ser. 1068, ed. R. Wijnands, D. Altamirano, P. Soleri, N. Degenaar, N. Rea, P. Casella, A. Patruno, & M. Linares (Melville, NY: AIP), 99–102

Kumar, P., Ao, C. O., & Quataert, E. J. 1995, ApJ, 449, 294

Landolt, A. U. 1968, ApJ, 153, 151

Lee, J. W., Kim, S.-L., Lee, C.-U., et al. 2013, ApJ, 763, 74

Lehmann, H., Southworth, J., Tkachenko, A., & Pavlovski, K. 2013, A&A, 557, A79

Lehmann, H., Zechmeister, M., Dreizler, S., Schuh, S., & Kanzler, R. 2012, A&A, 541, A105

Libralato, M., Nardiello, D., Bedin, L. R., et al. 2016, MNRAS, 463, 1780

Livingston, J. H., Dai, F., Hirano, T., et al. 2018, AJ, 155, 115

Livingston, J. H., Dai, F., Hirano, T., et al. 2019, MNRAS, 484, 8

Littlefield, C., Garnavich, P., Kennedy, M., Szkody, P., & Dai, Z. 2018, AJ, 155, 232

Littlefield, C., Garnavich, P., Szkody, P., et al. 2020, arXiv: 2004.08923

Lodieu, N., Alonso, R., González Hernández, J. I., et al. 2015, A&A, 584, A128

Loeb, A., & Gaudi, B. S. 2003, ApJL, 588, L117

Lohmann, A. W., Weigelt, G., & Wirnitzer, B. 1983, ApOpt, 22, 4028

Luan, J., & Goldreich, P. 2018, ApJ, 863, 82

Lund, M. N., Lundkvist, M., Silva Aguirre, V., et al. 2014, A&A, 570, A54

Lund, M. N., Chaplin, W. J., Casagrande, L., et al. 2016a, PASP, 128, 124204

Lund, M. N., Basu, S., Silva Aguirre, V., et al. 2016b, MNRAS, 463, 2600

Lundkvist, M. S., Kjeldsen, H., Albrecht, S., et al. 2016, NatCo, 7, 11201

Maceroni, C., Gandolfi, D., Montalbán, J., & Aerts, C. 2012, in IAU Symp. 282, From Interacting Binaries to Exoplanets: Essential Modeling Tools, ed. M. T. Richards, & I. Hubeny (Cambridge: Cambridge University Press), 41

Maceroni, C., Lehmann, H., da Silva, R., et al. 2014, A&A, 563, A59

Mann, A. W., Gaidos, E., Mace, G. N., et al. 2016a, ApJ, 818, 46

Mann, A. W., Newton, E. R., Rizzuto, A. C., et al. 2016b, AJ, 152, 61

Mann, A. W., Gaidos, E., Vanderburg, A., et al. 2017, AJ, 153, 64

Mann, A. W., Vanderburg, A., Rizzuto, A. C., et al. 2018, AJ, 155, 4

Marcy, G. W., Isaacson, H., Howard, A. W., et al. 2014, ApJ, 210, 20

Marsh, T. R., Armstrong, D. J., & Carter, P. J. 2014, MNRAS, 445, 309

Mathur, S., García, R. A., Huber, D., et al. 2016, ApJ, 827, 50

Matson, R. A., Gies, D. R., Guo, Z., et al. 2015, ApJ, 806, 155

Matson, R. A., Howell, S. B., Horch, E. P., & Everett, M. E. 2018, AJ, 156, 31

Matthews, J. M., Kuschnig, R., Guenther, D. B., et al. 2004, Natur, 430, 51

Maxted, P. F. L., & Hutcheon, R. J. 2018, A&A, 616, A38

Maxted, P. F. L., Bloemen, S., Heber, U., et al. 2014, MNRAS, 437, 1681

Mazeh, T., Nachmani, G., Sokol, G., Faigler, S., & Zucker, S. 2012, A&A, 541, A56

Mazumdar, A., Monteiro, M. J. P. F. G., Ballot, J., et al. 2014, ApJ, 782, 18

McQuillan, A., Mazeh, T., & Aigrain, S. 2014, ApJS, 211, 24

Meibom, S., Barnes, S. A., Platais, I., et al. 2015, Natur, 517, 589

Meibom, S., Mathieu, R. D., & Stassun, K. G. 2009, ApJ, 695, 679

Meibom, S., Torres, G., Fressin, F., et al. 2013, Natur, 499, 55

Meissner, F., & Weiss, A. 2006, A&A, 456, 1085

Metcalfe, T. S., Egeland, R., & van Saders, J. 2016, ApJL, 826, L2

Miglio, A., Girardi, L., Rodrigues, T. S., Stello, D., & Chaplin, W. J. 2015, in Asteroseismology of Stellar Populations in the Milky Way, Vol. 39, (Berlin: Springer) 11

Miglio, A., Chaplin, W. J., Brogaard, K., et al. 2016, MNRAS, 461, 760

Miszalski, B., Acker, A., Moffat, F. J., et al. 2009, A&A, 496, 813

Moe, M., & De Marco, O. 2006, ApJ, 650, 916

Molnár, L., Szabó, R., & Plachy, E. 2016, JAVSO, 44, 168

Montgomery, M. H., Hermes, J. J., & Winget, D. E. 2019, arXiv: 1902.05615

Montgomery, M. H., Provencal, J. L., Kanaan, A., et al. 2010, ApJ, 716, 84

Mosser, B., Goupil, M. J., Belkacem, K., et al. 2012, A&A, 548, A10

Murphy, S. J. 2018, arXiv: 1811.12659

Naoz, S., & Fabrycky, D. C. 2014, ApJ, 793, 137

Nardiello, D., Libralato, M., Bedin, L. R., et al. 2016, MNRAS, 463, 1831

Nather, R. E., Winget, D. E., Clemens, J. C., Hansen, C. J., & Hine, B. P. 1990, ApJ, 361, 309

Nowak, G., Palle, E., Gandolfi, D., et al. 2017, AJ, 153, 131

Obermeier, C., Henning, T., Schlieder, J. E., et al. 2016, AJ, 152, 223

O'Leary, R. M., & Burkart, J. 2014, MNRAS, 440, 3036

Orosz, J. A. 2015, in ASP Conf. Ser. 496, Living Together: Planets, Host Stars and Binaries, ed. S.-M. Rucinski, G. Torres, & M. Zejda (San Francisco, CA: ASP), 55

Orosz, J. A., & Hauschildt, P. H. 2000, A&A, 364, 265

Osaki, Y. 1989, PASP, 1, 1

Østensen, R. H., Bloemen, S., & Vučković, M. 2011, ApJL, 736, L39

Østensen, R. H., Silvotti, R., Charpinet, S., et al. 2010, MNRAS, 409, 1470

Pablo, H., Kawaler, S. D., & Green, E. M. 2011, ApJL, 740, L47

Pablo, H., Kawaler, S. D., Reed, M. D., et al. 2012, MNRAS, 422, 1343

Pala, A., Gaensicke, B. T., Marsh, T. R., et al. 2019, MNRAS, 483, 1080

Parker, Q. A., Acker, A., Frew, D. J., et al. 2006, MNRAS, 373, 79

Pecaut, M. J., Mamajek, E. E., & Bubar, E. J. 2012, ApJ, 746, 154

Pepper, J., Gillen, E., Parviainen, H., et al. 2017, AJ, 153, 177

Pinsonneault, M. H., Elsworth, Y. P., Tayar, J., et al. 2018, ApJ, 239, 32

Preibisch, T., Brown, A. G. A., Bridges, T., Guenther, E., & Zinnecker, H. 2002, AJ, 124, 404

Prisinzano, L., Damiani, F., Micela, G., & Sciortino, S. 2005, A&A, 430, 941

Prša, A., Guinan, E. F., Devinney, E. J., et al. 2008, ApJ, 687, 542

Prša, A., Guinan, E. F., Devinney, E. J., et al. 2012, in IAU Symp. 282, From Interacting Binaries to Exoplanets: Essential Modeling Tools, ed. M. T. Richards, & I. Hubeny (Cambridge: Cambridge University Press), 271–8

Prša, A., Robin, A., & Barclay, T. 2015, IJAsB, 14, 165

Prša, A., & Zwitter, T. 2005, ApJ, 628, 426

Raghavan, D., McAlister, H. A., Henry, T. J., et al. 2010, ApJS, 190, 1

Prša, A., Batalha, N., Slawson, R. W., et al. 2011, AJ, 141, 83

Raghavan, D., McAlister, H. A., Henry, T. J., et al. 2010, ApJS, 190, 1

Rampalli, R., Vanderburg, A., Bieryla, A., et al. 2019, AJ, 158, 62

Ramsay, G., Brooks, A., Hakala, P., et al. 2014, MNRAS, 437, 132

Ramsay, G., Wood, M. A., Cannizzo, J. K., Howell, S. B., & Smale, A. 2017, MNRAS, 469, 950

Rappaport, S., Deck, K., Levine, A., et al. 2013, ApJ, 768, 33

Rappaport, S., Nelson, L., Levine, A., et al. 2015, ApJ, 803, 82

Rappaport, S., Lehmann, H., Kalomeni, B., et al. 2016, MNRAS, 462, 1812

Rappaport, S., Vanderburg, A., Borkovits, T., et al. 2017, MNRAS, 467, 2160

Rawls, M. L., Gaulme, P., McKeever, J., et al. 2016, ApJ, 818, 108

Rebull, L. M., Stauffer, J. R., Cody, A. M., et al. 2018, AJ, 155, 196

Rebull, L. M., Stauffer, J. R., Hillenbrand, L. A., et al. 2017, ApJ, 839, 92

Rebull, L. M., Stauffer, J. R., Bouvier, J., et al. 2016a, AJ, 152, 113

Rebull, L. M., Stauffer, J. R., Bouvier, J., et al. 2016b, AJ, 152, 114

Reed, M. D., Baran, A., Østensen, R. H., Telting, J., & O'Toole, S. J. 2012, MNRAS, 427, 1245

Reed, M. D., Foster, H., Telting, J. H., et al. 2014, MNRAS, 440, 3809

Reed, M. D., Kawaler, S. D., Østensen, R. H., et al. 2010, MNRAS, 409, 1496

Reed, M. D., Baran, A., Quint, A. C., et al. 2011, MNRAS, 414, 2885

Reed, M. D., Baran, A. S., Telting, J. H., et al. 2018, OAst, 27, 157

Rendle, B. M., Miglio, A., Chiappini, C., et al. 2019, arXiv: 1906.07489

Rizzuto, A. C., Mann, A. W., Vanderburg, A., Kraus, A. L., & Covey, K. R. 2017, AJ, 154, 224

Rizzuto, A. C., Vanderburg, A., Mann, A. W., et al. 2018, AJ, 156, 195

Rodrigues, T. S., Bossini, D., Miglio, A., et al. 2017, MNRAS, 467, 1433

Romanova, M. M., Ustyugova, G. V., Koldoba, A. V., & Lovelace, R. V. E. 2013, MNRAS, 430, 699

Rowe, J. F., Borucki, W. J., Koch, D., et al. 2010, ApJL, 713, L150

Salabert, D., Régulo, C., Pérez Hernández, F., & García, R. A. 2018, A&A, 611, A84

Salabert, D., Régulo, C., García, R. A., et al. 2016, A&A, 589, A118

Sandquist, E. L., Mathieu, R. D., Quinn, S. N., et al. 2018, AJ, 155, 152

Santos, A. R. G., Campante, T. L., Chaplin, W. J., et al. 2018, ApJS, 237, 17

Scaringi, S., Manara, C. F., Barenfeld, S. A., et al. 2016, MNRAS, 463, 2265

Scaringi, S., Maccarone, T. J., D'Angelo, C., Knigge, C., & Groot, P. J. 2017, Natur, 552, 210

Schmid, V. S., Tkachenko, A., Aerts, C., et al. 2015, A&A, 584, A35

Serenelli, A., Johnson, J., Huber, D., et al. 2017, ApJ, 233, 23

Sharma, S., Stello, D., Bland-Hawthorn, J., Huber, D., & Bedding, T. R. 2016, ApJ, 822, 15

Shibahashi, H., & Kurtz, D. W. 2012, MNRAS, 422, 738

Shporer, A., Fuller, J., Isaacson, H., et al. 2016, ApJ, 829, 34

Silva Aguirre, V., Basu, S., Brandão, I. M., et al. 2013, ApJ, 769, 141

Silva Aguirre, V., Davies, G. R., Basu, S., et al. 2015, MNRAS, 452, 2127

Silva Aguirre, V., Lund, M. N., Antia, H. M., et al. 2017, ApJ, 835, 173

Silva Aguirre, V., Bojsen-Hansen, M., Slumstrup, D., et al. 2018, MNRAS, 475, 5487

Skumanich, A. 1972, ApJ, 171, 565

Slawson, R. W., Prša, A., Welsh, W. F., et al. 2011, AJ, 142, 160

Smullen, R. A., & Kobulnicky, H. A. 2015, ApJ, 808, 166

Southworth, J. 2015, in ASP Conf. Ser. 496, Living Together: Planets, Host Stars and Binaries, ed. S. M. Rucinski, G. Torres, & M. Zejda (San Francisco, CA: ASP), 164

Southworth, J., Zima, W., Aerts, C., et al. 2011, MNRAS, 414, 2413

Spitoni, E., Silva Aguirre, V., Matteucci, F., Calura, F., & Grisoni, V. 2019, A&A, 623, A60

Stauffer, J., Cody, A. M., Baglin, A., et al. 2014, AJ, 147, 83

Stauffer, J., Rebull, L., Bouvier, J., et al. 2016, AJ, 152, 115

Stauffer, J., Collier Cameron, A., Jardine, M., et al. 2017, AJ, 153, 152

Stello, D., Cantiello, M., Fuller, J., et al. 2016a, Natur, 529, 364

Stello, D., Basu, S., Bruntt, H., et al. 2010, ApJL, 713, L182

Stello, D., Vanderburg, A., Casagrande, L., et al. 2016b, ApJ, 832, 133

Stello, D., Zinn, J., Elsworth, Y., et al. 2017, ApJ, 835, 83

Tayar, J., & Pinsonneault, M. H. 2018, ApJ, 868, 150

Themeßl, N., Hekker, S., Southworth, J., et al. 2018, MNRAS, 478, 4669

Thomas, A. E. L., Chaplin, W. J., Davies, G. R., et al. 2019, MNRAS, 485, 3857

Thompson, S. E., Everett, M., Mullally, F., et al. 2012, ApJ, 753, 86

Timmes, F. X., Townsend, R. H. D., Bauer, E. B., et al. 2018, ApJL, 867, L30

Tokovinin, A. 2014, AJ, 147, 87

Torres, G., Andersen, J., & Giménez, A. 2010, A&ARv, 18, 67

Torres, G., Curtis, J. L., Vanderburg, A., Kraus, A. L., & Rizzuto, A. 2018, ApJ, 866, 67

Tovmassian, G., Szkody, P., Yarza, R., & Kennedy, M. 2018, ApJ, 863, 47

Tutchton, R. M., Wood, M. A., Still, M. D., et al. 2012, JSARA, 6, 21

Uytterhoeven, K., Moya, A., Grigahcène, A., et al. 2011, A&A, 534, A125

Van Eylen, V., Agentoft, C., Lundkvist, M. S., et al. 2018, MNRAS, 479, 4786

Van Eylen, V., & Albrecht, S. 2015, ApJ, 808, 126

Van Eylen, V., Albrecht, S., Huang, X., et al. 2019, AJ, 157, 61

Van Grootel, V., Charpinet, S., Fontaine, G., et al. 2010, ApJL, 718, L97

van Kerkwijk, M. H., Rappaport, S. A., Breton, R. P., et al. 2010, ApJ, 715, 51

van Saders, J. L., & Gaudi, B. S. 2011, ApJ, 729, 63

VandenBerg, D. A., Brogaard, K., Leaman, R., & Casagrand e, L. 2013, ApJ, 775, 134

Vanderburg, A., Mann, A. W., Rizzuto, A., et al. 2018, AJ, 156, 46

Vauclair, G., Moskalik, P., Pfeiffer, B., et al. 2002, A&A, 381, 122

Verma, K., Raodeo, K., Antia, H. M., et al. 2017, ApJ, 837, 47

Verma, K., Raodeo, K., Basu, S., et al. 2019, MNRAS, 483, 4678

Walker, G., Matthews, J., Kuschnig, R., et al. 2003, PASP, 115, 1023

Wang, J., Fischer, D. A., Xie, J.-W., & Ciardi, D. R. 2014, ApJ, 791, 111

Wang, J., Fischer, D. A., Horch, E. P., & Xie, J.-W. 2015, ApJ, 806, 248

Warner, B., & Robinson, E. L. 1972, NPhS, 239, 2

Welsh, W. F., Orosz, J. A., Aerts, C., et al. 2011, ApJS, 197, 4

White, T. R., Huber, D., Maestro, V., et al. 2013, MNRAS, 433, 1262

Wilking, B. A., Gagné, M., & Allen, L. E. 2008, Handbook of Star Forming Regions, Star Formation in the ρ Ophiuchi Molecular Cloud, Vol. 5, ed. B. Reipurth (San Francisco, CA: ASP), 351

Wilson, R. E. 1979, ApJ, 234, 1054

Winget, D. E., Nather, R. E., Clemens, J. C., et al. 1991, ApJ, 378, 326

Winget, D. E., Nather, R. E., Clemens, J. C., et al. 1994, ApJ, 430, 839

Wittrock, J. M., Kane, S. R., Horch, E. P., et al. 2016, AJ, 152, 149

Wood, M., Still, M. D., Howell, S. B., Cannizzo, J. K., & Smale, A. P. 2011, ApJ, 741, 105

Wu, Y., & Goldreich, P. 2001, ApJ, 546, 469

Zima, W. 1997, The Campaigns of the Delta Scuti Network Delta Scuti Star Newsletter, Issue 11, p37

Zinn, J. C., Pinsonneault, M. H., Huber, D., & Stello, D. 2019, ApJ, 878, 136

Zong, W., Charpinet, S., Vauclair, G., Giammichele, N., & Van Grootel, V. 2016, A&A, 585, A22

Zucker, S., Mazeh, T., & Alexander, T. 2007, ApJ, 670, 1326

Chapter 5

The Solar System as Observed by K2

Csaba Kiss, László Molnár, András Pál and Steve B Howell

5.1 Introduction

The original Kepler mission continuously observed a location in the sky near the Northern Ecliptic Pole. This sky location, being out of the plane of the Ecliptic, kept the number of solar system objects that wandered into the field-of-view at a minimum. Nevertheless, some asteroids and comets, having very inclined, near-polar orbits, were imaged. A search in the Kepler data for microlensing events caused by primordial black holes found none, but revealed correlated brightenings across the field of view which turned out to be three comets that passed in front of the Kepler spacecraft in quarters 5 and 9 (Griest et al. 2014).

The K2 mission, viewing in the ecliptic plane, opened up the possibility to make in depth observations of solar system objects. The Kepler space telescope, however, was never designed to image moving targets due to its primary exoplanet discovery mission. While stellar sources were observed using small postage stamp groups of pixels assigned to pre-selected sources, solar system objects required the use of specific large rectangular pixel masks made big enough to capture the moving object throughout a K2 campaign. Early pixel budget calculations suggested that only slowly moving distant objects were feasible. Trans-Neptunian objects (TNOs) spent several weeks near their stationary points with slow apparent motions, whereas Main Belt asteroids (MBAs) moved across the field-of-view within days.

5.2 Source Identification and Data Reduction

Imaged targets can be associated with known objects using the standard tools that provide ephemerides (e.g., JPL/Horizons). However, imaged unidentified asteroids which crossed through a K2 field-of-view caused two significant problems. First, public software tools useful to search for asteroids in large field-of-views, such as

5-1

MPChecker[1] provide sky locations for observers on Earth while the Kepler spacecraft orbited the Sun in an Earth-trailing orbit. Second, other software packages, such as virtual observatory (VO) extensions for SkyBoT (Berthier et al. 2016) were designed for identifying asteroids in a single epoch, one observation only, method. Thus, such software tools were not optimal for identifying and producing lists of objects passing through a K2 superstamp of pixels during a full campaign. Therefore, as we see in Section 5.2; new and efficient software tools were needed to be implemented in order to effectively handle these problems. One solution is the EPHEMD approach (Pál et al. 2018), which performs a cone search of moving objects from a pre-defined location (such as the Kepler spacecraft) where the time interval of the query is also specified. While specifying an interval, a reported position. This type of object query is rather relevant in the case of analyzing serendipitous detection of asteroids (Szabó et al. 2016; Molnár et al. 2018).

Following the identification of imaged objects—i.e., obtaining the pixel positions of the asteroids by combining the astrometric plate solutions of the individual CCD images with the individual per-object ephemerides retrieved by one of the aforementioned services (NASA/JPL Horizons, EPHEMD)—one can then extract the fluxes corresponding to the targets of interest. As it can be in Figure 5.1, this is done in mainly two steps: first, a pre-processing stage is executed providing differential images while in the next step, flux is extracted from the apparent position of the minor planets. Due to the lengthy integration time and the moving asteroid, the aperture for this flux extraction is going to be an oblong-shaped area on the grid of CCD pixels due to the large apparent proper motion. In the case of trans-Neptunian objects and Centarus, this motion is negligible but for main-belt objects, Hilda family members, and Trojans, one should indeed take into account this fast proper motion, usually in the range of a few (3–5) pixels per 30 minute long cadence observation. We note here that despite the employment of differential imaging analysis, background variations can also be present due to zodiacal light and stray light variations presented in the background annulus.

After extraction of the instrumental fluxes, observed magnitudes are derived using the USNO-B1 catalog (Monet et al. 2003). This is simply performed by shifting the zero points using a few to a dozen of reference stars. The accuracy of this transformation is good to ~0.1 magnitude (see, e.g., Pál et al. 2015) since the spectral response of Kepler is centered nearly at the same wavelength as the R photometric band. This post-processing of the extraction of magnitudes for minor planet light-curves is then followed by the time-domain analysis in order to retrieve the rotation and shape characteristics of the object. Via the example of the Hilda asteroid 46302 (2001 OG$_{13}$), displayed on Figure 5.1; we show the result of the residual spectrum analysis and the corresponding best-fit folded light-curve in Figure 5.2. The aforementioned examples and the light-curves presented in this section were generated by the various built-in tasks of the FITSH photometric pipeline package (Pál 2012). This procedure can then be repeated for all of the

---

[1] https://cgi.minorplanetcenter.net/cgi-bin/checkmp.cgi.

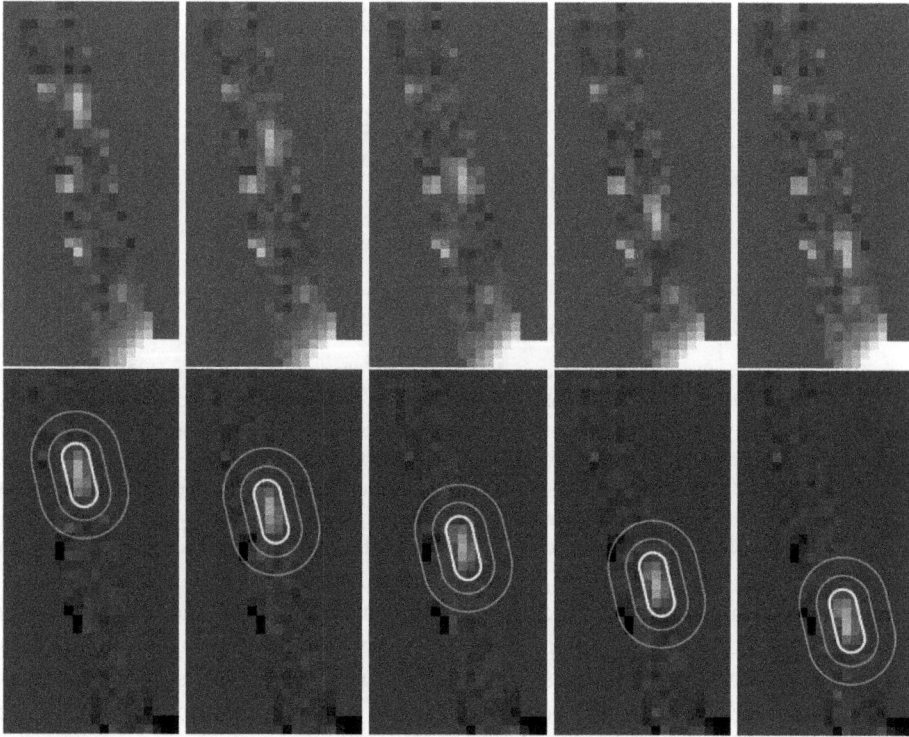

Figure 5.1. Plots showing the most relevant features of small solar system body observations via the example of the Hilda asteroid 46302 (2001 OG_{13}), measured during K2 Campaign 6, on Module/CCD 23.1. The upper panel displays five consecutive frames from the original data series while the lower panel shows these frames after the pre-processing for image subtraction-based photometry. The oblong-shaped aperture used for flux extraction is marked by the yellow curve while the annulus used for background estimation is the area within the two red oblong curves. These series of images clearly show that differential image processing effectively removes the background stars and structures, even the ones which are much brighter than our target asteroid (see, e.g., the bright star on the lower-right corner).

moving objects in the same Kepler imager CCD, assuming the same astrometric plate solution.

5.3 Minor Bodies in the Solar System

5.3.1 Main Belt Asteroids

Observing in the ecliptic plane presented new challenges for stellar photometry, with sharply increased contamination from solar system objects. Initial tests done by Szabó et al. (2015) using the very first K2 engineering campaign test data confirmed that this was to be expected, but it also showed that, in principle, light-curves of asteroids that crossed the stellar postage stamp pixels during an integration can be extracted too. This initial study concluded that the viewing geometry of the K2 mission meant that asteroid discoveries, however, were unlikely.

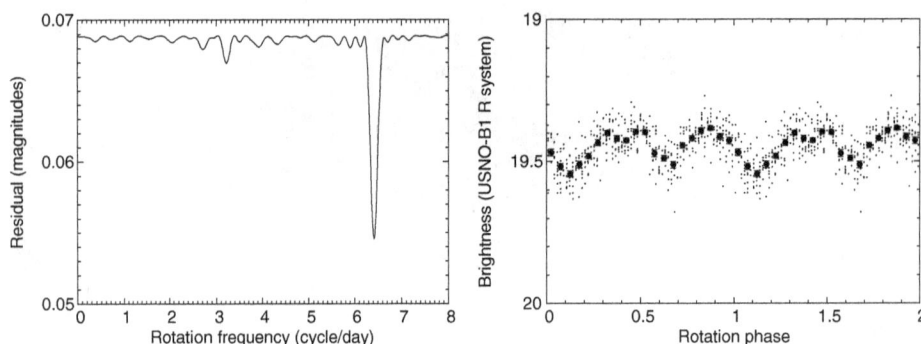

Figure 5.2. The final results of the photometric data analysis of the light-curve of the Hilda asteroid 46302 (2001 OG$_{13}$). The left panel shows the residual root mean squares spectrum as the function of possible rotation frequencies. The lower the residual, the more significant is the corresponding rotation frequency. The width of the prominent peak also yields the uncertainty of the possible period: for this one it is $f = 6.419 \pm 0.029$ cycles/day. The right panel shows the folded light-curve with the half of this frequency, due to the double-peaked nature of the observed periodic light variations. As in the case of other surveys, this feature is rather common in minor planet light-curves. For this Hilda object, therefore, the true rotation period is $P = 7.477 \pm 0.034$ hr.

As the mission progressed, and the pointing instabilities of the spacecraft were better understood, stellar pixel postage sizes got smaller and asteroid crossings became too short to provide meaningful light-curves. However, some areas in the sky were covered with larger, continuous pixel superstamp mosaics. Main Belt asteroids crossed over these pixel areas typically in 1–2 days, providing an opportunity to collect light-curves that were often longer than single-night measurements from the ground. Most of these mosaics, or superstamps, were placed on high-density stellar fields, open clusters and the galactic bulge, where detection efficiency was limited even with image subtraction employed. The Uranus and Neptune superstamps could be exploited much more effectively (Figure 5.3). Overall, Szabó et al. (2016) and Molnár et al. (2018) were able to determine rotation rates for 127 asteroids out of 1628 detected moving targets, most of which had no previously known periods.

Although most of the asteroid light-curves did not exceed three days in length, they still provided coverage that was superior to ground-based observations. Detailed studies found the median rotation rate of the K2 asteroid sample, 9.45 hr, to be significantly longer than the 7.00 hr value of the current ground-based light-curve database (Warner et al. 2009). This result clearly confirms that ground-based observations are biased toward shorter periods with long spin periods being underrepresented. This finding highlights the need for space-based photometry in solar system and planetary science.

Analysis of the individual asteroid light-curves revealed a wide variety of light-curve shape, from simple contact binaries to more complex, triple- and quadruple-peaked ones. Some targets suggested multiple periods or amplitude changes that could indicate tumbling asteroids, the best example being (94314) 2000 AO$_{165}$ (Molnár et al. 2018). Despite the initial skepticism, two new objects were discovered. 2015 BO$_{519}$ was identified in the Neptune superstamp, and later turned out to be a

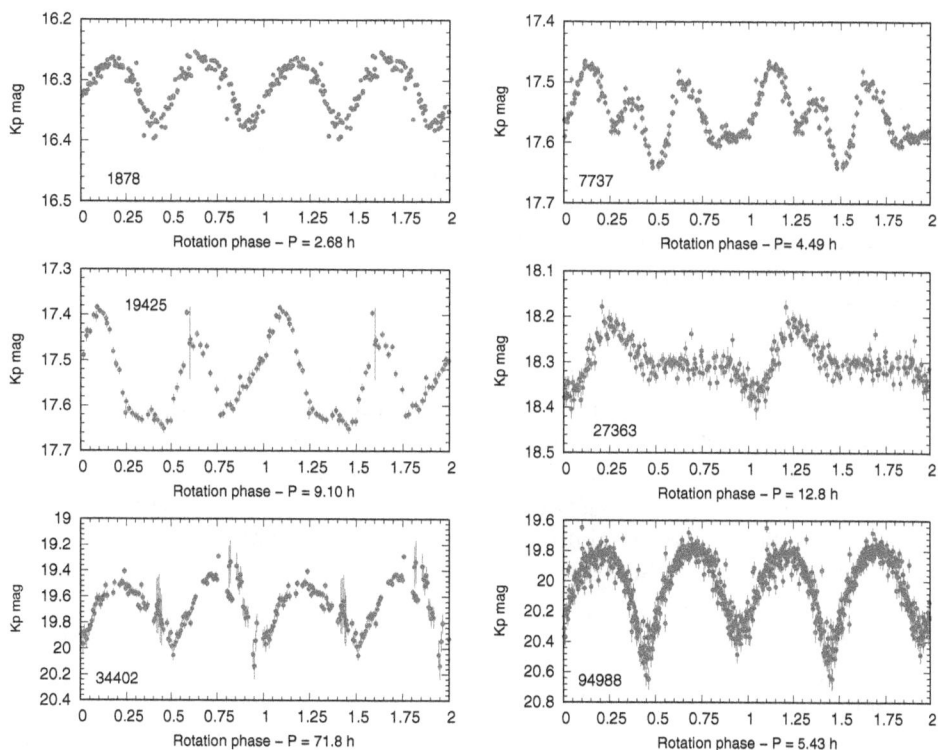

Figure 5.3. Six examples of MBA light-curves from the C8 Uranus superstamp. Note the wide range of rotation periods. Catalog numbers of each asteroid are indicated in the plots.

potentially hazardous asteroid (PHA). (506121) 2016 BP$_{81}$, a TNO from the cold classical group, was discovered in the Uranus field, and was independently identified by the outer solar system origins survey (Bannister et al. 2017).

Later in the mission, the Small Bodies Near and Far (SBNAF) group started to propose targeted observations for selected asteroids as well, spanning several days per object. These were typically bodies where thermal modeling was feasible but the photometric amplitudes or periods were ambiguous or poorly understood. Here K2 data was used in conjunction with prior and additional visual and infrared observations to derive accurate shape, size, rotation, and thermophysical properties. At the time of writing, spin axis direction and thermal inertia have been derived for the first target, (100) Hekate, by Marciniak et al. (2019).

K2 also observed two well-known PHAs, (99942) Apophis and (162173) Ryugu, the latter being the target of the Hayabusa-2 space probe (Tsuda et al. 2013). However, they were both faint, at the detection limit of the telescope, as their already low apparent brightness gets smeared out a bit over several pixels in each cadence due to their fast apparent motion. To date, no light-curve photometry has been published.

5.3.2 Jovian Trojans and Comets

Outside the Main Belt we find asteroid groups that are in mean motion resonances with Jupiter. These resonances maximize their distances from the gas giant, keeping them in their orbits. The two largest groups are the Hilda and Trojan families, at 3:2 and 1:1 orbital resonances. More than hundred members of each of the two groups were observed with special target masks that tiled their apparent motions near the stationary points throughout the mission. As of this writing, most of the observations are still being processed.

The first set of data were taken toward 56 members of the L4 (leading) swarm of the Trojan family, and were analyzed by two teams independently. Both Szabó et al. (2017) and Ryan et al. (2017) found that slow rotators, above 60–100 hr periods, are overabundant compared to other asteroid populations. Various tests done by the teams suggest that several members, roughly a fifth of the sample, could be binary objects. The Lucy space mission will visit a number of asteroids that include both fast and extremely slow rotators, with (11351) Leucus having a period of 445.7 hr, so an unbiased rotation rate distribution could tell how representative these targets are within the Trojan family (Buie et al. 2018). Out of the five primary Lucy targets, only Leucus was observed by K2, in Campaign 11.

All observed Trojans rotate slower than five hours, although the sample is large enough to detect faster ones that should be present if they were to follow the period distribution of the Main Belt. The lack of fast rotators in the sample suggests a low critical density, about 0.5 g cm^{-3}. This is much lower than the critical density in the Main Belt but consistent with that of cometary nuclei that have a lower spin barrier before they break apart. The K2 results confirm earlier findings that suggested icy, porous composition for these objects, and therefore favor formation scenarios where Trojans formed in the outer solar system, and were scattered inwards until they were captured and locked into resonance with Jupiter.

A number of comets were also observed during the K2 mission with dedicated pixel mosaics, analogous to the targeted asteroid observations. The most notable targets were 67P/Churyumov–Gerasimenko, the target of the *Rosetta* mission, C/2013 A1 (Siding Spring) that flew past very close to Mars, and 2P/Encke, one of the shortest-period comets known to us. However, these observations are still being processed, and only preliminary results have been presented. Brightness measurements of 67P, for example, agreed with the predicted rate of fading at a time when it was not observable any more from Earth.

5.3.3 Centaurs and TNOs

At the start of the mission trans-Neptunian objects (and Centaurs, their siblings) were assumed to be the possible primary targets for K2 moving object light-curve photometry. The feasibility of the concept was first proven by the observations of two Kuiper Belt objects, 2002 GV$_{31}$ and 2007 JJ$_{43}$ (Pál et al. 2015). As 2007 JJ$_{43}$ is relatively bright ($R \approx 20.5$ mag) a well-defined light-curve with a period of 12.10 hr could be obtained. This was also the case for the much fainter 2002 GV$_{31}$ ($R \approx 23.0$ mag, $P = 29.2$ hr), close to the limit that could be achieved with K2 observations. As

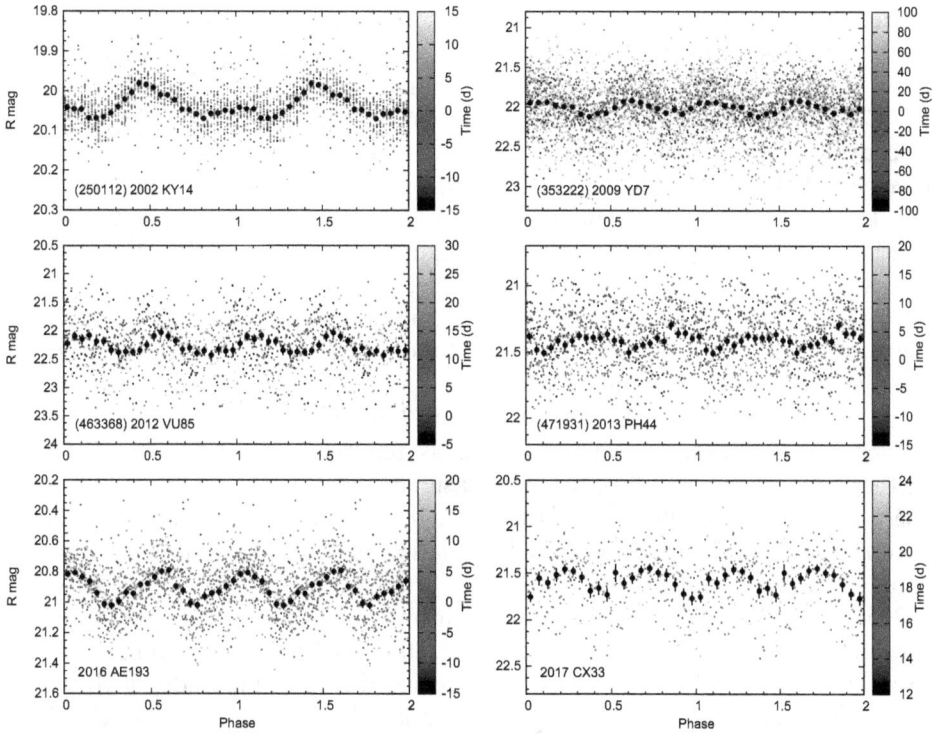

Figure 5.4. Folded light-curves for six centaurs observed in the K2 campaigns. The folding periods are: 2002 KY_{14}: 8.5004 hr; 2009 YD_7: 10.1592 hr; 2012 VU_{85}: 56.20 hr; 2013 PH_{44}: 22.2 hr; 2016 AE_{193}: 9.115 hr; 2017 CX_{33}: 21.51 hr. All targets are shown assuming double peak light-curves. Adapted from Marton et al. (2020). CC BY 4.0.

it was later seen in other Centaur and TNO observations the brightness of the target is not the only factor that influences detectability. Background features and close passes near fixed stars during the motion of the target were important as well—in many cases these are the prime limiting factors leading to unsuccessful light-curve detection even at $R \approx 21$ mag.

Centaur light-curves are rare; prior to K2 measurements only 19 Centaurs had reported light-curve periods (Peixinho et al. 2020). Light-curves from K2 observations of ten Centaurs have been obtained by (Marton et al. 2020), including (250112) 2002 KY_{14}, (353222) 2009 YD_7, 2010 GX_{34}, 2010 JJ_{124}, (499522) 2010 PL_{66}, (471931) 2013 PH_{44}, (463368) 2012 VU_{85}, (472760) 2015 FZ_{117}, (514312) 2016 AE_{193}, and (523798) 2017 CX_{33}. One target, 2002 KY_{14} had previous light-curve measurements, but its rotation period was uncertain. K2 measurements revealed a high-quality double-peak light-curve ($P = 8.4005 \pm 0.0010$hr) with two very different half-periods. Light-curve periods could be derived for five other K2 Centaurs: 2009 YD_7, 2013 PH_{44}, 2012 VU_{85}, 2016 AE_{193}, and 2017 CX_{33} (see Figure 5.4). For the other four targets only amplitude upper limits could be derived, and light-curve periods could not be obtained, mainly due to the large number of frames that were affected by stellar residuals and electronic cross-talk patterns, and had to be discarded. A target from this sample, 2012 VU_{85} has the longest known rotation period among Centaurs, $P = 56.20 \pm 0.16$ hr, and two other targets, 2013 PH_{44} and 2017 CX_{33} have rotation periods above 20 hr ($P = 22.2 \pm 0.8$ hr and 21.51 ± 0.17 hr, respectively), which are also rather long, compared with the mean pre-K2 light-curve period of ~8.1 hr (Peixinho et al. 2020). For these long rotation period targets, feasible equilibrium configurations representing a rotating distorted body approximated by a Jacobi ellipsoid, could not be obtained. The required densities would be too low, below ~$0.1 \mathrm{g} \, \mathrm{cm}^{-3}$, which is unfeasible, even for icy bodies in the outer parts of the solar system. In these cases the slow rotation could be explained by a tight binary where the components are tidally locked and we see the common orbital period as the light-curve period.

The light-curve of the prominent Centaur (2060) Chiron was also derived from K2 observations taken in Campaign 12, as reported by Marton et al. (2017), together with space based far-infrared measurements of the same target. The small observed amplitude confirms that the decadal light-curve amplitude variations of Chiron are caused by the aspect angle variation of the disk + nucleus system, as proposed by Ortiz et al. (2015).

(225088) 2007 OR_{10} is a large Kuiper Belt object with one of the reddest surfaces known in the solar system (Boehnhardt et al. 2014). K2 data obtained in Campaign 3 were analyzed by Pál et al. (2016), and revealed a low amplitude, likely double peak light-curve with $P = 44.81 \pm 0.37$ hr period, however, a single peak solution ($P = 22.40 \pm 0.18$ hr) could not have been ruled out either. Even the single peak period is unusually long among trans-Neptunian objects. Due to the large size (>1200 km) its shape is likely close to a Maclaurin spheroid, and consequently the light-curve is probably caused by surface albedo variations, rather than by shape effects.

The slow rotation observed by K2 for this object was supposed to be caused by tidal interactions with a relatively massive satellite, and a subsequent study by Kiss et al. (2017) indeed revealed the presence of a satellite on archival Hubble Space Telescope (HST) images. This finding lead to the conclusion that *all* objects in the Kuiper Belt with diameters above ~1000 km have satellites, an important constraint for planet formation theories. New observations with the HST in 2017 allowed the determination of the satellite orbit and the derivation of the system mass (Kiss et al. 2019). An analysis of the system's dynamical evolution suggests that the presently known satellite is probably too small to be responsible for the slow rotation of 2007 OR_{10}.

A general result from Centaur and TNO K2 observations is that they allowed the detection of rotation periods notably longer ($\geqslant$10 hr) than those obtained from ground-based observations. At the time this chapter is written, suitable statistics are available for Centaurs only. For these objects, the mean (Maxwellian distribution fitted) rotation period has increased from 8.1 hr to 8.8 hr by adding five objects (+20%) to the sample. The K2 TNOs with known rotation periods also rotate notably slower than the average Kuiper Belt objects. These long rotation periods could not have been detected easily by ground-based telescopes.

Revisiting Pluto

Contributed by László Molnár, Konkoly Observatory.

The most well-known TNO, (134340) Pluto, was observed by K2 in Campaign 7, only months after the New Horizons probe made its historic flyby (Stern et al. 2015). With the high-resolution images at hand, it was possible to match features of the light-curve to prominent dark or bright areas, such as the Chtulhu Macula and the Tombaugh Region, directly. As so little time passed between the two observations, the K2 integrated-light measurements did not reveal any surface changes, but were important nevertheless. They confirmed that the amplitude of the visual light-curve has been dropping faster since Pluto's perihelion passage than what a static frost model would predict. Instead, volatiles continue to move across the surface of Pluto through evaporation and re-deposition processes (Figure 5.5). With the high-resolution surface maps at hand, it is now possible to decipher whether further changes to the light-curve are caused by geologic activity at certain areas or by volatile deposition on the surface of the dwarf planet (Benecchi et al. 2018).

Figure 5.5. A composite image shows enhanced-color views of Pluto (lower right) and Charon (upper left). Credit: NASA/JHUAPL/SwRI.

5.3.4 Planetary Satellites

The major challenge in observing regular satellites (moons of the major planets) with K2 is their spatial proximity to the bright host planet. As the giant planets' scattered light is very strong, accurate photometry of the regular satellites can only be performed if the point-spread function of the planet is very well known. While K2 was observing Saturn, Uranus, and Neptune as well, with many potentially observable regular satellites, time series photometry was performed only on Titan, the largest moon of Saturn, and the only natural satellite in the solar system with a dense atmosphere. Parker et al. (2019) applied a k-means aperture optimization to the K2 data of Titan and obtained ⩽0.33% photometric scatter despite the typically very high background levels. At this level of accuracy, no clear signal of atmospheric variations could be found. However, in the same period, the Cassini spacecraft detected some cloud activity on Titan. The clouds observed by Cassini had low contrast and small hemispherical coverage, and their deduced detectability was close to that achieved by the k-means aperture optimization technique.

Unlike regular moons, irregular satellites orbit their host planet at large distances on highly eccentric and inclined orbits; their orbital periods often exceed many years (see, e.g., Jewitt & Haghighipour 2007; for a review). These moons were most likely captured from heliocentric orbits early after the main planetary accretion phase, and therefore represent one of the best probes of this important era. All four giant planets have extensive satellite swarms. Due to their larger distances from their host planet, time series photometric light-curves, in many cases, did not require special software tools, as in the case of Titan above. However, the detectability of the light-curve was limited by the faintness of the targets.

Nereid, the largest irregular satellite of Neptune, was observed in Campaign 3 of the K2 mission. From this data set Kiss et al. (2016) obtained a well-defined light-curve with a rotation period of $P = 11.594 \pm 0.017$ hr and an amplitude of $\Delta m = 0.0328 \pm 0.0018$ mag, confirming previous short rotation periods obtained in ground-based observations. The similarities of light-curve amplitudes between 2001 and 2015 show that Nereid is in a low-amplitude rotation state nowadays but it could have been in a high-amplitude rotation state in the mid-1960s. This difference is likely due to changes in the aspect angle of its spin axis. Another high-amplitude rotation period is expected in about 30 years. Based on the light-curve amplitudes observed in the last 15 years the shape of Nereid can be constrained to a maximum axial ratio of 1.3:1. This excludes the previously suggested very elongated shape of ≈1.9:1 and clearly shows that Nereid's spin axis cannot be in forced precession due to tidal forces. Additional thermal emission data from the Spitzer Space Telescope and the Herschel Space Observatory also indicate that Nereid's shape is close to the ratio limit of 1.3:1 obtained with the K2 data and it has a very rough, highly cratered surface.

Five Uranian irregular satellites were observed in Campaign 8 by K2. Caliban and Ferdinand fell with their full tracks on the same large mosaic that was set to follow the apparent motion of Uranus, while the motion of Setebos and Sycorax was covered partly by this mosaic and some additional pixels on the same CCD module. Prospero fell into an adjacent CCD module and its motion was covered with a narrow band of pixels. Light-curve analysis by Farkas-Takács et al. (2017) clarified the previously ambiguous rotation period obtained for the largest satellite, Sycorax (now 6.9190 ± 0.0082 hr considering a double-peaked light-curve), and provided rotation periods for the other four satellites. Previously, they had no firm light-curve periods measured in the literature. The main result from the K2 data is that irregular Uranian satellites typically rotate faster than those orbiting Jupiter and Saturn, suggesting a different collisional evolution. The mean spin rate of the Uranian irregulars is close to that of Centaurs and trans-Neptunian objects. The character-istic red color and moderately bright surfaces of Kuiper Belt objects appear only around Uranus, and not in the irregular satellite swarms of the other giant planets. This is in agreement with solar system formation and evolution theories in which Uranus formed at a heliocentric distance beyond that of Neptune, and should have captured more irregular satellites from a regions which was the closest to the Kuiper Belt.

5.4 Planets

Planets moving across the K2 field of view, such as Jupiter or the Earth, normally meant nuisance, creating intense scattered light, internal reflections, and video cross-talk in the electronics (Figure 5.6). A passage of the Earth was captured in a well-timed full frame image collected by K2 and planned awareness of the event was raised through the #waveatkepler social media campaign. But in scientific terms, solar system planets were mostly just noise sources for the mission—except for one.

Neptune is far away and dim enough to move slowly and be only moderately saturated in the K2 images. While a moving saturated image was still tricky to measure accurately, hopes were raised that global oscillations (planet seismology) might be detectable in the planet. This type of study would by analogous to asteroseismic studies of stars, and was successfully done for Jupiter in the past (Gaulme et al. 2011). A clear oscillation spectrum was indeed found in the light coming from Neptune. However, those oscillations came not from the planet but from the Sun, and were preserved in the Sun light reflected by the planet. Once those were removed, no other signals could be found: the oscillations generated within the ice giant remained elusive. Nevertheless, this gave an excellent opportunity to study the Sun as it would appear to Kepler as if from a distance exoplanet host star. Since many asteroseismic applications depend on scaling the solar seismic parameters, observing the Sun itself with the same instrument could provide important calibration data. Interestingly, two different asteroseismic approaches applied to the solar reflected light data came to different conclusions. The "peak-bagging" method, which fits each frequency component simultaneously, gave results that agreed generally with con-temporaneous solar data. In contrast, a global approach based on the $\nu_{\max}$ and $\Delta\nu$ values only, overestimated the radius and mass of the Sun. This discrepancy was traced back to the stochastic nature of the oscillations that may cause small differences in the value of $\nu_{\max,\odot}$ for a given time period compared to the canonical value as well as the higher level of noise in the K2 data (Gaulme et al. 2016; Rowe et al. 2017).

Although planetary oscillations were not detected, Neptune did show some variation thanks to its rotation, and the cloud features in its atmosphere. This observation also had broad implications. Photometric variations observed in brown dwarfs are connected to their atmospheric chemistry, cloud patterns, and weather,

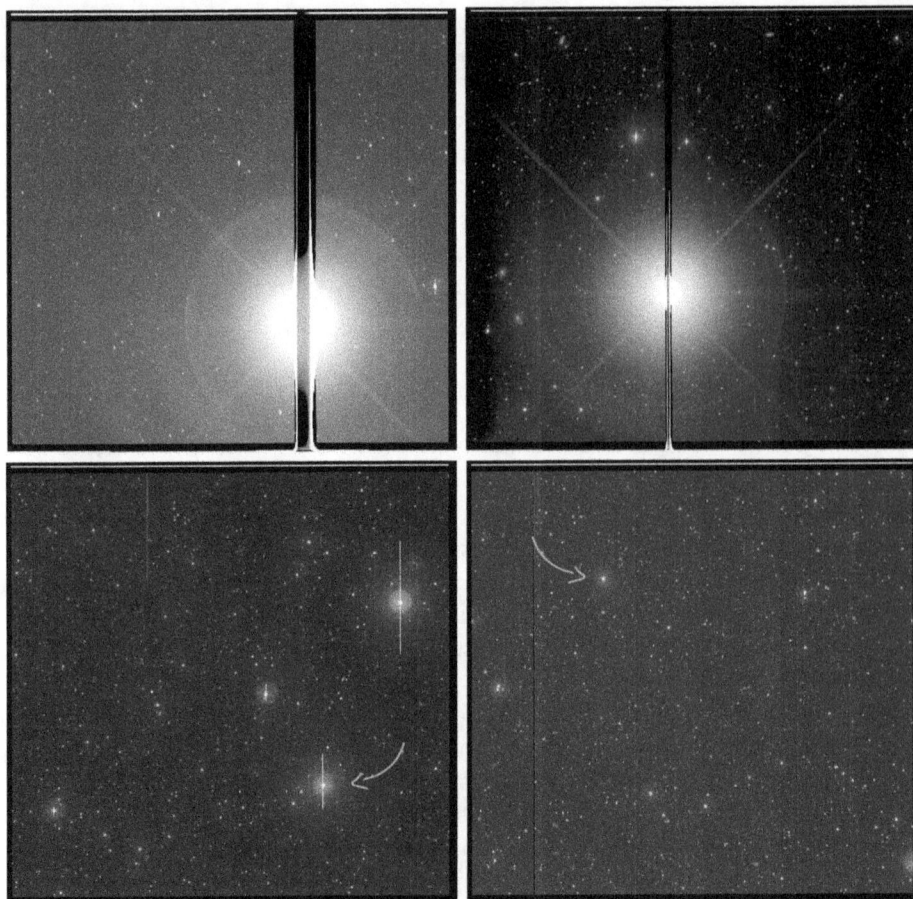

Figure 5.6. Solar system planets, as seen by *Kepler*: Earth and Moon (top left), Mars (top right), Uranus (bottom left), and Neptune (bottom right). Each image is an K2 full frame readout showing only a single CCD channel.

making them look more similar to hot gas giants than to stars (Marley et al. 2013; Biller 2017). As they rotate, the cloud features cause photometric modulations revealing weather patterns in their atmospheres. These modulations have been predicted and, in rare cases, have already been detected on exoplanets (Armstrong et al. 2016; Zhou et al. 2016). It is not easy, however, to directly relate light-curve variations to possible atmospheric or surface patterns on these objects. In contrast, detailed resolved images from large telescopes and space missions exist in abundance for solar system giant planets. However, producing disk-integrated, continuous photometric observations of our solar system giant planets would be of great interest as such data can be used for comparisons with far away, unresolved sources like brown dwarfs. This need made the K2 photometric data for Neptune especially valuable.

Contemporaneous Keck telescope and subsequent Hubble Space Telescope observations of Neptune showed that the K2 light-curve was dominated by a single large discrete storm, and the finer variations likely came from smaller, evolving cloud features (Simon 2016). The K2 light-curves were also compared with high-cadence Spitzer Space Telescope light-curves, which covered a single planet rotation, and sparse WISE IR data. The optical variation mirrored the infrared one but with a relative amplitude reduced by two orders of magnitude. This is because in the optical, reflected light is dominated by Rayleigh-scattering that mutes the signal that comes from high-altitude clouds. In the infrared, this scattering is reduced, and light contrast becomes especially high at the methane and other gas absorption bands, making surface features more visible (Stauffer et al. 2016).

In a later campaign, Uranus was also observed by K2. The more homogeneous appearance of the planet in the optical could, in principle, help in identifying planetary oscillation signals. But the planet is bright, and thus saturated much more severely in the K2 images. During the K2 observations, multiple bright moons of Uranus can be seen orbiting the planet, crossing the saturation columns regularly making the photometric data analysis much more challenging than it was in the case of Neptune.

References

Armstrong, D. J., de Mooij, E., Barstow, J., et al. 2016, NatAs, 1, 4

Bannister, M. T., Schwamb, M. E., Fraser, W. C., et al. 2017, ApJ, 851, L38

Benecchi, S. D., Lisse, C. M., Ryan, E. L., et al. 2018, Icar, 314, 265

Berthier, J., Carry, B., Vachier, F., Eggl, S., & Santerne, A. 2016, MNRAS, 438, 3394

Biller, B. 2017, AstRv, 13, 1

Boehnhardt, H., Schulz, D., Protopapa, S., & Götz, C. 2014, EM&P, 114, 35

Buie, M. W., Zangari, A. M., Marchi, S., Levison, H. F., & Mottola, S. 2018, AJ, 155, 245

Farkas-Takács, A., Kiss, C., Pál, A., et al. 2017, AJ, 154, 119

Gaulme, P., Schmider, F.-X., Gay, J., Guillot, T., & Jacob, C. 2011, A&A, 531, 104

Gaulme, P., Rowe, J. F., Bedding, T. R., et al. 2016, ApJ, 833, L13

Griest, K., Cieplak, A. M., & Lehner, M. J. 2014, ApJ, 786, 158

Jewitt, D., & Haghighipour, N. 2007, ARAA, 45, 261

Kiss, C., Pál, A., Farkas-Takács, A., et al. 2016, MNRAS, 457, 2908

Kiss, C., Marton, G., Farkas-Takács, A., et al. 2017, ApJL, 838, L1

Kiss, C., Marton, G., Parker, A. H., et al. 2019, Icar, 334, 3

Marciniak, A., Alí-Lagoa, V., Müller, T. G., et al. 2019, A&A, 625, 139

Marton, G., Kiss, C., Molnár, L., et al. 2020, Icar, 345, 113721

Marley, M. S., Ackerman, A. S., Cuzzi, J. N., & Kitzmann, D. 2013, in Comparative Climatology of Terrestrial Planets, ed. S. J. Mackwell, A. A. Simon-Miller, J. W. Harder, & M. A. Bullock (Tucson, AZ: University of Arizona Press), 367

Marton, G., Kiss, C., Müller, T., et al. 2017, K2 and Herschel/PACS light curves of the Centaur (2060) Chiron; 2017, EPSC abstracts, 11, EPSC2017-213

Molnár, L., Pál, A., Sárneczky, K., et al. 2018, ApJS, 234, 37

Monet, D. G., Levine, S. E., Canzian, B., et al. 2003, AJ, 125, 984

Ortiz, J. L., Duffard, R., Pinilla-Alonso, N., et al. 2015, A&A, 576, A18

Pál, A. 2012, MNRAS, 421, 1825

Pál, A., Szabó, R., Szabó, G. M., et al. 2015, ApJL, 804, 45

Pál, A., Kiss, C., Müller, T., et al. 2016, AJ, 151, 117

Pál, A., Molnár, L., & Kiss, C. 2018, PASP, 130, 114503

Parker, A. H., Hörst, S. M., Ryan, E. L., & Howett, C. J. A. 2019, PASP, 131, 084505

Peixinho, N., Thirouin, A., Tegler, S. C., et al. 2020, in The Trans-Neptunian Solar System, ed. D. Prialnik, M. A. Barucci, & L. Young (Amsterdam: Elsevier), 307

Rowe, J. F., Gaulme, P., Lissauer, J. J., et al. 2017, AJ, 153, 149

Ryan, E. L., Sharkey, B. N. L., & Woodward, C. E. 2017, AJ, 153, 116

Simon, A. A. 2016, ApJ, 817, 162

Stauffer, J., Marley, M. S., Gizis, J. E., et al. 2016, AJ, 152, 142

Stern, S. A., Bagenal, F., Ennico, K., et al. 2015, Sci, 350, 1815

Szabó, R., Sárneczky, K., Szabó, G. M., et al. 2015, AJ, 149, 112

Szabó, R., Pál, A., Sárneczky, K., et al. 2016, A&A, 596, 40

Szabó, G. M., Pál, A., Kiss, C., et al. 2017, A&A, 599, 44

Tsuda, Y., Yoshikawa, M., Abe, M., Minamino, H., & Nakazawa, S. 2013, AcAau, 91, 356

Warner, B. D., Harris, A. W., & Pravec, P. 2009, Icar, 202, 134

Zhou, Y., Apai, D., Schneider, G. H., Marley, M. S., & Showman, A. P. 2016, ApJ, 818, 176

Chapter 6

Extragalactic Studies from the Kepler/K2 Missions

Krista Lynne Smith, Armin Rest, Peter Garnavich and Steve B Howell

6.1 Introduction

While the primary focus of the Kepler and K2 missions was to discover transiting exoplanets, the unique design of the mission opened up a new discovery space for time-domain astrophysics well outside our own Galaxy. Important processes in spectacularly powerful events occurring in other galaxies, sometimes out to high redshifts, have long been obscured by the constraints of ground-based, trigger-oriented observing.

Active galactic nuclei (AGN), supermassive black holes surrounded by a hot, magnetized accretion flow, are variable objects across many wavelengths and on a wide range of timescales. Hidden within their stochastic variations are clues to the physics of accretion, detailed processes within relativistic jets, and useful probes of general relativity.

Supernovae, catastrophic explosions that occur at the deaths of massive stars or during thermonuclear runaway reactions on the surfaces of white dwarfs, have been observed and studied from the ground for centuries. However, even in the modern era, most supernovae light-curves are only observed after the explosion has occurred, since multi-wavelength follow-up programs are typically triggered by the appearance of the bright transient source. Much can be learned from the pre-explosion early time light-curves, a completely novel space that Kepler, and especially K2 exquisitely explored.

6.2 Studies of Active Galactic Nuclei (AGN)

6.2.1 Introduction

Active galactic nuclei (AGN) are the most luminous non-transient objects in the Universe. They consist of supermassive black holes (SMBHs) located in galactic centers that are actively accreting matter. Vast amounts of energy are produced in both the accretion flow, which forms a geometrically thin, optically thick disk in the standard paradigm, and in the jets and outflows launched by the interaction between the disk and the spinning black hole. The accretion disk is a dynamic object: it is viscous, turbulent, and magnetized. Despite being a valuable laboratory for high energy processes in relativistic environments, accretion disks cannot be directly imaged except in one or two nearby cases (e.g., the Event Horizon Telescope image of the accretion flow in M87).

AGN are variable across the electromagnetic spectrum, from low frequency radio observations to TeV gamma-rays. The physical mechanism responsible for the variability depends on the waveband, but in the optical and UV, the dominant source of the radiation is the accretion disk itself. Therefore, the ubiquitous variability of AGN in the rest-frame optical and UV is a very rare direct observational probe of accretion disk physics.

The situation is different for the class of AGN known as blazars: objects with very low inclination in which the relativistic jet is pointed nearly along our line of sight. For these AGN, the vast majority of the emission originates in the jet and observed light variations can be very rapid. Detailed light curves of blazars are sensitive probes of high-energy processes in relativistic jets, especially propagating shocks.

6.2.2 Key Observables in AGN Variability

There are several observables of interest in optical AGN variability, some directly from the light-curve, but most from the power spectral density (PSD) or similar functions. The even sampling cadence and low number of data gaps mean that the classical PSD is quite easy to calculate. Short gaps in the Kepler observations can be interpolated over, and after the light-curve is flattened by subtracting a simple linear fit, the PSD is simply the discrete Fourier transform of the treated light-curve, squared. If one does not wish to interpolate, equivalent methods of exploring the amount of power per frequency exist that can accommodate unevenly sampled behavior. The common tools used for AGN light-curve analysis are the Lomb–Scargle periodogram and the structure function, which are complementary regarding AGN temporal observables in many respects.

First, it is important to know a few things about AGN variability in general. In all observed wave-bands the variations are stochastic, best described by a red noise process that is potentially non-stationary. This means that the PSD is well-fit by a power law, with increasing power at low frequencies: spectral density S varies with the temporal frequency as $S \propto \nu^{\alpha}$. As a rule, the variations show no trace of periodicity except in a few rare cases, discussed further below.

Power Spectral Slope
Let's begin with the high frequency slope of the PSD: the slope of the logarithm of the best fitting power law in the range between the flat region dominated by white noise (in 30 minute Kepler observations, approximately 10^{-5} Hz) and any observed low-frequency turnover. One popular mathematical model for flux diffusion through

an accretion disk, the "damped random walk," predicts a slope of exactly $\alpha = -2$; this model has worked well to describe ground-based AGN light-curves in the past.

Characteristic Variability Timescale

Because the power in the variability rises toward lower frequencies, it must at some point turn over and flatten, or decrease. Otherwise, the power would diverge to infinity at long timescales. The location at which the power spectrum turns over or "breaks" corresponds to a characteristic variability timescale, τ_{char}. In X-ray observations, these breaks are frequently detected in AGN power spectra at timescales of $\sim 10^4$ Hz (González-Martín & Vaughan 2012). It is important to remember that most AGN have cosmological redshifts, so characteristic timescales must be corrected to the object's rest frame before comparisons to expected physical timescales.

Variability and Physical Parameters

The overall variability of a light-curve can be measured by many metrics. The most common used in the Kepler and K2 AGN studies are the standard deviation and the fractional rms variability. Ground-based AGN studies have found that the overall variability of optical light-curves is anti-correlated with the bolometric luminosity (e.g., the PanSTARRS result by Simm et al. 2015), and positively correlated to the redshift. The latter can be understood in the following context: an AGN accretion disk can be imagined as a superposition of many annuli, each of which acts as a thermal blackbody. The hottest annuli come from the smallest regions closest to the supermassive black hole, and radiate in the far-UV. These regions are expected to be the most variable on the shortest timescales. As we observe objects at higher and higher redshifts, more of this UV light is redshifted into the Kepler imager optical bandpass, and so objects with higher redshifts appear more variable.

Characteristic timescales from X-ray light-curves have shown a good correlation with black hole mass, most likely related to the size of the emitting region (Markowitz et al. 2003; McHardy et al. 2006). In addition to offering insight into the relative geometry of the X-ray corona and the black hole, such a correlation raises the tantalizing possibility of using power spectral breaks as a probe of black hole mass.

Measuring SMBH masses is a perilous task given present methods and data sets. The best masses are measured with dynamical observations of stars or gas in the direct gravitational influence of the black hole; however, this is only possible for a handful of very nearby AGN. Good mass estimates can also be obtained using water masers in edge-on systems, but such signals require fortuitous geometric alignment and are therefore rare. The most numerous sample of well-constrained black hole masses was obtained through reverberation mapping campaigns in which the time lag between continuum and broad emission line variations is used to estimate the size of the line emitting region and therefore the black hole mass (Bentz & Katz 2015). While the reverberation method generates good masses, it is observationally intensive, requiring many epochs of spectroscopic observations. To get single-epoch mass estimates for large samples, the most common mass proxies are the width of the broad emission lines in UV/optical spectra ($H\beta$, $H\alpha$, and Mg II $\lambda2799\text{Å}$), used for Type I AGN in which broad lines are visible, and the stellar velocity dispersion of the bulge via the $M - \sigma_*$ relation. Such methods provide accurate masses when compared to estimates from reverberation mapping and dynamical studies, but with large scatter.

Therefore, as we are perched at the beginning of an era in which vast databases of optical light-curves will be obtained (e.g., with LSST, but also existing observatories and surveys like PanSTARRS, PTF/ZTF, and CRTS), the possibility of a black hole mass probe from optical light-curves is very exciting. If such a correlation can be found and calibrated, it may become possible to measure millions of SMBH masses across a vast redshift space from these upcoming surveys.

RMS–Flux Relation and Flux Distributions
In X-ray light-curves of AGN, there is a remarkably consistent, tight correlation between the rms variability of a light-curve segment and that segment's mean flux; in other words, X-ray emission in AGN appears to be more variable when brighter (e.g., Uttley et al. 2005). This is called the rms-flux relation, and is often interpreted to indicate that the underlying process is multiplicative. Such a process may also give rise to a log-normal rather than gaussian flux distribution histogram. The model of propagating fluctuations within an accretion disk fulfills this requirement: the mass inflow rate to the accretion disk varies at large radii and these inhomogenieties propagate inwards through the disk (e.g., Kelly et al. 2011; Hogg & Reynolds 2016). Both of these characteristics are commonly observed in X-ray variability of many accreting systems, from young stellar objects and X-ray binaries to AGN. However, Scargle et al. (2020) posits that the rms-flux relation and flux distribution shape does not necessitate a multiplicatve underlying process. Since Kepler finally provided the same long-baseline (with respect to the timescales observed) and even sampling as X-ray light-curves have been doing for AGN for decades, Kepler observations offered the possibility to observe such signatures in optical AGN variability.

6.2.3 Challenges to AGN Science

The sample of AGN observed by the original Kepler mission was small. This is mainly due to the position of the field of view (FOV): as Kepler's goal was to discover exoplanets in our galaxy, the FOV was positioned near the galactic plane (see Chapter 1). This was unfortunate for AGN science, since this region did not overlap with any large extragalactic surveys, like SDSS, where the vast majority of AGN are cataloged. Only one or two AGN were known in the Kepler FOV before the mission. To find additional AGN in the Kepler FOV, two main techniques were undertaken. The first was AGN selection via infrared photometric cuts, based on the criteria developed by Edelson & Malkan (2012). The second was an X-ray survey of four modules of the Kepler FOV with the Swift satellite (Smith et al. 2015). Before requesting Kepler monitoring, any AGN selected by these techniques was first confirmed with optical spectroscopic follow-up. In the end, 41 AGN were monitored with 30 minute cadence in the original Kepler FOV; the reaction wheel failure in 2013 terminated plans to monitor an additional few dozen newly-discovered AGN.

Because the Kepler mission was primarily intended to monitor only point sources, the initial photometric extraction apertures provided by Kepler were too small to accommodate the extended emission from the AGN host galaxies. When an aperture does not leave sufficient "wiggle-room" for a source, spacecraft drift and thermal PSF effects can cause more or less of the host galaxy emission to wander in or out of the aperture, causing artificial increases and decreases in the light-curve. The stochastic nature of AGN variability makes it extremely difficult to distinguish between these artifacts and true variability. Furthermore, the persistent rolling band or "Moire pattern" electronic noise that ripples across the Kepler detector can look very similar to rapid AGN variability. The cotrending basis vectors (CBVs) provided by Kepler to assist observers in removing systematic trends can be of service, but often, even very conservative application of just a few CBVs can have dramatic overfitting effects, resulting in removal of real variability (Smith et al. 2018a). All of these systematics result in data products that require careful attention to how detrending affects the physical results, as pointed out by Kasliwal et al. (2015b). Indeed, Dobrotka et al. (2017) initially reported the discovery of multiple power spectral components in the Kepler light-curves of four AGN, which they later attributed to instrumental systematics (Dobrotka et al. 2019). The fact that these PSD components were especially convincing due to similar features seen in simultaneous X-ray data speaks to the insidiousness of these systematic effects in objects with intrinsic red-noise variability. For a more thorough description of the detailed systematics and how each one can affect AGN science with Kepler data, see Smith (2019).

The K2 observing campaigns did occasionally overlap with SDSS and other extragalactic surveys where large numbers of spectroscopically-confirmed AGN were already known. However, the short baselines of K2 monitoring (~85 days) restrict the science to only the most rapid Seyfert variability, preventing observable break timescales longer than ~10 days. Additionally, the systematics of K2 can have deleterious effects on AGN light-curves that are extremely difficult to characterize and remove (e.g., O'Brien et al. 2018), especially given the faintness and extended nature of AGN compared to most of Kepler's bright, point source targets.

6.3 Key Results from the Kepler/K2 AGN

Despite the difficulties posed for AGN by the complex intersection of their faint, stochastic nature and the design parameters and systematics of the Kepler space-craft, many important insights and phenomena were discovered. In this section, we offer an overview of the main results for each AGN sub-class: Seyfert galaxies, which are the most numerous AGN type observed by Kepler; radio-loud AGN and radio galaxies; and blazars.

6.3.1 Radio Quiet Seyfert Galaxies

While some of the most famous (and first discovered) AGN have powerful relativistic radio jets that extend well outside their host galaxies, such objects are the exception. The vast majority of AGN are radio-quiet, with only weak radio emission from an unresolved core. Such AGN typically have modest accretion rates, at a few to tens of percent of the Eddington limit. Accretion flows in this regime are generally modeled as geometrically thin, optically thick disks described by the standard model (Shakura & Sunyaev 1976). The inner regions of AGN accretion disks radiate in the ultraviolet. Therefore, the "white light" waveband of Kepler probes the outer regions of accretion disks in the local universe, and at higher redshifts, can access the inner regions. Which component of the accretion disk radiates the majority of its light into the Kepler imager bandpass also depends on the mass of the black hole: less massive holes have smaller disks with optically emitting regions closer in.

The best-studied AGN with Kepler data is Zw229-15 (KIC 6932990), mainly because it was the only bright, known Seyfert in the original Kepler FOV prior to the time when necessary search methods to find more were carried out (see Section 6.2.3). Its light-curve is given in Figure 6.1, demonstrating the stochastic (random) variability typical of Seyfert galaxies. A detailed analysis of this object was first performed by Carini & Ryle (2012), who reported a 44-day characteristic variability timescale and a steep power spectral slope inconsistent with the damped random walk. Later, Edelson et al. (2014) reported a characteristic timescale of five days, and a similarly steep PSD slope.

Of the few dozen AGN in the original Kepler field discovered by the methods described in Section 6.2.3, the great majority are radio-quiet Seyferts. Although ground-based variability studies of AGN tend to find power spectral slopes consistent with the damped random walk value of $\alpha = -2$, AGN in such samples are typically comprised of luminous quasars with more massive black holes with higher accretion rates than the common Kepler Seyferts. Every study of radio-quiet Seyfert galaxies using Kepler light-curves found steeper slopes regardless of the

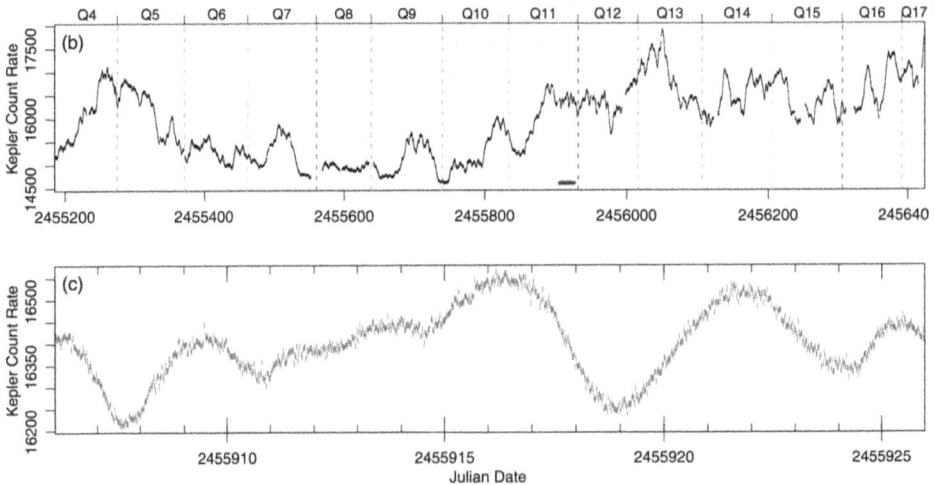

Figure 6.1. Kepler light-curve of Zw229-15 across the full monitoring period (top) and zoomed in to a short period in Quarter 11 (bottom; indicated in the top panel by the horizontal blue bar). Reproduced from Edelson et al. (2014). © 2014. The American Astronomical Society. All rights reserved.

details of the data correction methods used, sometimes even steeper than $\alpha = -3$ (Mushotzky et al. 2011; Kasliwal et al. 2015a; Smith et al. 2018a; Aranzana et al. 2018). The damped random walk is therefore an incomplete description of optical AGN variability in moderate luminosity Seyferts.

Smith et al. (2018a) also found curious flux distributions for a large number of the Kepler AGN. Some did indeed exhibit log-normal flux distributions consistent with the predictions for multiplicatively-interacting propagating fluctuations, but most had more unusual shapes, including some bi-modal distributions that may indicate obscuring material passing along the line of sight, and none had well-defined linear rms–flux relations. It is possible, however, that since the process giving rise to the optical variability acts over much larger distances, and longer timescales, than the X-ray variability, the Kepler baselines are simply insufficient to capture the full range of behavior necessary to give rise to the multiplicative signatures, as suggested by Alston (2019).

Smith et al. (2018a) reported significant detections of characteristic variability timescales in five of the 20 AGN they analyzed. The timescales ranged from 9 to 53 days. Although five objects are too few to search for a correlation of timescale with mass, the objects with higher black hole masses did indeed exhibit the longer timescales. Furthermore, when Smith et al. (2018a) compared the timescales to those expected given the putative masses and therefore disk sizes, the observed values of $\tau_{\rm char}$ were consistent with either the Keplerian orbital period or the freefall accretion time (depending on which equation is used to estimate the disk temperature profile); see Figure 6.2.

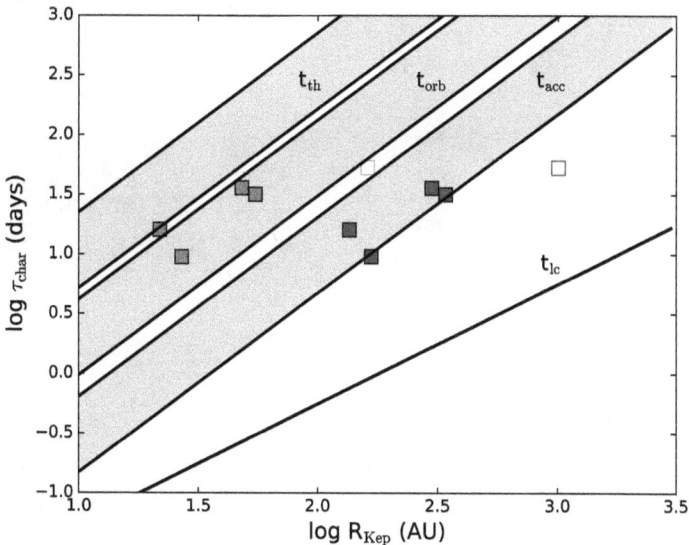

Figure 6.2. Characteristic timescales in Kepler AGN versus the radius in the accretion disk probed by the Kepler bandpass. The relevant disk radius depends on assumptions about disk temperature; here, blue symbols correspond to a treatment including more detailed inner disk contributions, and red symbols to treating the disk as a superposition of local blackbodies. Truncated accretion disks would act similarly to the red data points. The bands refer to the thermal ($t_{\rm th}$), Keplerian orbital ($t_{\rm orb}$), freefall accretion ($t_{\rm acc}$), and light-crossing ($t_{\rm lc}$) timescales. From Smith et al. (2018a).

A detailed analysis of individual flares in Kepler AGN was done by Chen & Wang (2015). These authors found that the flares of both AGN and quasars are highly symmetric, ruling out models for optical variability that require different rise and decay timescales, such as disk instabilities.

Finally, Smith et al. (2018a) and Aranzana et al. (2018) found weak anti-correlations between the variability and the bolometric luminosity, as expected based on ground-based results.

An Optical Quasi-periodic Oscillation

Contributed by Krista Smith, Stanford University, Stanford, CA.

Quasi-periodic oscillations (QPOs) are common features in the X-ray light-curves of accreting stellar mass black holes in X-ray binaries: peaks in the X-ray power spectrum that appear in most stellar mass black holes (Remillard & McClintock 2006). They come in two principle regimes: high-frequency and low-frequency. The cause of these remarkably coherent periodic features is still mysterious; many possible physical origin theories exist, including disk precession, warped accretion disks, and diskoseismology, just to name a few. The search to understand these features is largely motivated by the apparent correlation between the oscillation frequency and the black hole mass. X-ray QPOs appear to be far less common in AGN, having been found in only a small

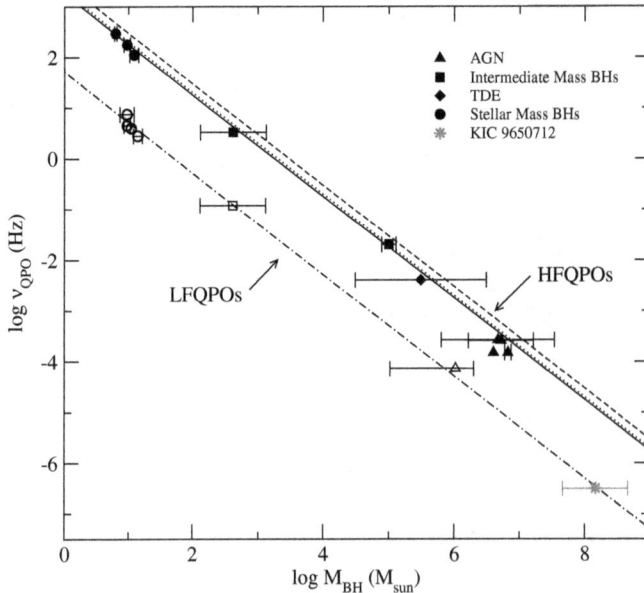

Figure 6.3. X-ray QPOs from the literature, and the newly-discovered Kepler optical QPO candidate. Solid and hollow symbols represent high- and low-frequency QPOs, respectively. The solid line shows the relationship for high frequency QPOs from stellar mass black holes only. Dashed and dotted lines indicate different possible physical origin models. The dot-dashed line is our independent linear regressive fit through all the low-frequency QPOs prior to the discovery of our candidate, KIC 9650172. Reproduced from Smith et al. (2018b). © 2018. The American Astronomical Society. All rights reserved.

handful of objects. These also follow the frequency-mass relation, despite vast differences in scale, implying that accretion processes may be quite similar across enormous mass ranges.

A question arises when pondering the very different apparent fraction of QPOs in stellar versus supermassive systems. Are AGN QPOs actually rare, implying a change in the accretion process at large scales, or have we simply not had the observational faculties to detect them? The oscillations in AGN occur on timescales of many hours to months, as opposed to the mHz-Hz oscillations in stellar X-ray binaries. Further, the red noise variability of AGN can often mimic periodicities quite convincingly (Vaughan et al. 2016), and so many cycles and very even sampling are typically required to confirm a quasi-periodic signal. We simply do not have large enough samples of light-curves with even sampling and months-to-years baselines.

Smith et al. (2018b) reported the discovery of a possible optical QPO in the Kepler light-curve of the Seyfert galaxy KIC 9650712. This is the first optical QPO in an AGN, and this galaxy has by far the largest mass black hole around which a QPO has been detected, at 1.5×10^8 $M_\odot$. Remarkably, it falls precisely upon the low-frequency QPO—black hole mass relationship extrapolated from stellar- and intermediate-mass black holes, as shown in Figure 6.3. Since it was discovered in only one out of a modest sample of 20 AGN with Kepler, this detection may indicate that QPOs are more common in AGN than previously thought, and that their apparent low occupation fraction is a consequence of insufficient observational capability and not intrinsic differences between stellar and supermassive black hole accretion.

6.3.2 Radio-loud AGN

Since radio-loud AGN make up only about 10% of the overall AGN population, it is not surprising that only four such objects were studied in the original Kepler field of view. These four objects are at higher redshifts than the radio-quiet sample ($0.51 < z < 1.52$), and are more luminous overall. Three are flat-spectrum radio quasars (FSRQs), in which the AGN's relativistic jet is within a few degrees, but not directly along, the line of sight, giving them properties similar to blazars. The variability in such objects may arise within the accretion disk or due to processes within the jet and may depend on the accretion state of the object.

Wehrle et al. (2013) studied these four targets in detail, followed by Revalski et al. (2014) with longer Kepler baselines. While Wehrle et al. (2013) did not find any fast flares, the longer baseline light-curves of two objects did flare on few-day timescales with amplitudes of a few to ~20% (Revalski et al. 2014). The power spectral slopes of the radio-loud targets were appreciably flatter than those in radio-quiet Seyferts, with $-2 < \alpha < -1.5$, as later confirmed by Smith et al. (2018a). The Wehrle and Revalski studies state that this behavior is consistent with variability arising from turbulence behind a shock, either in a jet or in the accretion disk, following the models of Mangalam & Wiita (1993) and Marscher & Travis (1991).

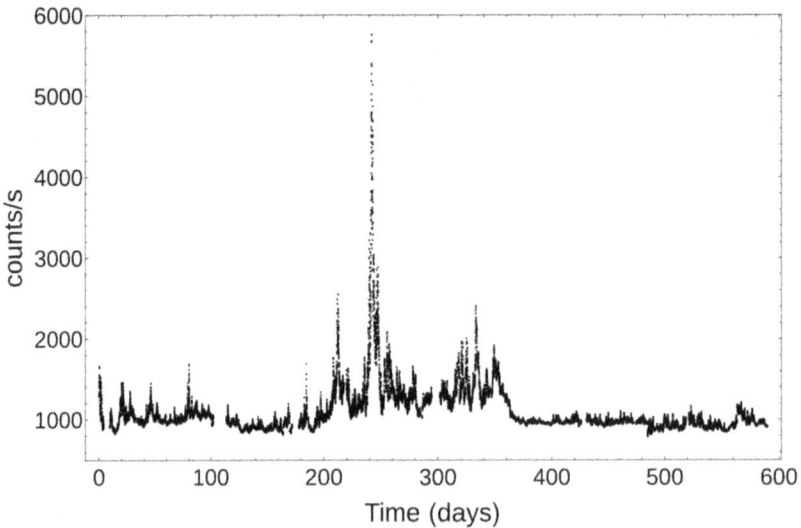

Figure 6.4. Kepler light-curve of the BL Lac object W2R 1926+42. Reproduced from Figure 1 of Mohan et al. (2016), by permission of Oxford University Press on behalf of the Royal Astronomical Society.

6.3.3 Blazars

Blazars are even rarer than radio-loud AGN: a narrow, collimated relativistic jet is not often fortuitously pointed along our line of sight. Only one blazar, the BL Lac object W2R 1926+42 (KIC 6690887) was known in the original Kepler field. This object has been heavily studied and several groups have attempted to draw insights into the jet physics using the new observation window opened by Kepler light-curve sampling. There are, in principle, dozens of blazars in the K2 ecliptic plane fields, but most are too faint for reliable photometry, and are beset by difficult systematics. Accordingly, there is currently only one published study of blazars in the K2 fields (see below).

The Kepler Blazar W2R 1926+42
The variability of W2R 1926+42, shown in Figure 6.4, is markedly different from that of Seyfert galaxies, consisting of a clear baseline with many superimposed flaring events. It has been studied by a number of groups, beginning with Edelson et al. (2013).

Edelson et al. (2013) found evidence for an approximately 4 hr characteristic timescale in their power spectral analysis. Translating this observed timescale into an intrinsic one depends on the model assumed for the optical variability mechanism within the jet. The most common assumption is that the variability arises due to shocks within the jet, with variability timescales subject to Doppler boosting factors of approximately 10; in this case, the 4 hr observed timescale corresponds to a couple of days in the intrinsic frame. Another possibility is the "jet-within-a-jet" scenario (e.g., Biteau & Giebels 2012): inside the larger jet, itself relativistic, small mini-jets form in which particles are accelerated to very high Lorentz factors, probably by magnetic reconnection. Here, the Lorentz factor of the emitting matter is the product of the factors of the jet and mini-jet, and thus the physical timescale corresponding to the 4 hr observed characteristic time is about 15 days. Edelson et al. (2013) also found that the flux distribution is best fit with a log-normal histogram, and a strong rms–flux relation. Both of these are consistent with jet-within-a-jet scenarios, as well as multiplicative propagating fluctuations within an accretion disk (e.g., Hogg & Reynolds 2016). A similar characteristic timescale of ~6 hr was reported by Mohan et al. (2016).

The Kepler light-curve of W2R 1926+42 was also studied by Bachev et al. (2015), who searched for evidence of low-dimensional chaotic behavior, which can be used to differentiate between variability driven by a purely stochastic process or by nonlinear dynamics. No evidence for such chaos was found in the Kepler light-curve, thus Bachev et al. (2015) suggests that variability arises in many independent zones within the jet.

From detailed analyses of the flaring behavior in W2R 1926+42, Mohan et al. (2016) calculated an average outburst lifetime of tens to hundreds of days, and concluded that the emitting regions are shocks within 2 parsecs of the jet base. They also attempt to measure the black hole mass using an empirical relation between mass and fractional variability (Bian & Zhao 2003), arriving at an estimate of $1.7–47.8 \times 10^7 \, M_\odot$. Li et al. (2018) investigated the asymmetry of the optical flares, finding that the flare durations and amplitude distributions are log-normal, indicating different acceleration and dissipation timescales for high energy particles in the emitting region. Such a finding is also consistent with the jet-within-a-jet scenario.

K2 Blazars

Wehrle et al. (2019) studied the K2 light-curves of nine AGN that were also detected in gamma-ray emission by the Fermi-LAT instrument. This sample including the possible binary black hole candidate OJ 287 as well as both FSRQs and BL Lac objects. Six objects in the sample had Fermi gamma-ray detections simultaneous with the K2 observations. The objects exhibit a wide range of power spectral slopes, from −1.7 to −3.8. Most of these slopes are consistent with the emitting region being turbulent within the relativistic jet, but the very steepest slopes ($\leqslant -3$) have large, low-frequency variations that are not well described by this model, and must arise from a separate component.

Wehrle et al. (2019) also report a characteristic timescale of 5.8 days in OJ 287. This result was also found in the large multi-wavelength study of OJ 287 by Goyal et al. (2018).

6.3.4 Selection of AGN by Optical Variability

There are a wide variety of selection techniques used to compile samples of AGN, most commonly X-ray and radio selection, and optical and infrared spectroscopic techniques; for a review, see Mushotzky (2004). Because most of these methods require that the AGN be significant in luminosity compared to its host galaxy, weak AGN can be missed even in the X-ray band. One remedy to this may be selection via optical variability: since the host galaxy light remains constant, even weak variations in galaxy light-curves may indicate a hidden active nucleus. Indeed, variability has been used to select AGN in dwarf galaxies with ground-based timing data (e.g., Baldassare et al. 2018). Because Kepler has such high photometric precision, it should be a highly capable instrument for finding weakly accreting AGN via optical variability selection. Shaya et al. (2015) searched the Kepler light-curves of ~500 galaxies, and found that 4% (19 objects) exhibited excess variability indicative of AGN activity. Several of these were not known to be AGN by any of the other selection methods used to search the Kepler field, a strong indication of the complementarity of this approach to AGN selection.

6.4 Studies of Supernovae and Transients

6.4.1 Introduction

Despite the growing volume of supernova (SN) data coming from wide-field surveys such as PS1, PTF, ATLAS, and ASAS-SN, fundamental questions remain as to the nature of these explosions and their progenitor systems. The early light-curves of SN probe the physics of the progenitor systems, e.g., specific features of the outer stellar layers and surrounding matter distributions reveal how, where, and why the explosions were triggered. However, it has been difficult to obtain the needed data from the ground. With Kepler, plus a focused observational campaign by K2, high-cadence early light-curves with exquisite photometric accuracy for every class of SNe were able to be collected.

6.4.2 Type Ia Supernovae

From both theoretical considerations as well as strong observational constraints, it is well understood that the thermonuclear explosion of C/O white dwarfs in a binary system is the physical mechanism behind Type Ia SN (e.g., Hoyle & Fowler 1960; Colgate & McKee 1969; Woosley et al. 1986; Bloom et al. 2012). However, despite the fact that SN Ia have been used to discover Dark Energy (Riess et al. 1998; Perlmutter et al. 1999) and constrain its equation of state (e.g., Garnavich et al. 1998; Scolnic et al. 2018), we still do not know whether they come from various progenitor systems and, if so, in what proportion.

The possible progenitor scenarios can be split into two categories: the single-degenerate (SD) channel, in which the explosion is triggered when material accreted from a companion star reaches the Chandrasekhar mass (e.g., Whelan & Iben 1973); and the double-degenerate (DD) channel, in which the explosion is triggered by the merger of two WDs (e.g.; Iben & Tutukov 1984). Both channels are consistent with the SN Ia population, and it is unclear how much or if each channel contributes to the SN Ia population. No progenitor system has yet been detected (e.g., Goobar et al. 2014; Kelly et al. 2014), and searches for surviving companion stars in Galactic supernova remnants have not been successful (e.g., Kerzendorf et al. 2012; Schaefer & Pagnotta 2012).

The early rise of SNe Ia may contain signatures of the progenitor superimposed on the flux emitted from the SN ejecta. In general, the initial flux produced by SN ejecta and ^{56}Ni decay can be approximated by the "fireball model" (Arnett 1982), in which the SN ejecta expands with nearly constant speed, and as the opacity drops, the more interior parts in the "expanding fireball" become visible. Arnett (1982) found analytically that the early light-curves of SNe Ia rise quadratically in time, which is in agreement with observations (Garg et al. 2007; Hayden et al. 2010; Nugent et al. 2011; Bianco et al. 2011). Kasen (2010) suggested that there is a way to directly detect the signature of a SD progenitor: if an accreting companion star is present, then there should be an early time excess from shocked emission as ejecta from the SN collides with it. This shock emission is viewing angle dependent and rapid, so obtaining early time, high-cadence light-curves of several SN Ia is crucial to detect this excess signal.

Even though this early excess flux is a robust prediction for the Roche-lobe-filling SD scenarios, there are also other mechanisms that may produce early heating and thus early excess flux. Piro & Nakar (2014) suggested that an overdensity of ^{56}Ni near the ejecta surface should also introduce flux in excess of the canonical "expanding fireball" model. This ^{56}Ni overdensity is possible in both channels and might resemble the signature of the shock interaction model. Piro & Morozova (2016) also found that interaction with circumstellar material expelled during a DD merger can modulate the early light-curve shape. A last possible mechanism to produce early excess flux is the double-detonation model. In the double-detonation model, the companion star can be either a helium star or a WD with a He layer on its surface (Shen & Bildsten 2009). Through accretion, a helium layer is formed and when this layer is sufficiently massive and degenerate, an explosive helium ignition can trigger a detonation. The detonation shock will propagate into the C/O WD core, then trigger a second carbon detonation at the center to produce a SN Ia. The first detonation on the surface will produce a remarkable amount of radioactive nuclei and produce an early bump which is red in color due to line blanketing of iron-group elements produced in outer layer (Noebauer et al. 2017; Bildsten et al. 2007).

It is difficult with ground-based telescopes to obtain early-light-curve observations of SN Ia, at faint magnitude limits and with fast enough cadence, to search for such excess flux. Only two normal SN Ia, SN 2012cg and SN 2017cbv, have shown excess flux at early times (Marion et al. 2016; Hosseinzadeh et al. 2017). Both find that a SD scenario is the favored scenario most consistent with the observations. However, this scenario has been challenged for both SNe, because the late-time spectra did not show any signs of Hα swept up by the interaction of the SN ejecta with the binary companion (Sand et al. 2018; Shappee et al. 2018).

The ideal survey to search for early excess flux as signature of the progenitor would have continuous, high-cadence coverage with high photometric precision. The Kepler telescope (Haas et al. 2010) is especially well-suited for this task with its wide field of view, 30 minute cadence, and exquisite light-curves. During the Kepler mission, Olling et al. (2015) discovered three likely SNe Ia with extraordinary light-curves from the onset of the explosion through the rise and decline of the SNe. These exquisite early light-curves showed no evidence of any excess flux, excluding companion stars above 2 $M_\odot$ (Figure 6.5). They are very well fit with power laws consistent with the expanding fireball model (Olling et al. 2015).

In contrast, an early blue flux excess was detected in SN2018oh with K2 in Campaign 16 (Figure 6.6, Dimitriadis et al. 2018; Shappee et al. 2019). While the blue color of the excess flux favors the shocked emission as ejecta from the SN collides with binary companion, the lack of hydrogen in the nebular phase spectra again is more consistent with an overdensity of ^{56}Ni on the surface of the ejecta as the source of the excess flux (Dimitriadis et al. 2019; Tucker et al. 2019).

Clearly, a large sample of SN Ia with high-cadence early rise light-curve observations is needed to solve this puzzle. With Kepler/K2, between 2010 and 2017, there now exists more than a dozen observed SN Ia's with good light-curves suitable for an analysis (see Figure 6.7 for a set of SN Ia from K2 Campaign 16, Villar et al. 2019). For each of these SNe, we run a detailed statistical analysis to calculate the efficiency of early bump detection as well as the false-alarm probability. This type of analysis allows robust constraints as to whether there is any excess flux, and which mechanism caused it (Villar et al. 2019).

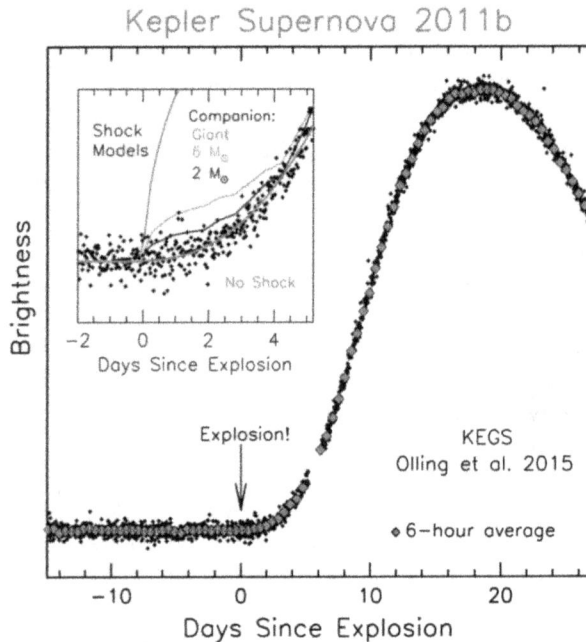

Figure 6.5. The SN Ia light-curve of KSN2011b observed with Kepler. The black points are the 30 minute cadence observations, while the red diamonds shows the data in 6 hr bins. The inset compares a range of shock models assuming different companion star radii. KEGS refers to the Kepler Extra-Galactic Survey. Reprinted by permission from Springer Nature: Nature, Olling et al. (2015).

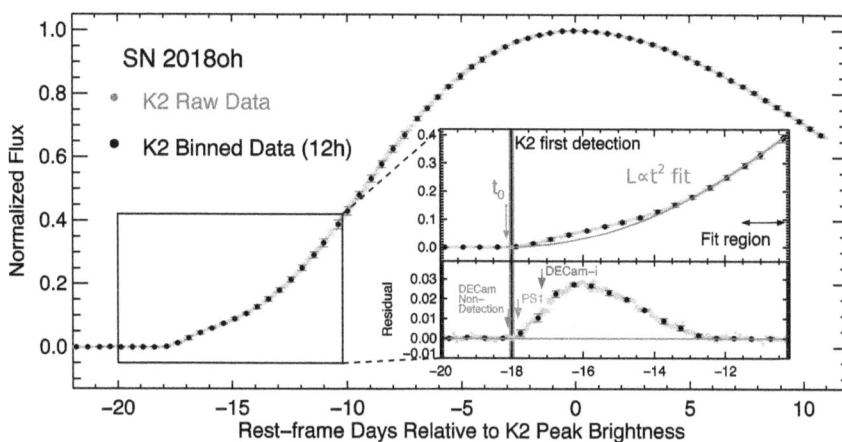

Figure 6.6. SN 2018oh K2 light-curve, normalized to peak flux, with respect to peak brightness. Unbinned K2 photometry and data averaged over 12 hr are shown as gray and black points, respectively. In the inset's upper panel, we show the zoomed light-curve from 20 to 10 days before peak brightness. A $L \propto t^2$ model (red line) fit to the data in the "Fit region" is displayed. The residual of the fit is shown in the lower panel. The time of our ground-based contemporaneous observations (last DECam non-detection, first PS1 and DECam) are marked with green, orange and red arrows, respectively. The black vertical line corresponds to the estimation of the onset of the K2 light-curve, as described in the text, with the blue-shaded region representing the 3-σ standard deviation. Reproduced from Dimitriadis et al. (2018). © 2018. The American Astronomical Society. All rights reserved.

6.4.3 Core-collapse Supernovae

SN IIP

The core-collapse of a massive star with a significant hydrogen envelope can result in a type IIP supernova. The shock generated in the creation of a neutron star takes about a day to propagate to the surface where it can finally "breakout" with a flash that reaches temperatures of $\sim 10^6$ K. This shock breakout is expected to rise and decay very rapidly ($t < 3$ hr) before being out shadowed by the expanding photosphere of the supernova. Models suggest that the visibility of the shock breakout depends strongly on the evolutionary state of the star. Nakar & Sari (2010) predict that the shock breakout in a red supergiant (RSG) will have a bright peak in optical flux just an hour after explosion while more compact progenitors (e.g., blue supergiants) will have a faint maximum minutes after breakout. In general, the luminosity and timing of the shock breakout will depend on the radius of the progenitor star.

Shock breakouts have been detected only rarely and often by accident. No breakout had ever been seen at optical wavelengths in a normal SN IIP until Kepler supernova KSN2011d (Garnavich et al. 2016). The paucity of detected shock breakouts could be attributed to the difficulties in catching such a short timescale event occurring at random locations on the sky. However, it could also be that mass loss during the RSG phase could result in sufficient circumstellar gas density to muffle the shock breakout (e.g., Förster et al. 2018). Limited mass-loss from a RSG will result in the energy of the breakout shock being spread out over time and

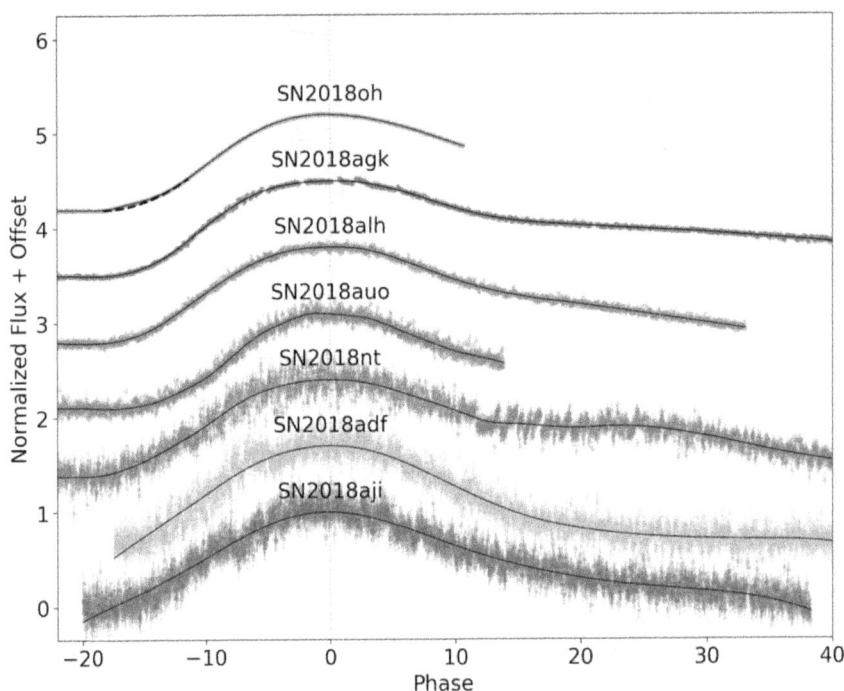

Figure 6.7. Light-curves of the seven high signal-to-noise ratio (SNR) Type Ia supernovae (SNe) observed during the K2 mission. Each light-curve is generated from the 30 minute cadence observations and normalized relative to maximum flux. The light-curves are vertically ordered by SNR and aligned relative to maximum light. The black solid lines are smoothed light-curves to guide the eye. The dashed line in SN2018oh follows a power-low profile to emphasize the early-time of flux excess.

appearing cooler than if the breakout occurred on a "bare" RSG. No flash would be visible, but the light-curve would rise rapidly before blending into the normal photospheric expansion phase.

Kepler recorded two likely type IIP events in 2011. Both KSN2011a and KSN2011d displayed long plateau phases lasting about four months, indicative of a type IIP supernova coming from a RSG. The continuous coverage and 30 minute cadence of Kepler provided an excellent test of supernova models during the early rise. KSN2011d showed a 13 day rest-frame rise to maximum light consistent with a nearly 500 $R_\odot$ RSG star (Garnavich et al. 2016). The rise was surprisingly well fit by a simple (Rabinak & Waxman 2011) photospheric model (see Figure 6.8) except in the earliest few hours after the explosion. The Kepler data showed an excess of flux just as the model predicted for the beginning of photospheric expansion and this is what is expected for a shock breakout. The 5-σ brightness excess amounts to an absolute Kepler magnitude of -15.6 ± 0.3 which is similar to the predictions of Nakar & Sari (2010) for a similarly sized RSG.

The excellent fit to the photospheric rise of KSN2011d by the basic models of Rabinak & Waxman (2011) was encouraging, but similar calculations did not do as good a job with the light-curve of KSN2011a (see Figure 6.8). Over the first few

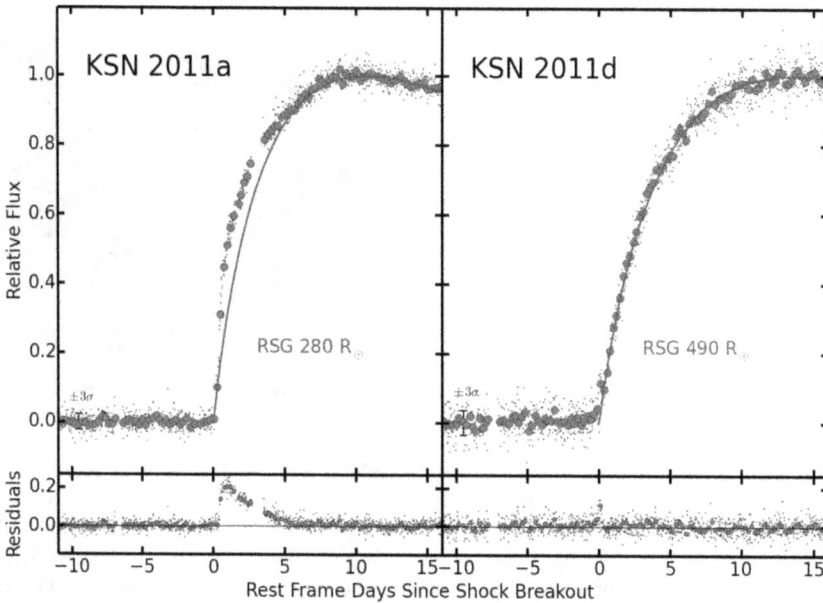

Figure 6.8. The top panels show the relative flux light-curves for the type IIP supernovae KSN2011a (left) and KSN2011d (right). Small blue points are observations at the 30 minute long cadence while large red points show 6 hr medians. The solid lines are best fit photospheric models (Rabinak & Waxman 2011) for exploding RSG stars with the indicated radii. The very fast rise seen in KSN2011a can be attributed to a weak circumstellar interaction. The photospheric model for KSN2011d is an excellent match to the data except for a small residual at the time of first light that may be attributed to a shock breakout. The lower panels show the residuals after the best-fit models are subtracted. Reproduced from Garnavich et al. (2016). © 2016. The American Astronomical Society. All rights reserved.

days, the brightness of KSN2011a quickly increased, but the rise was inconsistent with a shock breakout or a basic photospheric expansion model. The light-curve is best explained by a weak circumstellar interaction (Moriya et al. 2011). The interaction fades away in the optical after two weeks and prevents the detection of the expected shock breakout.

The two well-observed Kepler SN IIP events show a diversity in their early rise characteristics. One appears to have a circumstellar interaction that may be common in RSG due to episodic mass loss from the extended envelope. The other supernova, KSN2011d, appeared to have a cleaner local environment that allowed the detection of the shock breakout.

SN IIn

Explosions of stars that are embedded in a dense circumstellar medium result in type IIn supernovae (Schlegel 1990). The shock produced by such a supernova propagates through the dense medium converting kinetic energy of the explosion into heat and light. Because kinetic energy often dominates the energy budget of a supernova, these SN IIn can be both bright and long-lived. Indeed, depending on the

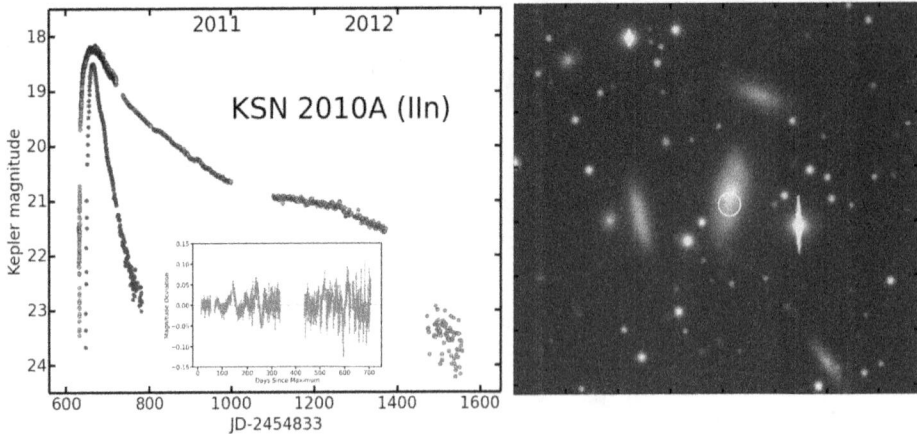

Figure 6.9. Left: The Kepler light-curve of the transient KSN2010a (red points) based on median bins of the long-cadence data. The transient was detectable for 3 yr and its variations are consistent with a type IIn supernova. For comparison, the Kepler light-curve of the type Ia supernova KSN2011b is displayed (blue points). The inset shows the deviations from a smooth light-curve decline. Right: An LBT/LBC false-color image at the location of the event (central circle). The transient appears to have been hosted by the brightest galaxy in the cluster Abell 2319. The host galaxy shows blue clumps near its center consistent with regions of recent star formation.

mass-loss history of the progenitor star, SN IIn light-curves can be very diverse and the star can remain luminous for years (e.g., Stritzinger et al. 2012).

Kepler detected a long-lived transient designated KSN2010a that remained visible for 1000 days. Unfortunately, the event had faded by the time it was identified, so no spectra were obtained. The light-curve is shown in Figure 6.9 and it is characterized by a month-long rise to maximum, a year-long decline followed by a plateau, and finally a rapid decay to beyond the detection limit of Kepler. The rise to maximum is linear in magnitudes over the first nine days but then the light-curve suddenly transitions to a parabolic rise before reaching maximum light. The light-curve is typical of a long-lived SN IIn, such as SN2005ip (Stritzinger et al. 2012), and thus a IIn type SN is probably the best classification for this transient.

The transient source KSN2010a exploded near the core of the brightest member of a cluster of galaxies, Abell 2319 at a redshift of 0.046. Typically, cluster members are elliptical galaxies that do not have on-going star formation, thus, core-collapse supernovae are rare in galaxy clusters. However, Large Binocular Telescope (LBT) imaging of the cluster shows blue knots near the core of the host indicating a star-burst or perhaps a merger with another star-forming galaxy (Figure 6.9).

The decay in the light-curve of KSN2010a shows variability on the timescale of about a month with an amplitude of ~5%. Detection of small variations on this scale is only possible from the continuous coverage and cadence of Kepler. These brightness variations may come from the supernova shock running through density fluctuations in the circumstellar gas. With an estimate of the speed of the shock, we can approximate the physical extent of the density variations. Chevalier (1998) found that SN IIn have expansion velocities between 1000 km s^{-1} and 3000 km s^{-1}, thus brightness variations on a timescale of a month correspond to density fluctuations with sizes between 20 and 50 au. The event remained bright for 2 yr suggesting that the circumstellar envelope extends out to about 10^{16} cm or about 800 au (Fransson et al. 2014).

6.4.4 Exotic Transients

The Fast-evolving Luminous Transient KSN2015K

For decades optical time-domain searches have been tuned to find ordinary supernovae, which rise and fall in brightness over a period of weeks. Recently, supernova searches have improved their cadences and a handful of fast-evolving luminous transients (FELTs) have been identified. FELTs have peak luminosities comparable to type Ia supernovae, but rise to maximum in <10 days and fade from view in <30 days. K2 observations have found the most extreme example of this class thus far, KSN2015K (Rest et al. 2018), with a rise time of only 2.2 days and a time above half-maximum brightness of only 6.8 days (Figure 6.10). Additional, ground-based photometry from DECam and SkyMapper show the color to be quite blue. These characteristics are similar to the fast blue transients discovered by the PS1 survey (Drout et al. 2014), which have rise time upper-limits of 3–5 days and peak luminosities similar to KSN2015K.

The progenitors of FELTs and the energy source that powers the light-curve variations are still under debate. Members of the class could originate from more

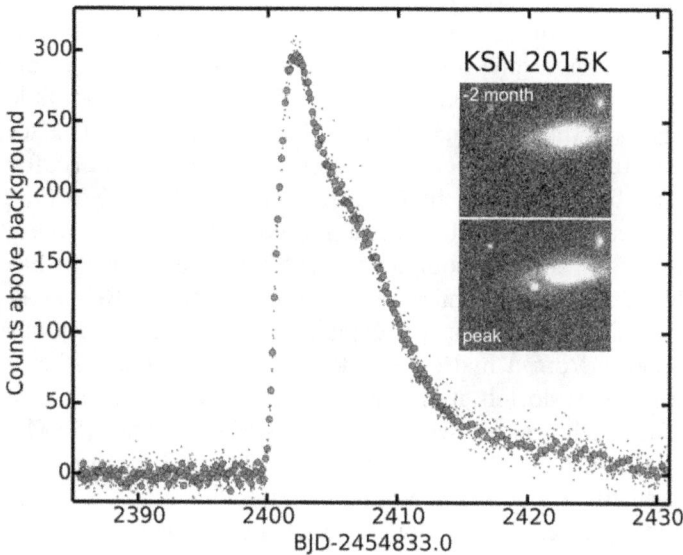

Figure 6.10. The K2 light-curve of KSN2015K. Blue dots are individual 30 minute cadence observations while the red points represent 3 hr median-value bins. The image cutouts in the inset show 60 s *i*-band DECam images from UT 2015 July 7 (2 months before peak) and 2015 August 1 (around peak) in the top and bottom panels, respectively. KSN2015K is marked with a red circle in the bottom panel. The photometric uncertainty is seen as the scatter of the K2 observations before the outburst. Reprinted by permission from Springer Nature: Nature, Rest et al. (2018).

than one type of progenitor. The high-cadence time-sampling of KSN2015K allows us to establish strong constraints on the origin of this particular event.

The decay of radioactive elements, like for SN Ia, is one of the possible energy sources of KSN2015K. However, the rise time of KSN2015k was ~8 times shorter than that of Type Ia supernovae, indicating that only a small amount of radioactive material could have been produced (10^{-4}–10^{-1} $M_\odot$). There are several scenarios that may lead to the ejection of such a small amount of radioactive material, e.g., the thermonuclear explosion of a shell of accreted helium on the surface of a white dwarf (Shen et al. 2010), the accretion induced collapse of a white dwarf to a neutron star (Dessart et al. 2006; Darbha et al. 2010), the merger of two neutron stars (i.e., kilonova; Abbott et al. 2017; Villar et al. 2017), or the core-collapse of massive stars if little ejecta is produced (Tauris et al. 2015; Moriya et al. 2010). Even though these scenarios can reproduce the timescales observed for the KSN2015K light-curve, they predict significantly fainter peak brightnesses on rather general physical grounds (e.g., Shen et al. 2010; Kasen et al. 2015): Figure 6.11 quantifies the allowed range of radioactive powered light-curves and shows that such a source can be ruled out for KSN2015K, which is above the so-called Arnett's rule for the range of synthesized ^{56}Ni masses (Arnett 1982).

A possible alternative power source for supernova light-curves is energy deposition from a central engine, such as a rotating magnetized neutron star (Maeda et al. 2007; Kasen & Bildsten 2010; a magnetar) or an accreting black hole (Dexter & Kasen 2013). The above constraints on the ejected mass apply to central engine heating, but the peak luminosity from a central engine can be substantially greater than is possible with radioactivity. However, explaining KSN2015K with a central engine implies extreme or fine-tuned parameters (Rest et al. 2018). For a black hole model, the small ejecta mass of KSN2015K would indicate a nearly failed supernova where all but $\lesssim 1\%$ of the star remained bound to the black hole. This scenario would also require fine tuning the fallback dynamics and/or accretion disk formation such that only a tiny fraction of the infalling material was tapped to power the light-curve. The light-curve of KSN2015K is a good match to orphan afterglow models, however, GRBs are very rare compared to SNe, so the chance of having found a GRB afterglow during the K2 mission is exceedingly small (Rest et al. 2018).

A final class of models that might explain the KSN2015K light-curve are those powered by energy from a hydrodynamical shock. This may either be the shock from a supernova explosion itself or one occurring post-explosion due to the interaction of the stellar ejecta with the circumstellar medium (CSM; Chevalier & Irwin 2011; Balberg & Loeb 2011; Ofek et al. 2010; Ginzburg & Balberg 2014; Kleiser & Kasen 2014). An explosion shock carries energy to the outer layers of the star and eventually vents in a shock breakout. To explain the rapid rise of KSN2015K as a shock breakout event requires a stellar radius larger than that of a red supergiant that are thought to be the progenitors of type IIP events. This issue can be solved by introducing an interaction with a dense and extended CSM at a radius of several times 10^{14} cm. The shock breakout energy is then absorbed in the extended CSM shell (Chevalier & Irwin 2011). Radiation-hydrodynamical simulations of a supernova running into a circumstellar shell show that the venting of the

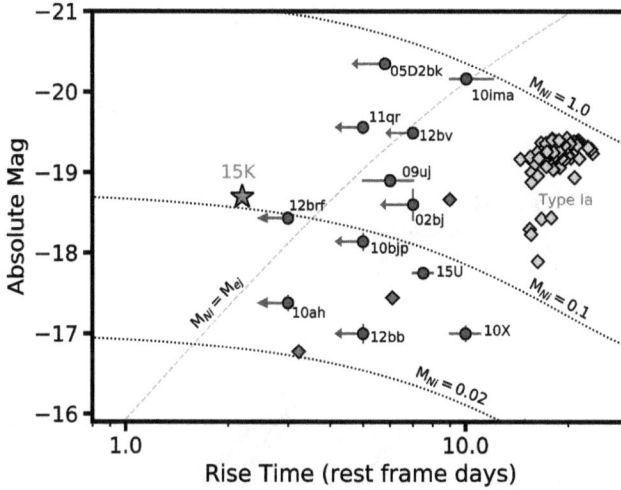

Figure 6.11. The peak luminosity versus rise time at optical wavelengths for fast transients (blue) and type Ia supernovae (green) from SDSS-II. The red star shows the location of KSN2015K and its extremely fast rise time. Purple diamonds indicate ".1a" models calculated by Shen et al. (2010). Dotted lines map out Arnett's rule for a range of synthesized ^{56}Ni masses. The dashed line is a thermonuclear scenario where a pure ^{56}Ni envelope is ejected at 10,000 km s^{-1} and this marks the limit of radioactivity-powered explosions. That is, events to the left of the dashed line cannot be fully powered with radioactive decay. The errors on the rise times are taken from the literature. For KSN2015K, the uncertainty is smaller than the symbol size. Reprinted by permission from Springer Nature: Nature, Rest et al. (2018).

post-shock energy at breakout can explain KSN2015K's very rapid rise to a luminous peak (Rest et al. 2018, Figure 6.12). The post-maximum luminosity derives from the diffusion of shock deposited energy from deeper layers. At later times ($t \gtrsim 10$ days) the decline of the KSN2015K light-curve becomes shallower and it is possible that radioactive ^{56}Ni decay contributes to the luminosity.

Fast transients are difficult to discover and follow-up, and sufficient numbers have been discovered only in recent years (e.g., Drout et al. 2014; Arcavi et al. 2016; Tanaka et al. 2016) due to surveys with improved cadence and depth such as Pan-STARRS1 (PS1) and Palomar Transient Factory (PTF). We find that KSN2015K, like most of the fast transients from the PS1 sample, is best described as a shock-breakout into a dense circumstellar shell. Such a model reproduces the significant characteristics of FELTs (fast, bright, blue) without much fine tuning.

The Unusual Transient of AGN AT2018qb, a tidal disruption event (TDE)?
One of the holy grails of AGN research is determining why quiescent supermassive black holes (SMBH) "turn on" to become strongly radiating active galactic nuclei (AGN). In the local universe, almost every massive galaxy appears to have a SMBH, but most of them are not active, and the processes that control their duty cycles may be either external, such as major or minor galaxy mergers, or internal and secular, such as gas streaming into the nucleus along a galactic bar. Observations of such

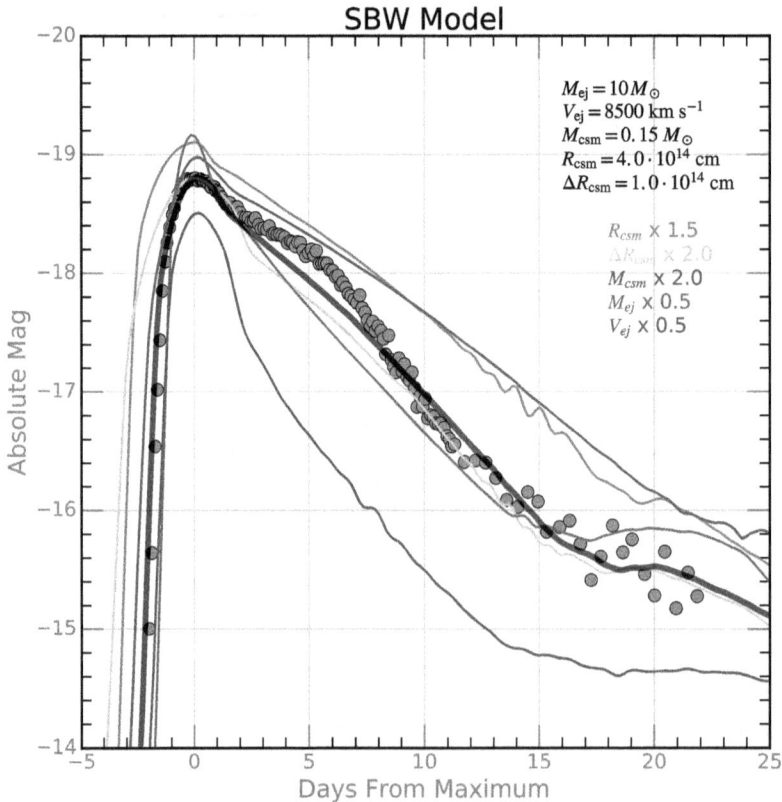

SBW Model

$M_{ej} = 10 M_\odot$
$V_{ej} = 8500 \ \text{km s}^{-1}$
$M_{csm} = 0.15 M_\odot$
$R_{csm} = 4.0 \cdot 10^{14} \ \text{cm}$
$\Delta R_{csm} = 1.0 \cdot 10^{14} \ \text{cm}$

$R_{csm} \times 1.5$
$\Delta R_{csm} \times 2.0$
$M_{csm} \times 2.0$
$M_{ej} \times 0.5$
$V_{ej} \times 0.5$

Figure 6.12. The KSN2015K data (red points) compared to numerical radiation hydrodynamics simulations of the CSM interaction models (lines). The best-fit model (black line, parameters listed on figure) is able to capture the fast rise and peak magnitude of KSN2015K as well as the rapid decline due to cooling of the shock-heated CSM and ejecta. Reprinted by permission from Springer Nature: Nature, Rest et al. (2018).

turn-ons/offs are very rare. To date, we know of only a few dozen AGN that have transitioned from having no broad emission lines, a hallmark of nuclear activity as the accretion disk irradiates nearby clouds, to having strong broad Balmer emission lines in their optical spectra (e.g., Yang et al. 2018). Astronomers have also observed the "light-echo" from several AGN (called voorwerps), caused by irradiated gas in tidal streams. Such a signature is taken to indicate that a now-inactive galaxy must have been active many thousands or millions of years in the past (e.g., Lintott et al. 2009). Therefore, we know for certain that the transition occurs. Despite about 30 light-curves of AGN covering multiple years, there has been only a handful of detections of an AGN "caught in the act" of turning on or off. The observations of these transitions are usually very poor sampled with cadences of months or years (e.g., the discovery in WISE data of J1657+2345 by Yang et al. 2019). Perhaps the best-sampled transition was found in SDSS J1554+36 with the intermediate Palomar Transient Factory (iPTF) by Gezari et al. (2017). This discovery was made using a light-curve with approximately daily sampling making it clear that the typical time

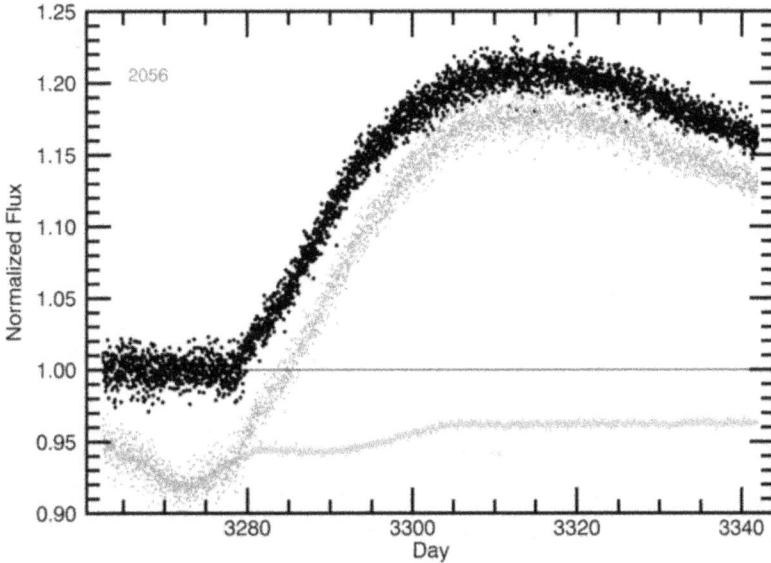

Figure 6.13. The K2 light-curve of the "changed look" AGN. The black points are the data after correcting for satellite motion and background variations. The blue points show the raw photometry while the orange points are an estimate of the background flux. K2 Campaign 16 began on 2017 December 7, and ended near the beginning of 2018 March.

resolution of months to years in ground-based surveys (e.g., SDDSS Stripe 82) is insufficient for this type of study.

Fortunately, K2 caught a galaxy in transition, providing us with the extremely rare opportunity to obtain photometric and spectroscopic observations of two AGN transitions along with K2's very high light sampling cadence (Shaya et al. 2020, in preparation).

AT2018qb was discovered in both Pan-STARRS and DECam monitoring of the Campaign 16 field, as part of the K2 Supernova Experiment. AT2018qb is a transient coincident with the nucleus of the Galaxy 2MASX J08565098+2107380, classified as a Narrow-line Seyfert 1 (NLSI) galaxy from SDSS observations. During the monitoring, AT2018qb brightened by more than 2 magnitudes in the g', r', and i' bandpasses over a timescale of only 30 days, reaching 18.5 mag in i'. The K2 light-curve, displayed in Figure 6.13 shows a quiescent phase, a smooth 35 day rise, and the beginning of a slow decline. We have continued to monitor AT2018qb from the ground, both photometrically and spectroscopically.

The smooth K2 light-curve of AT2018qb is not typical of AGN which show variations on all timescales. In addition, ground-based observations from PS1 from the last eight years show no variability. The transient event itself has a light-curve that is not consistent with a SN Ia, Ic/b, IIP, IIL, or IIb: The rise is too slow and, for all but the IIP, the decline is also too slow. The light-curve most closely resembles a tidal disruption event (TDE; Shaya et al. 2020, in preparation), even though the color is very red, unlike the blue colors of typical TDEs.

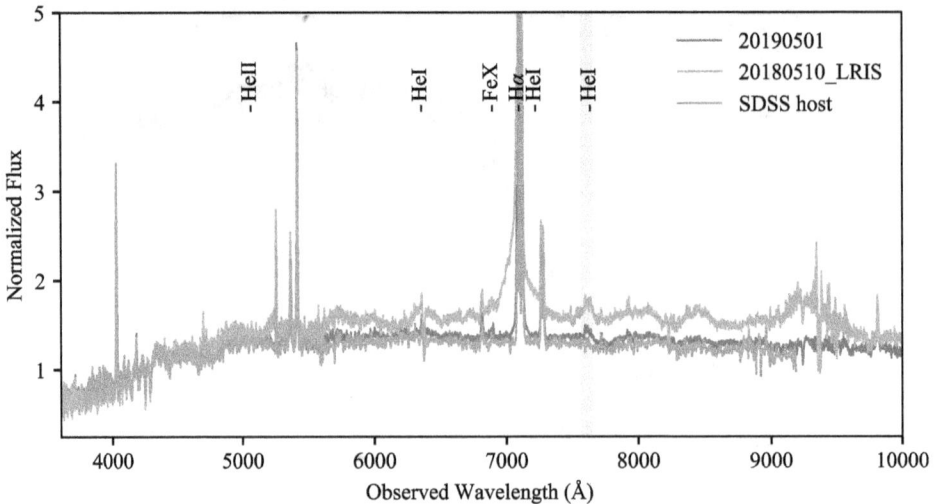

Figure 6.14. A series of spectra of AT2018qb obtained before, during, and after the "changing look" transient event. The gray spectrum shows the host galaxy observed by SDSS in 2005, long before the transient flared up. The spectrum taken in 2018 May (orange) is when the transient event was fading from maximum light, and the blue line shows the spectrum a year later. The pink stripe marks the region affected by A-band telluric absorption in the Earth's atmosphere. Atomic transitions, shifted to the observed wavelengths based on the Galaxy's redshift, are marked with black labels. As seen in the plot, several broad features, notably Hα and HeI, were observed only during the transient event. These broad features faded almost completely one year later.

Fortuitously, before the event, an archival spectrum had been obtained during the Sloan Digital Sky Survey in 2005 November and is shown in Figure 6.14. The pre-outburst spectrum is that of an Sc galaxy, combined with just a hint of broad Hα emission. The lack of significant broad emission and the ratios of the narrow lines both allow a clear classification of this galaxy core, in 2005, as an AGN. The lack of broad lines indicates that it was a Seyfert 1.9 (Shaya et al. 2020, in preparation). However, a spectrum taken with the Keck telescope in 2018 May (after the transient) shows a very different class of object. AT2018qb now presents very broad Hα and Helium lines (see Figure 6.14), the AGN has "changed looks" and become a Seyfert 1 nucleus. One year later, on 2019 May 1, the spectrum has evolved back to narrow lines, nearly indistinguishable from the SDSS spectrum from 2005! The discovery of the turn-on of AT2018qb, in which the optical and near-IR turn-on at the same time but for which the variability is larger in the red than the blue, is unprecedented and provides the first opportunity to observe the optical "turn-on" of an AGN (Rumbaugh et al. 2018), as well as its turn-off within one year. The K2 light-curve suggests that the turn-on event was likely triggered by the tidal disruption of a star.

References

Abbott, B. P., Abbott, R., Abbott, T. D., et al. 2017, ApJ, 848, L12
Alston, W. N. 2019, MNRAS, 485, 260
Aranzana, E., Körding, E., Uttley, P., Scaringi, S., & Bloemen, S. 2018, MNRAS, 476, 2501

Arcavi, I., Wolf, W. M., Howell, D. A., et al. 2016, ApJ, 819, 35

Arnett, W. D. 1982, ApJ, 253, 785

Bachev, R., Mukhopadhyay, B., & Strigachev, A. 2015, A&A, 576, A17

Balberg, S., & Loeb, A. 2011, MNRAS, 414, 1715

Baldassare, V. F., Geha, M., & Greene, J. 2018, ApJ, 868, 152

Bentz, M. C., & Katz, S. 2015, PASP, 127, 67

Bian, W., & Zhao, Y. 2003, MNRAS, 343, 164

Bianco, F. B., Howell, D. A., Sullivan, M., et al. 2011, ApJ, 741, 20

Bildsten, L., Shen, K. J., Weinberg, N. N., & Nelemans, G. 2007, ApJL, 662, L95

Biteau, J., & Giebels, B. 2012, A&A, 548, A123

Bloom, J. S., Kasen, D., Shen, K. J., et al. 2012, ApJ, 744, L17

Carini, M. T., & Ryle, W. T. 2012, ApJ, 749, 70

Chen, X.-Y., & Wang, J.-X. 2015, ApJ, 805, 80

Chevalier, R. A. 1998, ApJ, 499, 810

Chevalier, R. A., & Irwin, C. M. 2011, ApJ, 729, L6

Colgate, S. A., & McKee, C. 1969, ApJ, 157, 623

Darbha, S., Metzger, B. D., Quataert, E., et al. 2010, MNRAS, 409, 846

Dessart, L., Burrows, A., Ott, C. D., et al. 2006, ApJ, 644, 1063

Dexter, J., & Kasen, D. 2013, ApJ, 772, 30

Dimitriadis, G., Foley, R. J., Rest, A., et al. 2018, ApJ, 870, L1

Dimitriadis, G., Rojas-Bravo, C., Kilpatrick, C. D., et al. 2019, ApJ, 870, L14

Dobrotka, A., Antonuccio-Delogu, V., & Bajčičáková, I. 2017, MNRAS, 470, 2439

Dobrotka, A., Bezák, P., Revalski, M., & Strémy, M. 2019, MNRAS, 483, 38

Drout, M. R., Chornock, R., Soderberg, A. M., et al. 2014, ApJ, 794, 23

Edelson, R., & Malkan, M. 2012, ApJ, 751, 52

Edelson, R., Mushotzky, R., Vaughan, S., et al. 2013, ApJ, 766, 16

Edelson, R., Vaughan, S., Malkan, M., et al. 2014, ApJ, 795, 2

Förster, F., Moriya, T. J., Maureira, J. C., et al. 2018, NatAs, 2, 808

Fransson, C., Ergon, M., Challis, P. J., et al. 2014, ApJ, 797, 118

Garg, A., Stubbs, C. W., Challis, P., et al. 2007, AJ, 133, 403

Garnavich, P. M., Tucker, B. E., Rest, A., et al. 2016, ApJ, 820, 23

Garnavich, P. M., Jha, S., Challis, P., et al. 1998, ApJ, 509, 74

Gezari, S., Hung, T., Cenko, S. B., et al. 2017, ApJ, 835, 144

Ginzburg, S., & Balberg, S. 2014, ApJ, 780, 18

González-Martín, O., & Vaughan, S. 2012, A&A, 544, A80

Goobar, A., Johansson, J., Amanullah, R., et al. 2014, ApJ, 784, L12

Goyal, A., Stawarz, Ł., Zola, S., et al. 2018, ApJ, 863, 175

Haas, M. R., Batalha, N. M., Bryson, S. T., et al. 2010, ApJL, 713, L115

Hayden, B. T., Garnavich, P. M., Kessler, R., et al. 2010, ApJ, 712, 350

Hogg, J. D., & Reynolds, C. S. 2016, ApJ, 826, 40

Hosseinzadeh, G., Sand, D. J., Valenti, S., et al. 2017, ApJ, 845, L11

Hoyle, F., & Fowler, W. A. 1960, ApJ, 132, 565

Iben, I. Jr., & Tutukov, A. V. 1984, ApJS, 54, 335

Kasen, D. 2010, ApJ, 708, 1025

Kasen, D., & Bildsten, L. 2010, ApJ, 717, 245

Kasen, D., Fernández, R., & Metzger, B. D. 2015, MNRAS, 450, 1777

Kasliwal, V. P., Vogeley, M. S., & Richards, G. T. 2015a, MNRAS, 451, 4328

Kasliwal, V. P., Vogeley, M. S., Richards, G. T., Williams, J., & Carini, M. T. 2015b, MNRAS, 453, 2075

Kelly, B. C., Sobolewska, M., & Siemiginowska, A. 2011, ApJ, 730, 52

Kelly, P. L., Fox, O. D., Filippenko, A. V., et al. 2014, ApJ, 790, 3

Kerzendorf, W. E., Schmidt, B. P., Laird, J. B., Podsiadlowski, P., & Bessell, M. S. 2012, ApJ, 759, 7

Kleiser, I. K. W., & Kasen, D. 2014, MNRAS, 438, 318

Li, Y., Hu, S., Wiita, P. J., & Gupta, A. C. 2018, MNRAS, 478, 172

Lintott, C. J., Schawinski, K., Keel, W., et al. 2009, MNRAS, 399, 129

Maeda, K., Tanaka, M., Nomoto, K., et al. 2007, ApJ, 666, 1069

Mangalam, A. V., & Wiita, P. J. 1993, ApJ, 406, 420

Marion, G. H., Brown, P. J., Vinkó, J., et al. 2016, ApJ, 820, 92

Markowitz, A., Edelson, R., Vaughan, S., et al. 2003, ApJ, 593, 96

Marscher, A. P., & Travis, J. P. 1991, NASA STI/Recon Technical Report A, 93, 52913

McHardy, I. M., Koerding, E., Knigge, C., Uttley, P., & Fender, R. P. 2006, Natur, 444, 730

Mohan, P., Gupta, A. C., Bachev, R., & Strigachev, A. 2016, MNRAS, 456, 654

Moriya, T., Tominaga, N., Blinnikov, S. I., Baklanov, P. V., & Sorokina, E. I. 2011, MNRAS, 415, 199

Moriya, T., Tominaga, N., Tanaka, M., et al. 2010, ApJ, 719, 1445

Mushotzky, R. 2004, Supermassive Black Holes in the Distant Universe, in Astrophysics and Space Science Library, Vol. 308, ed. A. J. Barger (Berlin: Springer), 53

Mushotzky, R. F., Edelson, R., Baumgartner, W., & Gandhi, P. 2011, ApJ, 743, L12

Nakar, E., & Sari, R. 2010, ApJ, 725, 904

Noebauer, U., Kromer, M., Taubenberger, S., et al. 2017, MNRAS, 472, 2787

Nugent, P. E., Sullivan, M., Cenko, S. B., et al. 2011, Natur, 480, 344

O'Brien, J. T., Moreno, J., Richards, G. T., & Vogeley, M. S. 2018, RNAAS, 2, 127

Ofek, E. O., Rabinak, I., Neill, J. D., et al. 2010, ApJ, 724, 1396

Olling, R. P., Mushotzky, R., Shaya, E. J., et al. 2015, Natur, 521, 332

Perlmutter, S., Aldering, G., Goldhaber, G., et al. 1999, ApJ, 517, 565

Piro, A. L., & Morozova, V. S. 2016, ApJ, 826, 96

Piro, A. L., & Nakar, E. 2014, ApJ, 784, 85

Rabinak, I., & Waxman, E. 2011, ApJ, 728, 63

Remillard, R. A., & McClintock, J. E. 2006, ARA&A, 44, 49

Rest, A., Garnavich, P. M., Khatami, D., et al. 2018, NatAs, 2, 307

Revalski, M., Nowak, D., Wiita, P. J., Wehrle, A. E., & Unwin, S. C. 2014, ApJ, 785, 60

Riess, A. G., Filippenko, A. V., Challis, P., et al. 1998, AJ, 116, 1009

Rumbaugh, N., Shen, Y., Morganson, E., et al. 2018, ApJ, 854, 160

Sand, D. J., Graham, M. L., Botyánszki, J., et al. 2018, ApJ, 863, 24

Scargle, J. D. 2020, ApJ, 895, 90

Schaefer, B. E., & Pagnotta, A. 2012, Natur, 481, 164

Schlegel, E. M. 1990, MNRAS, 244, 269

Scolnic, D. M., Jones, D. O., Rest, A., et al. 2018, ApJ, 859, 101

Shakura, N. I., & Sunyaev, R. A. 1976, MNRAS, 175, 613

Shappee, B. J., Piro, A. L., Stanek, K. Z., et al. 2018, ApJ, 855, 6

Shappee, B. J., Holoien, T. W.-S., Drout, M. R., et al. 2019, ApJ, 870, 13

Shaya, E. J., Olling, R., & Mushotzky, R. 2015, AJ, 150, 188

Shen, K. J., & Bildsten, L. 2009, ApJ, 699, 1365

Shen, K. J., Kasen, D., Weinberg, N. N., Bildsten, L., & Scannapieco, E. 2010, ApJ, 715, 767

Simm, T., Saglia, R., Salvato, M., et al. 2015, A&A, 584, 106

Smith, K. L. 2019, AN, 340, 308

Smith, K. L., Mushotzky, R. F., Boyd, P. T., et al. 2018a, ApJ, 857, 141

Smith, K. L., Mushotzky, R. F., Boyd, P. T., & Wagoner, R. V. 2018b, ApJ, 860, L10

Smith, K. L., Boyd, P. T., Mushotzky, R. F., et al. 2015, AJ, 150, 126

Stritzinger, M., Taddia, F., Fransson, C., et al. 2012, ApJ, 756, 173

Tanaka, M., Tominaga, N., Morokuma, T., et al. 2016, ApJ, 819, 5

Tauris, T. M., Langer, N., & Podsiadlowski, P. 2015, MNRAS, 451, 2123

Tucker, M. A., Shappee, B. J., & Wisniewski, J. P. 2019, ApJ, 872, L22

Uttley, P., McHardy, I. M., & Vaughan, S. 2005, MNRAS, 359, 345

Vaughan, S., Uttley, P., Markowitz, A. G., et al. 2016, MNRAS, 461, 3145

Villar, V. A., Berger, E., & Miller, G. 2019, ApJ, 884, 83

Villar, V. A., Guillochon, J., Berger, E., et al. 2017, ApJ, 851, L21

Wehrle, A. E., Carini, M., & Wiita, P. J. 2019, ApJ, 877, 151

Wehrle, A. E., Wiita, P. J., Unwin, S. C., et al. 2013, ApJ, 773, 89

Whelan, J., & Iben, I. Jr. 1973, ApJ, 186, 1007

Woosley, S. E., Taam, R. E., & Weaver, T. A. 1986, ApJ, 301, 601

Yang, Q., Shen, Y., Liu, X., et al. 2019, ApJ, 885, 110

Yang, Q., Wu, X.-B., Fan, X., et al. 2018, ApJ, 862, 109

www.ingramcontent.com/pod-product-compliance
Lightning Source LLC
Chambersburg PA
CBHW080523220326
41599CB00032B/6178